DAVID O. MCKAY LIBRARY
3 1404 00748 0830

D1737621

The Dynamics of Hired Farm Labour

Constraints and Community Responses

The Dynamics of Hired Farm Labour

Constraints and Community Responses

Edited by

Jill L. Findeis

Department of Agricultural Economics and Rural Sociology and
Population Research Institute
The Pennsylvania State University
University Park, Pennsylvania, USA

Ann M. Vandeman

US Department of Agriculture
Economic Research Service
Washington, DC, USA

Janelle M. Larson

Department of Agricultural Economics and Rural Sociology
The Pennsylvania State University, University Park
Pennsylvania, USA

and

Jack L. Runyan

US Department of Agriculture
Economic Research Service
Washington, DC, USA

CABI *Publishing*

CABI *Publishing* is a division of CAB *International*

CABI Publishing
CAB International
Wallingford
Oxon OX10 8DE
UK

Tel: +44 (0)1491 832111
Fax: +44 (0)1491 833508
E-mail: cabi@cabi.org
Web site: www.cabi-publishing.org

CABI Publishing
10 E 40th Street
Suite 3203
New York, NY 10016
USA

Tel: +1 212 481 7018
Fax: +1 212 686 7993
E-mail: cabi-nao@cabi.org

A catalogue record for this book is available from the British Library, London, UK.

Library of Congress Cataloging-in-Publication Data
The dynamics of hired farm labour : constraints and community responses / edited by Jill L. Findeis, Ann M. Vandeman, Janelle M. Larson.
p. cm.
Includes bibliographical references and index.
ISBN 0-85199-603-5
1. Agricultural laborers- -United States. 2. Migrant agricultural labor- -United States. 3. Agricultural laborers- -Canada. 4. Migrant agricultural labor- -Canada. 5. Agricultural laborers- -Australia. 6. Migrant agricultural labor- -Australia. I. Findeis, Jill Leslie. II. Vandeman, Ann M. III. Larson, Janelle M.
HD1525 .D95 2002
331.7′63- -dc21

2002001334

ISBN 0 85199 603 5

Typeset in Adobe Palatino by Wyvern 21 Ltd, Bristol.
Printed and bound in the UK by Biddles Ltd, Guildford and King's Lynn.

Contents

Contributors

Daniel Carroll, *US Department of Labor, 200 Constitution Avenue, NW, Washington, DC 20210, USA.*

Beckie Mullin Denton, Correspondence address: *231 High Reeves Road, Richmond, KY 40475, USA.*

Eleanor Pepi Downey, *School of Social Work, Colorado State University, 116 Education Building, Fort Collins, CO 80523-1586, USA.*

Robert D. Emerson, *Food and Resource Economics Department, University of Florida, PO Box 110240, Gainesville, FL 32611-0240, USA.*

Alicia Fernandez-Mott, *US Department of Labor, 200 Constitution Avenue, NW, Washington, DC 20210, USA.*

Jill L. Findeis, *Department of Agricultural Economics and Rural Sociology, The Pennsylvania State University, 112E Armsby Building, University Park, PA 16802, USA.*

Susan Gabbard, *Aguirre International, 480 E 4th Avenue, Unit A, San Mateo, CA 94401-3349, USA.*

Michael J. Greenwood, *Department of Economics, University of Colorado at Boulder, Boulder, CO 80309, USA.*

Patricia M. Hennessy, Correspondence address: *433 13th Avenue E. #306, Seattle, WA 98102, USA.*

Wallace E. Huffman, *Department of Economics, Iowa State University, 478 Heady Hall, Ames, IA 50014, USA.*

Fred Krissman, *Center for US–Mexican Studies, University of California at San Diego, 9500 Gilman Drive, 0510, La Jolla, CA 92093-0510, USA.*

Janelle M. Larson, *Department of Agricultural Economics and Rural Sociology, The Pennsylvania State University, Berks Campus, University Park, 19610-6009, USA.*

Thomas R. Maloney, *Department of Applied Economics and Management, Cornell University, 306 Warren Hall, Ithaca, NY 14853-7801, USA.*

Jim McAllister, *School of Psychology and Sociology, Building 32, Room 2.4, Central Queensland University, Rockhampton, Qld 4702, Australia.*

Paul E. McNamara, *Department of Agricultural and Consumer Economics, University of Illinois at Urbana-Champaign, 437 Mumford Hall, 1301 W. Gregory Drive, Urbana, IL 61801-3605, USA.*

Michael D. Miller, *Department of Agricultural and Resource Economics, Colorado State University, Fort Collins, CO 80523-1172, USA.*

Richard Mines, *California Institute for Rural Studies, 221 G. Street, Suite 204, Davis, CA 95616, USA.*
Avis Mysyk, *Department of Anthropology, University of Manitoba, 435 Fletcher Argue Building, Winnipeg, Manitoba R3T 5V5, Canada.*
Christine K. Ranney, *Department of Applied Economics and Management, Cornell University, 315 Warren Hall, Ithaca, NY 14853, USA.*
Fritz Roka, *Southwest Florida Research and Education Center, University of Florida, 2686 State Road 29 North, Immokalee, FL 34142, USA.*
Rene P. Rosenbaum, *Julian Samora Research Institute and Department of Resource Development, 320 Natural Resources Building, Michigan State University, East Lansing, MI 48824-1222, USA.*
Jack L. Runyan, *US Department of Agriculture, Economic Research Service, 1800 M Street NW, Washington, DC 20036-5831, USA.*
Robert C. Seiz, *School of Social Work, Colorado State University, 127 Education Building, Fort Collins, CO 80523-1586, USA.*
Hema Swaminathan, *Department of Agricultural Economics and Rural Sociology, The Pennsylvania State University, 312 Armsby Building, University Park, PA 16802, USA.*
Dawn Thilmany, *Department of Agricultural and Resource Economics, Colorado State University, Fort Collins, CO 80523-1172, USA.*
Ann M. Vandeman, *c/o Jack Runyan, US Department of Agriculture, Economic Research Service, 1800 M Street NW, Washington, DC 20036-5831, USA.*
Qiuyan Wang, *Department of Agricultural Economics and Rural Sociology, The Pennsylvania State University, 308 Armsby Building, University Park, PA 16802, USA.*
Steven S. Zahniser, *US Department of Agriculture, Economic Research Service, 1800 M Street NW, Washington, DC 20036-5831, USA.*

Acknowledgements

The chapters in *The Dynamics of Hired Farm Labour: Constraints and Community Responses* are an outgrowth of a conference on hired farm labour and rural communities held in October 1999 near Kennett Square (Philadelphia), Pennsylvania. The conference was co-sponsored by the Economic Research Service at the US Department of Agriculture and The Pennsylvania State University, and sought to provide a forum for expressing the diverse views of scholars and practitioners concerned with the well-being of hired farm workers, the communities in which they live and work, and agriculture. The conference was multidisciplinary, bringing together social scientists who often have very different views of the farm worker problem. A word of thanks is extended to Miley Gonzalez, former Undersecretary of Agriculture (US Department of Agriculture); Susan Offutt, Economic Research Service Administrator; Al French, USDA Coordinator of Agricultural Labor Affairs; Betsey Kuhn, Head of the Food and Rural Economics Division, ERS/USDA; Juan Martinez, USDA National Farm Worker Coordinator; and David Blandford, Head, Department of Agricultural Economics and Rural Sociology at Penn State.

The editors of this volume wish to thank all participants in the conference as well as those authors who contributed their time to this book. Like the conference, this volume expresses different dimensions of a problem that concerns us all. The editors wish especially to thank Rose Ann Alters in the Department of Agricultural Economics and Rural Sociology at Penn State, for her skilled assistance and patience in pulling the entire volume together, and Rachel Ritz, also at Penn State, for helping to edit the book. Without their help, this volume would not have become a reality. There are many others who have helped to prepare individual chapters – we thank them for making our job easier.

Finally, the editors would like to thank Francisco Guajardo, a teacher of migrant farm worker children and a conference participant, who will always remind us that our work has the potential to have significant impacts on farm workers, rural communities and agriculture. For this reason, this volume is dedicated to the students and teachers at Edcouch-Elsa High School.

Preface

Labour remains a critical input in agriculture, despite current trends toward less total employment in farm production. *The Dynamics of Hired Farm Labour: Constraints and Community Responses* focuses on understanding current trends in labour use in agriculture, recent changes in the agricultural workforce and the conditions of work, and the interrelationships between local communities, employers and farm workers who live and work in these communities. The growing dependence on hired farm labour as a proportion of the total workforce in agricultural production in most developed countries makes issues surrounding a hired workforce in agriculture a timely topic. Technological change in agriculture and recent changes in the direction of farm policy in many countries may further influence the balance of reliance on family versus hired labour to provide an adequate food supply to society.

The challenge now is to better understand how changes in agriculture are affecting the work done by hired farm labour, the economic returns to farm work, and the interrelationships between agricultural employers, employees and local communities. This book seeks to provide insight into these questions. Although the chapters in this volume report on research undertaken in several different countries, the main focus is on the USA, where agriculture is undergoing significant changes in its agricultural workforce. Unlike some countries where agriculture is absorbing hired labour that would otherwise be unemployed, the USA faced very tight labour markets in the 1990s. As a result, long-term concerns surrounding the farm workforce (e.g. low economic returns to work, poor work conditions, occupational safety issues) have been further complicated to a very significant degree by a greater reliance on immigrant labour. Issues of race/ethnicity, culture and language become increasingly important in this context.

The Dynamics of Hired Farm Labour is divided into three sections. Section I (The Farm Workforce – Trends, Adjustments and Technical Change) provides an overview of changes in agricultural (farm) employment in developed countries, trends in the farm workforce, and the effects of technology on labour use in production agriculture. Chapter 1 (Jill Findeis) provides an overview of changes in employment in agriculture, documenting labour adjustments that have occurred in this sector. Chapter 2 (Susan Gabbard, Alicia Fernandez-Mott and Daniel Carroll) then documents the dynamic changes in the US farm workforce in the 1990s, changes that mean that many of the long-held images or stereotypes of farm workers in the USA are now outdated. These papers are followed by two chapters that focus on the issue of agricultural technology and labour. Chapter 3 (Wallace Huffman) analyses changes in technology and its effects on hired farm labour under different conditions, and Chapter 4 (Richard Mines) assesses the joint issues of technology and the composition of the hired labour force – solo male labour versus the settlement of family units. In both cases, the authors argue that technology could be developed to further reduce agriculture's reliance on hired labour and improve the conditions of farm work but that the current availability of low-cost (immigrant) labour in countries such as the USA reduces incentives for technological change.

Section II (Hired Farm Labour, Employers and Community Response) presents a series of studies from different disciplinary perspectives that focus on the issues faced by farm labourers, their employers and local communities. Chapter 5 (Fritz Roka and Robert Emerson), Chapter 6 (Thomas Maloney) and Chapter 7 (Dawn Thilmany and Michael Miller) examine the economic returns to farm workers and the issues faced by farm employers who require a stable supply of farm labour.

Chapter 8 (Rene Rosenbaum) takes a community development viewpoint, arguing that farm workers have direct and indirect positive economic impacts on local communities or regional economies that would be lost if hired labourers were not employed in the agricultural sector. Despite the positive impacts outlined by Rosenbaum, Chapter 9 (Beckie Mullin Denton) shows that immigrant farm workers can be met with 'hostility' by local communities. The chapter by Denton assesses the introduction of Hispanic farm workers into eastern Kentucky, examining the importance of community acceptance and employer–employee–community relationships. Chapter 10 (Steven Zahniser and Michael Greenwood) and Chapter 11 (Robert Emerson and Fritz Roka) approach the labour issue from a broad perspective; Zahniser and Greenwood explore the question of transferability of farm worker knowledge gained in the USA to their origin communities in Mexico, while Emerson and Roka examine whether changes in the income distribution in the USA have affected the returns to farm workers there. Finally, Section II closes with three chapters that explore issues of labour, capital and class. Chapter 12 (Jim McAllister), Chapter 13 (Avis Mysyk) and Chapter 14 (Fred Krissman) draw on literature in sociology and anthropology that, again, focuses on the low returns to farm labour but also on issues of race, ethnicity and class. The perspective is different from the earlier papers but the conclusion is the same – the returns paid to farm workers continue to be low.

Finally, Section III (Farm Worker Health and Safety) examines an issue that continues to be a major problem: the health and safety of the farm worker population. Safety issues are well known in production agriculture. Chapter 15 (Robert Seiz and Eleanor Pepi Downey) reports on a recent face-to-face survey of farm workers that asked about the safety of their work environment and other health care issues, while Chapter 16 (Paul McNamara and Christine Ranney) reports on research on health insurance for farm workers and farm families. Chapter 17 (Patricia Hennessy) takes a practitioner's view of the difficulties of providing health care to migrant workers.

The book concludes with a final chapter (Chapter 18 by Janelle Larson, Jill Findeis, Hema Swaminathan and Qiuyan Wang) that compares two of the major data sources used throughout this volume, i.e. the National Agricultural Workers Survey (NAWS) and the Current Population Survey (CPS). The two surveys provide very different 'pictures' of the farm worker population because of their different sampling frames and approaches. The two surveys include different sets of questions that help to provide data to gain added insight into the issues discussed above.

Section I

The Farm Workforce – Trends, Adjustments and Technical Change

In the past century and particularly since the Second World War, there have been significant declines in the labour employed on farms in most developed countries. Labour is no longer being absorbed by agriculture, as technological change has resulted in the substitution of other production inputs for labour. Adjustments in the farm sector have been substantial: fewer farms, fewer farm operators, fewer farm families, and more off-farm employment among those who remain in agricultural production. The labour that is hired to perform farm work has also experienced marked adjustments.

What has remained constant is the fact that most farm workers today continue to earn low wages, receive few benefits, and work concentrated hours in jobs that are often unsafe. Who is willing to do this work varies, depending on country, region and commodity. Workers in the farm production sector are typically at the bottom of the job ladder, and whether they can move upward to better jobs in agriculture or elsewhere in the economy very often depends on barriers imposed by immigration policy, cultural differences that may impose constraints to integration within the broader society, access to education and other means of enhancing human capital, among other barriers to upward mobility. The issues are complex and often involve race, ethnicity, differences in religion, and other societal barriers.

A question worth addressing is whether technologies could be developed to enhance the economic returns to hired farm labour, improve the conditions of work, and/or reduce the need for low-skill labour in this sector. The problem is that technological change is often induced by incentives to reduce costs, but in the case of hired farm labour, incentives to change are reduced by low wages coupled with low benefits. As long as there are workers willing to supply labour at low wages, it is unlikely that the situation will change. The only changes that may occur are in terms of *who* is willing to work under these conditions.

The four chapters included in this section outline trends in hired farm labour and the issue of technological change. Chapter 1 (Findeis) reviews trends in employment in agriculture in the last half of the 20th century, showing that the adjustments that have occurred in labour utilization in production agriculture are due to changes in technology, first and foremost, but also due to demographic changes, changes in farm policy, and the effects of labour markets external to agriculture. Adjustments in the farm labour

(eds J.L. Findeis, A.M. Vandeman, J.M. Larson and J.L. Runyan)

force have been dramatic in some instances and have had effects not only on farms but also on the local communities in which farm labourers live and work. In some cases, the community impacts may be beneficial, particularly if local labour is absorbed into the agricultural sector. If immigrant labour is needed, the impacts may also create community conflicts or at least challenges in terms of service provision – housing, health care and education, among other services – to a workforce that is often culturally different from the local population.

Chapter 2 (Gabbard, Fernandez-Mott and Carroll) assesses trends that have occurred in the US agricultural workforce in the past decade. Because the farm labour market is so dynamic, previous 'images' of farm workers often no longer apply. Unlike the past, when farm workers in much of the USA worked as family units, a significant shift toward young Hispanic solo male workers from Mexico is observed by Gabbard *et al.* Over half of farm workers in the USA are now unauthorized. They also are more likely to see themselves as temporary workers in agriculture and in the USA. As was also true in the past, the economic returns to this new workforce are, by and large, low.

Chapters 3 (Huffman) and 4 (Mines) document the use and effects of new technology adopted in agricultural production. Both authors argue that the potential exists for the development of better technologies to reduce the dependence and hazards of the work for which farm labourers are usually employed. Clearly, much more could be done to utilize technology to reduce the negative aspects of farm work. Yet the pace of technology development is slower than might be optimal from the farm workers' perspective, simply because hired labour continues to be paid a low wage. It can be argued that if the wages for farm labour would increase in real terms, the pace of technology development would increase correspondingly. In this case, it is immigration policy that is likely the cause for the continued reliance on low-wage, hired farm workers.

Finally, Mines in Chapter 4 also reflects on the trend towards solo male labour on farms in the USA. An issue faced by governments concerned about labour shortages (perceived or real) is the appropriate immigration policy to adopt, i.e. whether to open borders to families that will eventually become assimiliated, at least to some extent, into the destination country, or to encourage the use of 'guestworkers' or labourers who only temporarily fill a labour need. This issue runs throughout the four chapters in this section as they explore issues of technology, changes in the labour force and alternative immigration policies.

1 Hired Farm Labour Adjustments and Constraints

Jill L. Findeis
Department of Agricultural Economics and Rural Sociology, The Pennsylvania State University, University Park, Pennsylvania, USA

Abstract

This chapter examines trends in the agricultural workforce in developed countries, including both hired and family labour used in farm production. Trends show that while agricultural employment has declined significantly over time, this trend has been coupled with an increasing proportion of the agricultural workforce that is hired. Adjustments in the workforce are attributable to changes in labour supply, demand for labour, external labour markets, and agricultural and labour policies. Employers are affected as are communities adjacent to farms that help to provide services for farm workers. The extent to which communities are affected is related to labour demand and external labour market conditions, specifically the tightness or looseness of the labour market for local labour. If local labour cannot be hired to fulfil the need for labour in farming, local communities are likely to be affected by an influx of workers from outside the community. The impacts on local communities will vary in part depending on the size and demographic characteristics of the immigrant population.

Introduction

Employment in farm production declined significantly in most developed countries in the past century. Labour productivity has increased, farm households and farm operators have declined in numbers, and average farm size has increased (Barkley, 1990; Gardner, 1992; OECD, 1994, 2001; von Meyer, 1997; see also Schmitt, 1991). Some farm operators have left farming altogether, with the land being purchased by other farmers or used for development outside of agriculture. On other farms, labour has been increasingly absorbed by external (off-farm) labour markets, as off-farm employment and 'pluriactivity' (multiple job-holding) have increased (Fuller, 1990; Hallberg *et al.*, 1991; Bryden *et al.*, 1992; Post and Terluin, 1997; Weersink *et al.*, 1998; Oluwole, 2000). Part-time farming is now prevalent (Gasson 1986; Pfeffer, 1989; Gasson *et al.*, 1995) and a more concentrated number of large farms produces a significant share of farm output. These adjustments have had important implications for farm family labour utilization as well as the labour hired to support the agricultural production sector.

The shift in the structure of the farm production sector has affected the balance

(eds J.L. Findeis, A.M. Vandeman, J.M. Larson and J.L. Runyan)

between hired labour and farm family labour in the aggregate. Although employment on farms has declined as a result of the substitution of capital for labour and the adoption of other new labour-saving technologies, many farms today depend on hired labour in some form. This is particularly true for large farms. This may mean workers hired to produce labour-intensive fruits and vegetables, workers employed on a full-time basis throughout the year to work on farms such as dairy and mushroom operations, or casual workers who supply labour on a 'casual' or part-time basis. For farms hiring labour, labour is an input that is essential for production. For this reason, maintaining a stable supply of workers is important to producers.

There would not be a problem maintaining this supply if farm work was full-time year-round or could readily be combined with other jobs to provide year-round work, and if it paid wages and provided job benefits that would keep a family out of poverty, was physically less demanding and safe, and was flexible in terms of hours of work per day, then there would not be a problem with maintaining this supply. This has not been the case. In the past century, the problem of a low-income farm worker population has still not gone away. Changes in farm structure, technology and policies that directly affect farm labour (e.g. immigration and minimum wage policies) have not solved the problem.

Since it is unlikely that the demand for farm workers will decline to any significant extent in the short run and employers will continue to require substantial numbers of workers to support farm production in the aggregate, the question of how to deal with the labour supply problem becomes important. How changes affecting the agricultural sector and external labour markets (e.g. changes in farm structure and technology, the growth of low-skill service employment in external labour markets, greater use of farm labour contractors, and policy) are influencing the well-being of farm workers, and the interrelationships between employers, workers and local communities, are also important in this context. Multiple perspectives on the farm worker problem exist, but in general there is agreement that improvements in the well-being of farm workers is an important goal – for agriculture and its workforce.

Greater Reliance on a Hired Farm Workforce

Three important trends related to agricultural employment have been witnessed in most developed countries since the Second World War (OECD, 2001). Firstly, virtually all countries have witnessed a decline in the workforce engaged in agricultural production, an outgrowth of technological change that has enhanced capital and other non-labour inputs in agriculture (Gardner, 1992). New agricultural technologies have been largely labour-saving (Binswanger, 1974; Antle, 1986; Huffman and Evenson, 1989; Kang and Maruyama, 1992). The new technologies have led to higher rates of labour productivity and a reduced need for labour inputs in the farm sector (Gardner, 1992; Ahearn *et al.*, 1997). Declines in labour demand attributable to technological change in the farm sector have reduced employment of both farm family members and hired farm workers. Scale economies have also resulted in larger (and correspondingly fewer) farms in many countries.

The declines have been significant, consistent and widespread across developed countries over this period. In the past decade, OECD labour force statistics for a number of developed countries show that the percentage of agricultural employment in total civilian employment today is low in most nations, and declining (Table 1.1).

Secondly, there have also been simultaneous increases in the proportion of the total farm workforce that is hired in many countries (Table 1.2). Again, this trend is consistent and widespread. For example, in the OECD-22 countries (the European Union and Australia, Canada, Iceland, Japan, New Zealand, Norway and the USA), hired or paid labour comprised an average 28.1% of the agricultural workforce over the period 1986–1990, but by 1996–1997 the hired workforce was 34.3% of total labour (OECD, 2001). This

Table 1.1. Agricultural employment as a percentage of total civilian employment. (Source: *OECD Labour Force Statistics*.)

	Percentage of civilian employment		
Country	1986–1990	1991–1995	1996–1997
Austria	8.2	7.2	7.0
Belgium	2.8	2.5	2.4
Denmark	5.6	5.0	3.8
Finland	9.6	8.2	7.0
France	6.5	5.0	4.4
Germany[a]	3.9	3.6	3.2
Greece	26.2	21.3	20.3
Ireland	15.1	12.7	10.4
Italy	9.6	7.8	6.7
Luxembourg	3.7	1.8	na
The Netherlands	4.7	4.0	3.8
Portugal	19.9	12.4	12.8
Spain	13.6	9.8	8.4
Sweden	3.8	3.3	2.8
UK	2.2	2.1	1.9
European Union	7.1	5.6	4.9
Australia	5.7	5.2	5.1
Canada	5.9	5.5	5.2
Iceland	12.1	9.7	8.8
Japan	7.9	6.1	5.4
New Zealand	10.4	10.4	8.8
Norway	6.6	5.4	4.9
USA	2.9	2.8	2.7
Czech Republic	12.0	7.8	6.0
Hungary	na	na	8.1
Korea	20.7	14.7	11.3
Mexico	na	25.3	23.3
Poland	na	na	21.2
Switzerland	5.7	3.6	4.6
Turkey	45.9	44.3	42.4

The agricultural sector includes not only agriculture but also hunting, forestry and fishing (see OECD, 2001).
[a]Unified Germany, since 1991.
na = not available.

trend was observed in almost all countries: Austria (12.5–14.5%), Belgium (16.3–23.6%), France (19.1–27.8%), The Netherlands (33.5–41.6%), Australia (31.8–40.2%) and Canada (57.6–61.1%), among others. The dual trends of less reliance on labour but more reliance on hired farm labour as a proportion of the farm workforce has, in general, been the case in most developed countries. In some cases, there has been an absolute increase in the number of hired workers. Barthelemy (2001) noted that there had been large increases in the hired workforce in Greece and Denmark and that smaller increases had been observed in Belgium, Spain, Luxembourg and The Netherlands (see also discussion in Fasterding, 1997).

Finally, there is now more off-farm employment among farm households, in part an adjustment to the declines in farm labour requirements. Many farm households are now engaged in multiple job-holding, both

Table 1.2. Hired farm labour as a percentage of agricultural employment. (Source: *OECD Labour Force Statistics.*)

	Percentage of agricultural employment		
Country	1986–1990	1991–1995	1996–1997
Austria	12.5	13.7	14.5
Belgium	16.3	19.7	23.6
Denmark	35.3	42.4	48.5
Finland	23.1	25.6	25.2
France	19.1	24.3	27.8
Germany[a]	21.4	46.5	45.8
Greece	3.9	4.1	4.5
Ireland	13.6	15.1	15.2
Italy	38.6	38.8	36.9
Luxembourg	21.4	43.1	na
The Netherlands	33.5	39.6	41.6
Portugal	17.3	15.9	14.2
Spain	31.3	32.4	35.2
Sweden	39.3	37.3	34.4
UK	46.7	42.4	45.1
European Union	26.2	30.9	31.5
Australia	31.8	35.3	40.2
Canada	57.6	59.0	61.1
Iceland	45.5	46.4	51.9
Japan	9.3	11.2	12.3
New Zealand	41.6	38.3	41.7
Norway	26.0	28.8	35.7
USA	54.4	53.7	55.5
Czech Republic	na	na	85.7
Hungary	na	na	75.2
Korea	na	7.0	6.7
Mexico	na	22.1	30.4
Poland	na	na	9.7
Switzerland	na	na	na
Turkey	na	4.6	6.3

The agricultural sector includes not only agriculture but also hunting, forestry and fishing (see OECD, 2001).

on and off the farm – in the farm business, in off-farm employment and/or in non-farm businesses (Hallberg *et al.*, 1991). Employment of farm women in off-farm jobs has grown substantially in many countries (Bryden *et al.*, 1992; Findeis *et al.*, 2002), but even farm men have increasingly gone to work in non-farm jobs (Kada, 1980; Hallberg *et al.*, 1991). Off-farm labour supply has become increasingly important (Tokle and Huffman, 1991; Zweimuller *et al.*, 1992; Olfert, 1993; Mishra and Goodwin, 1997; Weiss, 1997; Corsi and Findeis, 2000; Oluwole, 2000; Woldehanna *et al.*, 2000). This trend has been observed in Europe, North America, Australia and Japan, among other industrialized countries. Off-farm employment may be related to hired farm labour decisions on some types of farms (Findeis and Lass, 1992; Benjamin *et al.*, 1996).

There is now a greater dependence of farm families on income earned in non-farm labour markets instead of principally from farming and other farm-related sources of income (e.g. land rental; value-added processing of farm products; custom work on other farms). The result is that farms are increasingly affected by changes in the broader economy unrelated to agriculture, and farm households are now more directly tied to non-farm labour markets. Agricultural policy reform (e.g. the Common Agricultural Policy in the EU and FAIR 96 in the USA) emphasizing direct payments over price supports may be influencing this trend (Frohberg, 1994; Findeis, 1998; Weiss, 1998; Weyerbrock, 1998; OECD, 2001).

The greater prevalence of off-farm work may also have implications for hired farm labour that needs other (potentially off-farm) employment for part of the year to supplement seasonal employment on farms. The issues of combining farm and non-farm work for farm households and farm workers have similarities because of the seasonality of agricultural production that affects the 'other employment' prospects of those working on farms.

The three trends outlined above together influence the labour that is needed on farms today and the ability of workers – both hired and farm family – to make a living in the agricultural sector. Since many farms are still located in regions and areas that have lower population densities, relevant questions become: (i) whether or not local labour markets are able to provide the labour needed to support farms; and (ii) whether farm production can adequately support a farm workforce that may only be able to work in farming for part of the year. There is also the question of large seasonal demands for particular farm enterprises – especially those situations where the agriculture that has developed is highly labour-intensive (e.g. fruits, vegetables and horticultural crops), requiring large numbers of workers.

Forms of Hired Labour

Labour is by no means a homogeneous input in agriculture (Mines *et al.*, 1993; see also Chapter 2, this volume) and the demands for labour vary considerably across the sector (Emerson, 1984). Labour demand depends on enterprise mix and on specific growing conditions that determine when and how much labour is required. Three types of labour needs – seasonal, full-time year-round, and casual – are discussed below.

Seasonal workers

In some regions, the labour problem faced by farm operators is largely seasonal, with potentially large requirements for labour for particular tasks in a production environment where timing of need is not exact (due, for example, to weather conditions and climate) (Martin, 1988; Martin *et al.*, 1995). Issues related to the significant need for seasonal hired or paid labour are observed in many developed countries, including Australia, New Zealand, Canada, the USA and parts of Europe. In some regions, seasonal labour is hired but agriculture is not so densely concentrated in the region as to require large numbers of workers to support agricultural production. In other regions, however, the agriculture that has developed is heavily labour-intensive. For example, in states such as California, Florida and Washington there is a significant need for seasonal labour in labour-intensive fruit and vegetable production. Large numbers of workers are needed on farms in a concentrated time period to perform a variety of tasks, such as planting, tying, trimming, pruning and harvesting. An inability to attract workers in time carries with it a high economic cost.

Whether hired or paid workers are needed from outside the local or regional labour market depends on: (i) the number of workers needed, which may be very significant (as in labour-intensive fruit and vegetable production); and (ii) the tightness of the labour market. Where unemployment rates are high and involuntary part-time work is prevalent, absorbing seasonal labour from the local or regional labour market is less of a problem than when labour markets are tight. However, even if unemployment and underemployment rates are high locally, the seasonal labour problem may still exist if the agricultural systems that have developed put a significant demand on labour resources. In these cases, in-migration of seasonal labour is required.

In the USA, both the need for a *large* number of seasonal workers and the tightness of labour markets (until recently) have resulted in a clear need to find solutions to the seasonal labour problem in agriculture. Several potential solutions are possible: (i) developing new technologies to reduce the need for labour in this industry; (ii) improving the economic returns to farm work (wages and job benefits) to attract labour into the hired farm workforce; (iii) improving the conditions of work and timing/time requirements of work on farms, again to attract workers into the sector; and (iv) supporting an immigration policy that allows for the influx of workers. The first solution – the development and adoption of new labour-saving technologies – has been an ongoing process in the past century in agriculture, but it is not clear that enough is being done in terms of developing new technologies that will ease the negative aspects of hired farm work, as discussed in Chapters 3 and 4. The rate of technology development depends on incentives for development, and Chapters 3 and 4 both argue that a low-wage immigration policy in the USA has reduced the development of labour-saving technologies, including some with particular promise (for reducing labour needs and worker injuries, for example).

Improvement in the economic returns to hired farm work and the physical conditions of work are a means of improving the attractiveness of the farm production sector to workers, but many producers argue that, to remain competitive, they are not able to pay higher wages. Whether or not this assertion is true has long been a point of debate (and sometimes heated disagreement), even in the literature on farm workers. Tight labour markets can lead to higher wages, unless there is an avenue that allows producers to access workers willing to work for less. For countries such as the USA, this avenue is immigration policy (Martin *et al.*, 1995; Taylor and Martin, 1997; Rothenberg, 1998).

Producers can also attract workers by offering better job benefits (e.g. paying quality bonuses, providing health insurance/health care, transportation), improving the physical conditions of work, or improving the timing of the work itself. Mines *et al.* (1993) and Larson *et al.* (Chapter 18, this volume) document that the benefits provided by farm employers in the USA are limited. Some producers have tried to improve the job benefits provided to their workers in attempts to attract and retain workers, but benefit levels in the aggregate remain extremely low. There are also possibilities in terms of improving the conditions of farm work, again through the development of new technologies to reduce some of the most onerous (and physically demanding and often debilitating) aspects of farm work. Finally, improvements in the time dimension of farm work may be possible. This strategy has at least two components: dealing with the problem of a heavy time-intensity of work and either lengthening the amount of time needed on the farm to year-round or coupling seasonal farm work with another form of (non-farm) work to allow farm workers the opportunity to earn a sufficient income. In the USA, the service industry hires workers away from agriculture. Reasons given for this are that service-industry jobs are less demanding in terms of physical work, more 'reasonable' in terms of on-the-job time requirements, and more typically year-round (Findeis and Chitose, 1994).

Technological change could reduce the physical demands of farm work, and growers should actively support the development of these technologies. Improving on-the-job time requirements is difficult, since even more workers must be found and hired. Perhaps the most promising (in the short run at least) is the better integration of seasonal farm work with other work, either off-farm or on the farm itself or in integrated food-processing facilities, a strategy that is already used by a number of producers. The problem here is one of integration of the work time requirements of multiple jobs, but under tight labour market conditions employers are more likely to provide greater cooperation in such efforts.

Finally, public policy provides a means of dealing with the seasonal labour issue in agriculture. While policy can help to support the long-term development of technologies to reduce dependence on labour, the more immediate policy prescription usually focuses on immigration policy. From an economic perspective, as long as differentials exist across labour markets between countries and even between regions, the economic incentive will be there for migration of labour to the market yielding the higher economic return. However, the root causes of migration, and particularly international migration, are an issue highly debated in the social science literature (Massey *et al.*, 1993). What is clear is that an immigration policy that allows in-migration of farm workers will continue to depress wages, likely retard technology development (that would benefit both agriculture and farm workers), may negatively influence farm production in the *origin* country or region (Ortega-Sanchez, 2001), and potentially lead to other undesirable consequences, depending on the specific dimensions of the policy (see Taylor *et al.*, 1997). In the USA, for example, many employers of seasonal workers support legislation for a guestworker programme to support agriculture. Many others oppose this type of programme, again because of the (negative) impacts on the returns to work, the rate of development of technology and the integration of immigrant families into local communities (versus mobile solo-male workers who are only temporarily in the USA to work on farms), among other concerns voiced throughout this book.

Full-time year-round workers

Another group of farm workers that warrant attention are those who work for hire throughout the year. This particular segment of the farm worker population is becoming more important in some regions and countries, in part because of the growth in the average size of farms.

The issues for the full-time hired farm workforce have some similarities to those for the seasonal labour force, although in many respects the overall situation is very different, at least in the USA. Firstly, the very fact that full-time workers are employed throughout the year means that they have continuous employment and therefore have higher overall annual earnings. One caveat is in order here, however, and that is in cases where the in-migrants fill 'full-time positions' but workers leave their jobs for part of the year to return to the origin country or region to be with family. Individual workers may not work full-year, despite the full-time year-round nature of the 'position'.

A second very significant difference between the full-time and seasonal workforces is the number of workers that are hired. Many farms hiring seasonal workers (for example, those in California and Florida) require large workforces. The number of full-time labourers needed in agriculture is smaller in the USA, although the seasonal/full-time breakdown is clearly dependent on the type of agriculture undertaken in a country or region.

Employers of full-time year-round workers face similar issues in terms of attracting and maintaining a workforce, i.e. low wages, long working hours, weak job benefits. Therefore, it is not surprising that farmers hiring full-time workers are also looking for ways to ensure a stable workforce. Again, where unemployed and under-employed labour exists outside of the farm production sector, agriculture can absorb local labour and therefore benefit communities. Conversely, under tight labour markets, employers need to develop strategies for recruitment and retention.

Finally, in the USA, reliance on immigrant labour has been a long-time trend for seasonal labour in states such as California, Florida, Texas, Washington and similar states and production regions. What has recently changed is that other areas of the USA that are not as directly connected to the major migrant worker streams and that also have a dependence on full-time year-round farm workers (e.g. dairy) are now hiring Hispanic farm workers in greater numbers than ever befo

Casual labour

A third category of labour is 'casual' hired labour, i.e. workers employed only on a casual basis, either part-time throughout the year or seasonally. The amount of labour time is typically small but for smaller and even some mid-size farms this type of labour helps to meet farm labour needs. Farms have utilized labour of this type to help out when labour requirements on the farm are higher than can be supplied by farm family labour alone, or to allow some time flexibility for the farm family. Small and often mid-size farms have been particularly dependent on casual labour, especially those farms that do not hire full-time year-round workers.

Farm families in the USA have found that this particular labour source is becoming increasingly scarce. In a study of labour use on Pennsylvania farms, farm operators reported that they were finding it increasingly difficult to find part-time or 'casual' workers – the farm hand who could help out on a part-time basis, sons of neighbouring farm families who wanted work to acquire farm skills (or money) useful for later operating their own farms, and custom workers willing to work when needed. Common reasons for these problems included the general lack of young people wanting to go into farming as an occupation, in part due to smaller family sizes but principally due to concerns over the future profitability of farming (Findeis and Chitose, 1994). The rapid growth of the service industries in the USA that provide alternative job opportunities to unskilled workers (including local high school students and college students during the summer months) was also 'blamed' for the decline in workers willing to provide the supplementary part-time labour that some farms need.

...ponse to a Changing ...n Work force

...e of the emerging issues ... workers (as well as to workers in the agricultural processing or agri-food sector) was the relationship between workers and local communities. Changes in the composition of the hired farm workforce can affect the communities in which farm workers live and work. When unemployed local labour is hired by farms, employment in the farm sector can result in very significant benefits for the local community, since the incidence of unemployment is reduced. Alternatively, locally supplied labour may not be available to do farm work, due to the attractiveness of alternative job opportunities relative to farming or due to the large labour needs of some forms of agriculture. In this case, labour from outside the local labour market is required. This is where problems have arisen in the USA, where immigrant labour has been widely used to work in agricultural production. Although labour from Latin America (principally Mexico) has long been the reality in a number of the dominant states in the southwest and southeast USA, these issues intensified in the 1990s due to local labour shortages in other areas across the USA. Food processors and farmers in areas of the country that typically hired from the domestic labour pool increasingly turned to immigrant labour to supply their labour needs. These actions in some cases created conflicts between local communities, agricultural employers and immigrant farm workers, particularly when communities viewed the change with suspicion or even hostility. Even when communities were accepting of the change and no overt problems arose, issues of how local communities could provide public and private services to the new workers, and in some cases their families, were of concern.

The characteristics of workers hired from outside the local labour market as well as the labour needs of farms hiring these workers can importantly influence the impacts on local communities (see Chapter 6). Several dimensions of this issue are worth mentioning. Firstly, farms can attract a permanent labour force that will settle in the region, or can attract a labour force that is temporary, typically migrant. This depends on the farm work that is being done but again also depends on the ability of local labour markets to provide off-season employment for farm workers or provide employment opportunities for other family members. To the extent that farm workers and their families from outside the local community take what are perceived as local jobs needed by local residents, they may be viewed as a threat. Viewed from another perspective, many farm workers and family members are willing to take jobs that others in the community may not be willing to take. When a more migratory workforce is attracted, issues that arise include education for children who live in the local community only part of the year, the continuity of health care, and the provision of temporary housing, among other issues. For a permanent workforce, issues of the provision of services are also present, but vary depending on the extent to which the work force is poor and illegal. Being migrant further complicates this problem.

Secondly, the size of the farm workforce may make a difference. That is, if the in-migrant population needed to support agriculture is relatively small, the impacts on the local community can be quite different than if a large workforce is hired. For example, Chapter 17 describes the migrant and seasonal workforce in the Alaska seafood fishing and processing industry, noting that small rural Alaskan communities increase very substantially in size when the migrant workforce arrives each year to support this industry. This influx of a large number of workers into otherwise small, geographically remote communities that are generally without well-developed community services further burdens local service providers. For health care provision, this issue is particularly critical because of the hazardous nature of the work being done (Chapter 16).

Finally, the demographics of the workforce hired from outside the local community will likely affect community acceptance and community effects. Issues of race/ethnicity may exist, a point raised in Chapter 9 (see also Luchok and Rosenberg, 1997). Community response will also differ depending on whether solo male farm workers are hired, with their families residing elsewhere (an increasingly frequent situation in the USA), or if worker families also reside locally. Chapter 4 contends that family settlement is more stable, resulting in fewer negative effects. The impacts may also be more or less severe depending on the extent to which farm workers are in poverty.

The Critical Nature of Policy

Under the conditions described above, the question becomes: what labour policies make the most sense? What policies support agriculture while improving the well-being of farm workers? That is, can policy help to provide farm employers with the workforce that they require while simultaneously making changes in the conditions of farm work and the economic returns to this work to improve the lives of hired farm labour?

One approach is to view the farm worker problem as a problem inherent in entry-level unskilled work – it is a starting place. The view that farm workers should be helped to move out of farm work eventually into other better employment opportunities has helped to guide policy in some countries. This means that the basic assumption is that the conditions and economic returns of farm work cannot be improved but rather that the worker's ability to move out of agriculture is the path to follow. The hope is that, over time, former farm workers and their families will become assimilated into the larger economy and leave poverty behind (Emerson, 1984).

A second approach is based on the observation that many farm workers are never able to move out of agriculture. Thus, more liberal immigration policies that allow the entry of more workers for agriculture, such as guestworker programmes, will simply result, under this view, in simply more poverty (Taylor *et al.*, 1997). A guestworker programme serves to keep wages and benefits low, and puts less pressure on employers to improve the economic returns to and conditions of farm work. The maintenance of a low-wage workforce in agriculture also has the indirect effect of slowing the development of new technologies that could reduce the need for hired farm workers or at least improve the safety of farm work.

Finally, other policies are possible, including changes in the minimum wage or relaxing restrictions on welfare or public support receipt, among a range of other potential policies to ameliorate poverty. The policy choices exist and the effects are critical for a population that supports the agricultural sector upon which society depends.

References

Ahearn, M., Yee, J., Ball, E. and Nehring, R. (1997) Agricultural productivity in the United States. *Agricultural Information Bulletin No. 740.* US Department of Agriculture, Washington, DC.

Antle, J. (1986) Aggregation, expectation, and the explanation of technological change. *Journal of Econometrics* 33, 213–236.

Barkley, A. (1990) The determinants of the migration of labor out of agriculture in the United States, 1940–85. *American Journal of Agricultural Economics* 72, 567–588.

Barthelemy, P.A. (2001) Changes in agricultural employment. In: *The European Commission, Agriculture, Environment, Rural Development – Facts and Figures.* http://europa.eu.int/comm/agriculture/envir/report/en/emplo_en/eport_en.h

Benjamin, C., Corsi, A. and Guyomard, H. (1996) Modelling labour decisions of French agricultural households. *Applied Economics* 28, 1577–1589.

Binswanger, H. (1974) The measurement of technical change biases with many factors of production. *American Economic Review* 64, 964–976.

Bollman, R. (1994) Agriculture's Revolving Door. In: *Canadian Agriculture At a Glance* (Catalog 96-301). Statistics Canada, Ottawa.

Bryden, J., Bell, C., Gilliatt, J., Hawkins, E. and MacKinnon, N. (1992) *Farm Household Adjustment in Western Europe 1987–1991.* Final Report to the Arkleton Trust (I-200,1750), Oxford, UK.

Corsi, A. and Findeis, J. (2000) True state dependence and heterogeneity in off-farm labour participation. *European Review of Agricultural Economics* 27, 127–151.

Emerson, R.D. (ed.) (1984) *Seasonal Agricultural Labor Markets in the United States.* The Iowa State University Press, Ames, Iowa.

Fasterding, F. (1997) Projection of the structure of labour input in Germany's agriculture. *Landbauforschung Volkenrode* 47, 135–145.

Findeis, J. (1998) *Labor Adjustment in Agriculture: Implications of Policy Reform in the NAFTA Countries.* Report to the Organization for Economic Cooperation and Development OECD, Paris.

Findeis, J. (2002) Penn State Survey of U.S. farm women. Paper presented at the USDA Agricultural Outlook Forum 2002, February, Washington, DC.

Findeis, J. and Chitose, Y. (1994) *Hired Farm Labor: US Trends and Survey Results for Pennsylvania.* AE & RS 244. Department of Agricultural Economics and Rural Sociology, The Pennsylvania State University, University Park, Pennsylvania.

Findeis, J. and Lass, D.A. (1992) Farm operator off-farm labor supply and hired labor use on Pennsylvania farms. Paper presented at the 1992 Annual Meeting of the American Agricultural Economics Association, Baltimore, Maryland.

Frohberg, K. (1994) Assessment of the effects of a reform of the common agricultural policy on labor income and outflow. *European Economy* 5, 179–206.

Fuller, A. (1990) From part-time farming to pluriactivity: a decade of change in rural Europe. *Journal of Rural Studies* 6, 361–373.

Gardner, B. (1992) Changing economic perspectives on the farm problem. *Journal of Economic Literature,* 62–101.

Gasson, R. (1986) Part-time farming: strategy for survival? *Sociologia Ruralis* 24, 364–376.

Gasson, R., Crow, G., Errington, A., Hutson, J., Marsden, T. and Winter, D. (1995) The farm as a family business: a review. Reprint from *Journal of Agricultural Economics* (1988). In: Peters, G.H. (ed.) *Agricultural Economics, International Library of Critical Writings in Economics.* Edward Elgar Publishing, Aldershot, UK, pp. 3–43.

Gunter, L.F., Jarrett, J.C. and Duffield, J.A. (1992) Effects of US immigration reform on labor-intensive agricultural commodities. *American Journal of Agricultural Economics* 74, 897–906.

Hallberg, M., Findeis, J. and Lass, D. (eds) (1991) *Multiple Job-Holding Among Farm Families.* Iowa State University Press, Ames, Iowa.

Huffman, W. and Evenson, R. (1989) Supply and demand functions for multiproduct US cash grain farms: biases caused by research and other policies. *American Journal of Agricultural Economics* 71, 761–773.

Kada, R. (1980) *Part-Time Farming: Off-farm Employment and Farm Adjustments in the United States and Japan*. Center for Academic Publications, Tokyo, Japan.

Kang, W. and Maruyama, Y. (1992) The effects of mechanization on the time allocation of farm households: a case study in Hokuriku, Japan. *Journal of Rural Economics* 64, 119–126.

Luchok, K.J. and Rosenberg, G. (1997) Steps in meeting the needs of Kentucky's migrant farmworkers. *Journal of Agromedicine* 4, 381–386.

Martin, P.L. (1988) *Harvest of Confusion: Migrant Workers in US Agriculture*. Westview Press, Boulder, Colorado.

Martin, P.L., Huffman, W.E., Emerson, R., Taylor, J.E. and Rochin, R.I. (1995) *Immigration Reform and US Agriculture*. Publication No. 3358. University of California, Division of Agriculture, Oakland, California.

Massey, D., Arango, J., Hugo, G., Kouaouci, A., Pellegrino, A. and Taylor, J.E. (1993) Theories of international migration: a review and appraisal. *Population and Development Review* 19, 431–466.

Mines, R., Gabbard, S. and Samardick, R. (1993) *US Farmworkers in the Post-IRCA Period: Based on Data from the National Agricultural Workers Survey (NAWS)*. Research Report No. 4. US Department of Labor, Office of the Assistant Secretary for Policy, Office of Program Economics, Washington, DC.

Mishra, A. and Goodwin, B. (1997) Farm income variability and the supply of off-farm labor. *American Journal of Agricultural Economics* 79, 880–887.

OECD (1994) *Farm Employment and Economic Adjustment in OECD Countries*. Organization for Economic Co-operation and Development, Paris.

OECD (2001) *Agricultural Policy Reform and Farm Employment*. Paper AGR/CA/APM (2001) 10/Final. Organization for Economic Co-operation and Development, Paris (paper by J.L. Findeis, with contributions by C. Weiss, K. Saito and J. Anton).

Olfert, M.R. (1993) Off-farm labour supply with productivity increases, peak period production and farm structure impacts. *Canadian Journal of Agricultural Economics* 41, 491–501.

Oluwole, T. (2000) An econometric analysis of off-farm labor participation and farm exit decisions among US farm families, 1977–1998. MS Thesis, Department of Agricultural Economics and Rural Sociology, The Pennsylvania State University, University Park, Pennsylvania.

Ortega-Sanchez, I. (2001) Labor out-migration impacts on agrarian economies: the case of Central and Southern Mexican communities. PhD Dissertation, Department of Agricultural Economics and Rural Sociology, The Pennsylvania State University, University Park, Pennsylvania.

Pfeffer, M. (1989) Part-time farming and the stability of family farms in the Federal Republic of Germany. *European Review of Agricultural Economics* 16, 425–444.

Post, J. and Terluin, I. (1997) The changing role of agriculture in rural employment. In: Bollman, R. and Bryden, J. (eds) *Rural Employment: an International Perspective*. CAB International, Wallingford, UK, pp. 305–326.

Rothenberg, D. (1998) *With These Hands: the Hidden World of Migrant Farm Workers Today*. Harcourt Brace, New York.

Schmitt, G. (1991) Why is the agriculture of advanced western economies still organised by family farms? Will this continue to be in the future? *European Review of Agricultural Economics* 18, 443–458.

Taylor, J.E. and Martin, P. (1997) The immigrant subsidy in US agriculture: farm employment, poverty and welfare. *Population and Development Review* 23, 855–874.

Taylor, J.E., Martin, P. and Fix, M. (1997) *Poverty Amid Prosperity: Immigration and the Changing Face of Rural California*. Urban Institute Press, Washington, DC.

Tokle, J. and Huffman, W. (1991) Local economic conditions and wage labor decisions of farm and non-farm couples. *American Journal of Agricultural Economics* 73, 652–670.

von Meyer, H. (1997) Rural employment in OECD countries: structure and dynamics of regional labour markets. In: Bollman, R.D. and Bryden, J.M. (eds) *Rural Employment: an International Perspective*. CAB International, Wallingford, UK, pp. 3–21.

Weersink, A., Nicholson, C. and Weerhewa, I. (1998) Multiple job-holding among dairy farm families in New York and Ontario. *Agricultural Economics* 18, 127–143.

Weiss, C. (1997) Do they come back again? The symmetry and reversibility of off-farm employment. *European Review of Agricultural Economics* 24, 65–84.

Weiss, C. (1998) *Farm Employment Adjustment and Recent Policy Reform: a Review of Empirical Evidence for Europe*. Report to the Organization for Economic Co-operation and Development. OECD, Paris.

Weyerbrock, S. (1998) Reform of the European Union's common agricultural policy: how to reach GATT-compatibility? *European Economic Review* 42, 375–411.

Woldehanna, J., Lansink, A. and Peerlings, J. (2000)

Off-farm work decisions on Dutch cash crop farms and the 1992 and Agenda 2000 CAP reforms. *Agricultural Economics* 22, 163–171.

Zweimuller, J., Pfaffermayr, M. and Weiss, C. (1992) Farm income, market wages and off-farm labour supply. *Empirica* 18, 221–235.

2 Examining Farm Worker Images

Susan Gabbard, Alicia Fernandez-Mott and Daniel Carroll[2]
[1]Aguirre International, San Mateo, California, USA; [2]US Department of Labor, Washington, DC, USA

Abstract

Popular images of farm workers are often outdated or correspond to only a small portion of the farm labour force. This is not surprising, given the fast turnover of the farm workforce and the periodic changes in its make-up. For the USA, images include families following the crops, parents and children working in the fields together and, more recently, large numbers of people entering the USA illegally. This chapter presents a snapshot of farm workers based on findings from the National Agricultural Workers Survey (NAWS). Using data from 12,375 interviews conducted during fiscal years 1993–1998, this chapter describes farm worker characteristics by family type and migrant status. In doing so, it examines the validity of popular perceptions and explores the implications of the current findings for farm work service programmes. A central challenge to administrators of both public and private programmes may be how to adjust service delivery activities to a farm worker population that consists increasingly of young, single males, many of whom are recent immigrants and first-time farm workers.

Why Discuss Farm Worker Images?

Popular images frequently play an important role in shaping public opinion on important issues, especially when facts, based on defensible data, are not well known. In some cases, popular images partially or completely substitute for facts. Such may be the case regarding knowledge about farm workers generally and about migrant farm workers specifically. For many people, images or stereotypes are a shorthand way to think about farm workers. Some images are carried forth and perhaps enhanced in popular culture and the media. For the USA, they include: families following the crops; parents and children working in the fields together; and farm workers illegally crossing the border.

While most of these images have or had some truth in the past, some are now outdated or they currently correspond to only a small portion of the farm labour force. This is not surprising given the rapid turnover in the farm labour force that leads to regular changes in the make-up of those who do farm work in the USA. Whether they are outdated or misplaced, images are not easily changed. Even when new images emerge, others linger as we cling to familiar perceptions.

This chapter has two purposes. The first is to review some of these commonly and strongly-held perceptions about farm workers in the USA and to examine their validity by comparing them with findings from the National Agricultural Workers Survey (NAWS). This analysis shows that some prevalent images do not match recent findings, while others more accurately reflect the data. The chapter's second purpose is to examine the implications of changing farm worker demographics for research direction and programme development. Here the discussion draws upon two major distinctions that have relevance for farm worker programmes: migrant status and family type.

Data and Methodology

The NAWS data presented here come from 12,375 interviews conducted during federal fiscal years 1993–1998 with a nationally representative sample of hired crop workers. The NAWS samples the population of all hired crop farm workers, including those who perform seasonal services within year-round employment. Each year between 1993 and 1998, more than 2000 crop workers across the USA were randomly selected and interviewed. (In the fiscal year 1999, NAWS began to interview approximately 4000 workers per year.) A multistage sampling procedure is designed to account for seasonal and regional fluctuations in the level of farm employment.

Seasonal fluctuations in agricultural employment are captured by three interviewing cycles that last for 10–12 weeks each. Cycles begin in February, June and October. The number of interviews conducted during a cycle is proportional to the amount of crop activity at that time of the year, which is approximated using administrative data from the Bureau of Labor Statistics and the Census of Agriculture. All states in the continental USA are divided into 12 regions, aggregated from the 17 agricultural regions used by the US Department of Agriculture (USDA). Within these regions, a roster of 47 Crop Reporting Districts (CRDs) containing 288 counties was selected. For each cycle, no fewer than two CRDs were selected randomly for each region.

Multistage sampling is used to choose respondents in each cycle. The number of sites selected is also proportional to the amount of farm work being done during the cycle. The likelihood of a site being selected varies with the size of its seasonal agricultural payroll. Because states such as California and Florida have relatively high agricultural payrolls throughout the year, even a random selection process results in several CRDs in these states being selected for interviews during each cycle. Within each CRD, a county is selected at random. Farm employers within each of the selected counties are chosen randomly from public agency records. Principal among these are unemployment insurance files, Agricultural Commissioners' pesticide registrations, and lists maintained by the Bureau of Labor Statistics and various state agencies. The availability of these data varies by state. NAWS staff annually review and update these lists in the field.

Once the sample is drawn, NAWS interviewers contact the selected agricultural employers, explain the purpose of the survey, and obtain access to the work site to schedule interviews. Interviewers then go to the farm, ranch, or nursery, explain the purpose of the survey to workers, and ask a random sample of them to participate. Interviews are conducted in the worker's home or at another location of the worker's choice (US Department of Labor, 2000).

Migrant Status

The NAWS defines migrants as individuals who travel over 75 miles to obtain a farm job. This broad definition encompasses international and domestic migrants and these groups, in turn, comprise several subgroups that are defined based on the number of farm jobs held, place of residence while working, and date of first entry into the USA. International migrants include: (i) shuttle migrants (workers who move back and forth from their foreign country home base – usually Mexico – where they live for about half of each year, to an area of US farm employment); (ii) follow-the-crop migrants (those who cross an international border and then travel to at least two farm jobs that are more than 75 miles apart); and (iii) newcomers (foreign-born workers whose first entry into the USA occurred less than 12 months prior to the NAWS interview). Some newcomers will settle in the USA, while others will become shuttle or follow-the-crop migrants. Domestic migrants may also be shuttle or follow-the-crop migrants. A worker who resides in a location that is more than 75 miles from any of their US farm jobs would be a domestic shuttle migrant; a worker who had at least two farm jobs, each more than 75 miles apart, would be a follow-the-crop migrant.

A review of NAWS data shows that the number of migrants increased between 1989 and 1998. Migrants comprised 43% of the farm labour force in 1989/90, compared with 51% in 1993–1998. The increase is attributed to the change in the share of crop workers who are international newcomers. In 1989/90, international newcomers comprised 4% of the crop farm labour force, compared with 16% in 1993–1998. Contrary to what was popularly believed 10 years ago, the 1986 Immigration Reform and Control Act (IRCA) did not cause a settling of the farm labour force, despite the fact that nearly 1.1 million undocumented persons were legalized under the Special Agricultural Worker (SAW) provision of IRCA. Rather, events since then seem to have favoured increased migration, particularly international migration. Table 2.1 shows the shares of farm workers by farm worker type for the period 1993–1998. The designation 'settled' means non-migrant.

Family Type

Two family types – those with no dependents (singles) and those who were parents of minor children – accounted for 82% of the US farm labour force in 1993–1998. The remainder of the farm labour force was made up of a variety of family types, mostly couples without dependent children (Table 2.1).

Table 2.1. Farm worker and family types in the National Agricultural Workers Survey, fiscal years 1993–1998. (Source: National Agricultural Workers Survey (NAWS) data, fiscal years 1993–1998.)

Characteristics	Per cent
Type of farm worker	
Settled	49
International newcomer	16
International shuttle	15
International follow-the-crop	4
Domestic shuttle	8
Domestic follow-the-crop	8
Type of family	
Single	42
Parent of minor(s)	40
Other	18

The most interesting finding here was that 42%, or approximately two in five farm workers, were single. This figure has doubled since 1989, when single workers made up 21% of the farm labour force in the USA. Parents of minor children accounted for the second largest group, comprising 40% of all crop workers. The share of parents, however, declined from the 1989 level, when they comprised 51% of all crop workers (US Department of Labor, 1991). The remaining 18% of farm workers encompassed a variety of family types. Most of these workers were part of a couple with no dependent children.

Another interesting finding was the increase in the share of farm workers who, at any one time, were not residing with their families while working. The NAWS classifies these workers as 'unaccompanied'. In 1990, 43% of all farm workers were unaccompanied, compared with 63% in 1993–1998. This represents nearly a 50% increase between the two time periods.

Combining migrant status with the three family types – singles, parents of minor children, and others – resulted in six groups that seemed to represent large numbers of farm workers with similar characteristics. The six groups are: migrant parents (22%); migrant singles (24%); migrant other (5%); settled parents (18%); settled singles (18%); and settled other (13%).

Four of the six groups – 'migrant parents', 'migrant singles', 'settled parents' and 'settled singles' – each comprised approximately one-fifth of the crop labour force. The groups 'migrant other' and 'settled other' comprised the remaining one-fifth. As these latter groups included a variety of family types, this discussion focuses on singles and parents. The demographic and employment variables for all groups are included in Table 2.2 for migrants and Table 2.3 for settled workers.

After tentatively defining these groups, a review of various characteristics of farm

Table 2.2. Characteristics of all farm workers and migrant farm workers based on the National Agricultural Workers Survey, fiscal years 1993–1998. (Source: National Agricultural Workers Survey (NAWS) data.)

Variable	All respondents	Migrant		
		Parents	Singles	Other
Frequency	12,375	2,701	3,010	563
Percentage of respondents	100	22	24	5
Years				
Median age	30	33	22	33
Mean highest grade completed	7	6	7	6
Median highest grade completed	6	6	6	6
Percentage of respondents				
Female	20	13	7	17
Foreign born	76	93	90	89
Hispanic	83	98	96	93
Non-native English speaker	82	98	97	96
US citizen	26	9	10	14
New farm worker	30	32	55	29
US dollars (nominal $)				
Mean wage	5.73	5.69	5.49	5.61
Median wage	5.25	5.15	5.14	5.15
Percentage of respondents				
Paid below minimum wage	7	9	11	8
Paid by the hour	76	70	71	72
Paid by the piece	21	28	27	27
Provided health insurance	7	4	4	3
Paid vacation	10	3	2	5
Employed by contractor	22	27	29	30
Weeks during last 12 months				
Mean farm work weeks	25	21	20	21
Mean weeks abroad	11	21	23	19
Mean non-work weeks	10	6	5	8
Mean non-farm work weeks	5	4	4	4
US dollars (nominal $)				
Median personal income	5,000–7,499	5,000–7,499	2,500–4,999	5,000–7,499
Median farm work income	5,000–7,499	5,000–7,499	2,500–4,999	5,000–7,499
Median family income	7,500–9,999	7,500–9,999	2,500–4,999	7,500–9,999
Percentage of respondents				
Income below US poverty level	57	75	75	60
Used public aid	22	24	63	17
Received AFDC[a]	2	2	0	1
Received Food Stamps	13	17	5	12
Received WIC[b]	10	12	1	3

[a]Received Aid to Families with Dependent Children (AFDC) programme support.
[b]Received Women, Infants and Children (WIC) programme support.

Table 2.3. Characteristics of all farm workers and settled farm workers based on the National Agricultural Workers Survey, fiscal years 1993–1998. (Source: National Agricultural Workers Survey (NAWS) data.)

Variable	All respondents	Settled Parents	Settled Singles	Settled Other
Frequency	12,375	2,287	2,193	1,621
Percentage of respondents	100	18	18	13
Years				
Median age	30	34	25	35
Mean highest grade completed	7	7	9	8
Median highest grade completed	6	6	9	7
Percentage of respondents				
Female	20	40	20	26
Foreign born	76	71	53	52
Hispanic	83	78	67	59
Non-native English speaker	82	76	64	58
US citizen	26	33	48	50
New farm worker	30	8	24	20
US dollars (nominal $)				
Mean wage	5.73	6.16	5.65	5.77
Median wage	5.25	5.75	5.25	5.37
Percentage of respondents				
Paid below minimum wage	7	4	5	4
Paid by the hour	76	79	82	84
Paid by the piece	21	16	15	13
Provided health insurance	7	13	10	10
Paid vacation	10	23	13	17
Employed by contractor	22	15	18	14
Weeks during last 12 months				
Mean farm work weeks	25	31	27	27
Mean weeks abroad	11	0	0	0
Mean non-work weeks	10	14	15	17
Mean non-farm work weeks	5	5	9	6
US dollars (nominal $)				
Median personal income	5,000–7,499	10,000–12,499	7,500–9,999	7,500–9,999
Median farm work income	5,000–7,499	7,500–9,999	5,000–7,499	7,500–9,999
Median family income	7,500–9,999	12,500–14,999	7,500–9,999	12,500–14,999
Percentage of respondents				
Income below US poverty level	57	50	44	29
Used public aid	22	55	13	15
Received AFDC[a]	2	5	1	1
Received Food Stamps	13	26	10	8
Received WIC[b]	10	32	2	2

[a]Received Aid to Families with Dependent Children (AFDC) programme support.
[b]Received Women, Infants and Children (WIC) programme support.

workers and their jobs was undertaken to see if these groups did indeed represent distinct subpopulations of farm workers. For most characteristics, the groups varied significantly from each other. The values for these subpopulations were then compared with the mean values for all farm workers. In some cases, the mean value for 'all farm workers' was not significantly different from the subpopulation means. In many cases, however, the means for subpopulations and 'all farm workers' differed significantly.

Single Farm Workers

Single migrants

Single migrants comprised the largest group of farm workers: they made up almost one-fourth of all farm workers and almost half of the migrants. This group was almost exclusively male (93%), Hispanic (96%), foreign-born (90%) and non-native English speaking (97%). Like most farm workers, they had a median of 6 years of formal education. More than half (55%) of the single migrants were new to US farm work (less than 1 year of experience), comprising 45% of all newly recruited farm workers in 1993–1998. The majority did not come from a farm background: only one in three single migrants had a parent who worked in US agriculture. This was the highest percentage of first-generation crop workers among the four categories. Overall, this group was the youngest. Half were less than 22 years old and three-quarters were less than 28 years of age. Approximately 8% of single migrants were minors who migrated without their families. The NAWS classifies these minor workers as *de facto* emancipated youth.

Overall, given that they were young and inexperienced and did not have a farm work background, it was not surprising that this group had the lowest wages. Their median hourly wage was US$5.14 per hour and their average wage was US$5.49 per hour. Eleven per cent reported being paid below the minimum wage. Like other migrants, more than one-quarter (27%) were paid by the piece and fewer than 5% received job benefits such as health insurance or paid days off. Like other migrants, about three in ten (29%) worked for farm labour contractors.

Single migrants worked the fewest number of weeks per year in the USA – finding just 20 weeks of farm work and 4 weeks of non-farm work. When not working, single migrants lived abroad, spending only 5 weeks in the USA looking for work or between jobs. Low wages combined with few weeks of work resulted in the lowest earnings of any group. Half of the single migrants earned less than US$5000 in US wages, or an average of US$208 per week.

Settled singles

While migrant singles were a fairly homogeneous group, settled singles were much more diverse. This group made up nearly one-fifth (18%) of the farm workforce. In general, members of this group were not newcomers to farm work; only 24% were in their first year of farm work and half had a parent who had worked in US agriculture.

Settled singles were predominantly male (80%) and somewhat older than migrant singles. Half were below the age of 25 and another one-fourth were between 28 and 40 years of age. Twenty-five per cent of settled singles were over 40 years of age, compared with just 8% of migrant singles. This group also included a small percentage of minors (6.5%) who did not live with their parents.

This group contained the highest share of US-born farm workers (47%). Among the foreign-born, one-third were naturalized citizens and one-sixth were lawful residents. About two-thirds of settled singles were Hispanic (67%) and a similar share were non-native English speakers (64%). This group had the highest education levels, with a median of 9 years of education, compared with 6 years for most other groups.

Settled singles were somewhat better paid and worked more than migrant singles. Their median hourly wage was US$5.25 and their mean hourly wage was US$5.65, compared with US$5.14 and US$5.49, respectively, for migrant singles. This group was half as likely as migrant singles to report being paid less than the minimum wage. Like other settled workers, they were paid by the piece about 15% of the time. Even though they had few job benefits, settled singles were more likely to receive benefits than migrants; about one in ten had health insurance or paid time off. Only 18% of settled singles worked for farm labour contractors, compared with 29% of migrant singles.

Settled singles worked and earned more than migrant singles. On average, they were employed 27 weeks in farm work. Furthermore, settled singles were the most successful at combining farm work and non-farm work; they worked in non-farm jobs for 9 weeks, giving them a total of 36 weeks of employment. Settled singles were out of work for about 15 weeks and spent less than 1 week abroad. As a result of higher wages and more weeks of employment, settled singles earned more than migrant singles; median farm earnings were less than US$7500 and median off-farm earnings were less than US$2500 for a total median income of less than US$10,000. While 44% of settled single farm workers had incomes below poverty-level, only 13% used needs-based public assistance – mostly food stamps.

Parent Farm Workers

Migrant parents

Migrant parents made up 22% of the farm labour force and 43% of the migrants. This group accounted for 23% of all new entrants and about four in ten had parents who also worked in US agriculture. Migrant parents tend to be older than single farm workers, with a median age of 33 years. They were mostly fathers, although one in six was a mother. Like other migrants, migrant parents were predominantly foreign-born (93%), Hispanic (98%) and non-native English speakers (98%).

Three-fifths of migrant parents reported having three or fewer children, 83% of which were under 14 years of age. Only 6% of the children of migrant parents were employed in agriculture and 56% of these were between 16 and 17 years of age. Over half (53%) of the children of migrants were left behind when their migrating parent, usually the father, moved. A significant share of the school-age children of migrants did not attend school. While 82% were attending school and two-thirds were performing at grade level, 18% had not attended school within the last 12 months.

Migrant parents were similar to single migrants in terms of wages and working conditions, although migrant parents had slightly higher earnings. Half of migrant parents made less than US$7,500 in US wages, and the median family income was less than US$10,000. While about 75% of migrant parent households had incomes below the poverty level, less than one-quarter (24%) of these families received federal assistance, mostly in the form of food stamps or Women, Infants and Children (WIC) programme support.

Until recently, the NAWS did not have a clear picture of how and when children travelled with their migrant parents. When the NAWS found international migrants with their children, it was clear that those children had migrated with their parents. On the other hand, when domestic migrants were residing with their children, it was harder to determine whether they were in a home-base location or if they had taken their children with them when they travelled to do farm work.

In fiscal 1998, the NAWS asked farm workers whether their spouse and/or children had migrated with them when they changed location. Analysis of these data revealed that while 29% of migrant parents were with their children at the time of the interview, only 14% of migrant parents said that they took their children with them when they migrated. Therefore, 86% of the time, children were left behind when a parent (usually the father) migrated. While migrant parents accounted for 22% of hired crop workers, migrant parents who took their children on the road with them accounted for just 3%. Moreover, migrant parents whose children worked with them in the fields accounted for less than 1% of hired crop workers.

Settled parents

Apart from age, settled parents were very different from migrant parents in terms of their demographics and working conditions (Table 2.3). The median age of settled parents was 34 years, compared with 33 years of age for migrant parents. About three-fourths of settled parents were foreign-born (71%), Hispanic (78%) and non-native English speakers (76%). As a group, settled parents were most likely to come from a farm work background; six in ten had a parent who had worked in agriculture in the USA. Settled parents were generally experienced farm workers; only a few had recently begun working in agriculture (8%). Many of these workers entered farm work before they became parents.

Female farm workers were more likely to be settled parents. Within the settled parent group, 40% were mothers and 60% were fathers. The relatively high percentage of women in this group resulted from two factors: (i) among both migrants and settled farm workers there were almost twice as many women parents as single women; and (ii) women were more likely to be settled than to be migrant.

Most settled parents (71%) reported having three or fewer children residing in the USA, four-fifths of which were under the age of 14 years old. Only 2% of the children of settled parents worked in agriculture and the majority of these (78%) were above the age of 14. Both school enrolment and performance were higher for the children of settled parents compared with the children of migrants; 93% attended school, with four-fifths performing at grade level. Still, 7% of the school-age children of settled farm workers had not attended school during the 12 months preceding the interview.

Settled parents had the highest wages of all the groups, with a median wage of US$5.75 and an average wage of US$6.16 per hour. They were less likely than migrant parents to be paid by the piece (16% versus 28%) and the most likely to have health insurance and paid days off. Still, only 13% of these workers had employer-provided health insurance and fewer than one-fourth (23%) had paid time off.

Settled parents had the steadiest incomes, averaging 31 weeks of farm work, 5 weeks of non-farm work and 14 weeks out of work. They also had the highest farm work earnings, with a median farm work income of less than US$12,500 per year. However, even with the earnings of additional household members, half of all settled parent households had total incomes of less than US$15,000, placing them below the poverty level. Correspondingly, about half of the settled parent families received federal assistance. Food stamps and WIC were the most common forms of assistance; about one-fourth and one-third of families, respectively, participated in these programmes.

Revisiting Popular Images and Exploring New Ones

Analyses of the NAWS data reveal that some farm worker images are inaccurate, while others are supported by survey findings. This section revisits the popular images presented at the beginning of this chapter and then presents new ones.

Families following the crops

This popular image corresponds to a relatively small share of crop workers in the USA. In 1993–1998, follow-the-crop migrants represented less than 12% of the hired crop workforce, while parents who took their children with them when they migrated accounted for just 3% of all hired crop workers. As discussed below, many more farm workers in 1993–1998 were international back-and-forth migrants, compared with follow-the-crop migrants.

Parents and children working together in the fields

Relatively few parent–children families work in the fields in the USA. Just 6% of migrant parents had children who worked, with the majority of the children being 14–17 years old. Likewise, only 7% of settled parents had children who worked and, again, most of these were 14–17 years old. As discussed above, parents today (particularly migrant fathers) are more likely to leave their children behind when they migrate. Those children who do work are more likely to be *de facto* emancipated teenagers than dependents of farm workers. Teenagers living away from their families made up half the youth aged 14–17 working in the fields.

Farm workers illegally crossing the border

In 1993–1998, 46% of all crop workers were unauthorized, compared with just 8% in 1989. As such, an image of unauthorized immigration is supported by the data. The popular image of illegal immigration, however, could be characterized as unidirectional – the familiar TV news clip portrays people running across the highway or, more recently, navigating the desert to get to the USA. A more accurate portrayal of immigration would include images of other migration patterns, such as back-and-forth migration and return migration. These other migration patterns are discussed below.

'New' farm worker images

An instructive farm worker image would convey important characteristics, such as gender, age, tenure working in US agriculture and household structure. An image that would correspond to nearly a quarter of all crop workers would be that of a 22-year-old solo male who recently immigrated to the USA (the characterization 'solo male migrant' was discussed in Mines and Kearney, 1982). In 1993–1998, single workers made up 42% of the hired crop workforce – double what they did in 1989. Among single farm workers, 54% were migrant males, meaning that single male migrants made up 23% of all crop workers in 1993–1998. When unaccompanied parents (i.e. those who leave their children behind when they migrate) are combined with the group of single farm workers, the share of farm workers who are unaccompanied at any one time rises to 64%. Taken as a whole, these findings show that many farm workers are men who are separated from family.

Periods of family separation, of course, vary depending on migration patterns. The NAWS was not designed to measure return migration rates: subjects are interviewed only once, and it is not possible to report on moves following the interview, although a study on return migration (Reyes, 1995) found that, on average, migrants from traditional sending states in West Central Mexico who worked in agriculture in the USA returned to their homes in Mexico after just 2 years; the same study found that only 2% of all migrants from West Central Mexico remained in the USA longer than 10 years. However, the NAWS does record the respondent's travel for the 12-month period *preceding* the interview. As indicated in Table 2.1, a significant share (15%) of crop workers in 1993–1998 were international shuttle migrants. While the share of this group remained stable at around 15% over the 6-year period examined here, the share of international newcomers increased by 45%, from 11% in 1993/94 to 16% in 1993–1998. It is not known how many international newcomers will become back-and-forth or return migrants.

An accurate and familiar image of the average farm worker is that of a low-paid, seasonally employed individual who experiences long periods when they are out of work. Today, much of the out-of-work time experienced by crop workers is invisible, as many of them are international shuttle migrants who spend much of their non-work time abroad. This may change if it continues to be difficult for workers to cross the US borders and NAWS findings of more unauthorized solo male migrants may be related to increased border enforcement. Persons who would normally return to their home country may be choosing to remain in the USA or to stay longer than normal as the risk and cost of entering the USA increases. The findings may also be simultaneously related to 'push' factors in sending countries.

In some ways, even the poverty resulting from low wages, unemployment and underemployment is invisible, as the vast majority of farm worker families do not receive public assistance. In 1993–1998, 75% of migrant parent households had incomes below the poverty level, but only one in three of these families received public assistance in the USA.

Implications and Additional Questions

The changing demographics of farm workers raises research and programme design questions. For example, how well are farm worker programmes serving current farm workers, and how should US federal government programmes respond to changes in the makeup of the farm labour force?

Many federal farm worker programmes focus on families. Migrant Head Start and Migrant Education, for example, both target services to meet the needs of migrant and seasonal farm worker families and their dependents. The budgets for these programmes are based in part on the number of persons who qualify for services. In addition to knowing the overall number of persons who qualify for these and similar federal programmes, it is important to understand the demographic characteristics of those who are eligible. The success of education programmes that target children, for example, will depend in part on how well existing information on the ages of children and on family work and migration patterns is used.

For families that experience separation, information about how long migrating parents are away from their children, and how families cope when parents are away, may also be important for guiding programme design. Similarly, for those parents whose children work with them, it is important to know whether the primary reason for their working is insufficient household income, lack of day care or adult supervision at home, the parent's and/or child's desire for the child to acquire work experience, or a combination of these and other factors. Programmes geared to serve youth, on the other hand, may need to be reviewed and altered to target the increasing share of teenage farm workers who live away from their families.

The findings presented here may also have implications for health delivery systems and workplace safety training. As the crop labour force is increasingly made up of young, single males, health clinics may need to modify their array of services to match the changes in the demographic and health characteristics of farm workers. Similarly, if the characteristics of this population pose special risk factors for occupational injury, then safety training for employees (particularly new hires) may become more important.

Historically, the farm labour force in the USA has had high turnover and agricultural employers' need for workers has been filled by a succession of groups. More and more, it is the task of researchers and programme staff to identify and respond to changes in the demographics of farm workers. Clinging to outdated images of farm workers can lead to out-of-step information and programmes. Accurate information about farm workers as a whole, as well as information about important subgroups within the farm labour force, is increasingly important in formulating appropriate policies and designing useful programmes.

References

Mines, R. and Kearney, M. (1982) *The Health of Tulare County Farm Workers*. State of California Health Department, California.

Reyes, B. (1995) *Return Migration from West Central Mexico*. California Institute for Public Policy, California.

US Department of Labor (1991) *Findings from the National Agricultural Workers Survey (NAWS) 1989*. Research Report No. 2. Office of the Assistant Secretary for Policy, Office of Program Economics, Washington, DC.

US Department of Labor (2000) *Findings from the National Agricultural Workers Survey (NAWS) 1997–1998*. Research Report No. 8. Office of the Assistant Secretary for Policy, Office of Program Economics, Washington, DC.

3 Changes in the Labour Intensity of Agriculture: a Comparison of California, Florida and the USA

Wallace E. Huffman
Department of Economics, Iowa State University, Ames, Iowa, USA

Abstract

This chapter examines changes in farm technology, farm labour (including hired and contract labour) and other inputs, and aggregated agricultural output and productivity for California, Florida and US agriculture from 1960 to 1996. Summaries for subperiods on either side of the US Immigration Reform and Control Act of 1986 are also presented. A major conclusion is that the relatively abundant and reliable supply of immigrant farm workers in the USA has undoubtedly slowed the discovery, development and adoption of mechanized technologies for agriculture, especially in vegetable and fruit production.

Introduction

The post Second World War period has brought dramatic reductions in labour use and in the labour intensity of US agriculture. Because of the heterogeneity across the USA of climate, soils and location relative to major urban markets, the changes have not been uniform. The states of California and Florida stand out because of their relatively high labour intensity, which is associated with large fresh fruit and vegetable production. The objective of this chapter is to examine changes in farm technology, farm labour (including hired and contract labour) and other inputs, and in aggregated agricultural output and productivity for California, Florida and US agriculture. Comparisons are also made with Iowa, which is the largest producer of agricultural products in the interior of the USA.

The chapter focuses on the period after 1960. It briefly reviews the development and adoption of new labour-saving technologies, biotechnology and innovations in organizational structures affecting the composition of farm labour, such as the shift from direct hire to farm labour contractors after the Immigration Reform and Control Act of 1986 (IRCA). The data are from a new US Department of Agriculture (USDA) data set on agriculture of the 50 states. They are used to summarize: growth in aggregate output, crop output and livestock output; changes in hired labour (including contract labour), capital services and intermediate input (materials) usage and in factor intensities; and total factor productivity (TFP) over the whole period 1960–1996. The chapter shows that different trends exist in farm labour use and labour intensity in agriculture in California and Florida relative to the whole USA and Iowa and that the differences are especially dramatic for 1986–1996 relative to 1976–1986.

Technology

Agriculture in the USA has undergone steady and seemingly relentless technological change during the 20th century, especially since 1950 (Gardner, 1992; Huffman and Moschini, 1999; Huffman and Evenson, 2001). In most of US agriculture, the adoption of increasingly mechanized techniques, the use of new chemical inputs (such as herbicides, insecticides and fertilizers), the availability of genetically improved crops and animals, and countless other technical and organization improvements have dramatically changed the way agriculture is practised. The coming of the new biotechnology era with Bt genes for selective insect resistance, selective herbicide-tolerant crop varieties, enhanced-trait maize and oilseed crops, new crops and pharmaceutical-producing farm animals has greatly changed the potential for biological advances and development of new products from agriculture. The widespread diffusion of new information technologies has accentuated the changing structure of agriculture (Huffman, 2002).

Research and development in the public and private sectors have been the sources of most of the new technologies for agriculture, although innovative farmers and local machine shops still make some important mechanical discoveries for agriculture (Huffman and Evenson, 1993; Fuglie *et al.*, 1996). A broad range of discoveries is required to sustain technical change in agriculture. Some discoveries are in the public sector, especially discoveries in the basic/general and pre-technology sciences; others, largely discoveries from applied research and development, are concentrated in the private sector.

The private sector has long been an important source in the production and marketing of innovations of a mechanical and chemical type, and it has been a major factor in shaping the structure of US agriculture. Mechanical and chemical innovations have allowed production techniques to be developed that economize on labour. In fact, the migration of labour from the agricultural sector generally to the rest of the economy has been the most striking feature of post-war US agriculture, especially over 1950–1970 (Gale, 1996). The trends in California have been somewhat different (Olmstead and Rhode, 1993).

Crop production

Agriculture is production by biological processes. Plant growth and development are very sensitive to day length, and crop production is land-surface area intensive. For non-greenhouse plants, day length and temperature trigger plant stages. Although the completion of any phase of crop production can sometimes be accelerated by using new technology, the timing from planting to harvest is largely unaffected. Furthermore, because of the use of large amounts of land surface area, mechanization where it occurs must be largely through the use of mobile power or machines that move through the fields (e.g. tractor-drawn maize planters, self-propelled combines). Packing and processing operations can be completed in the field or in packing/processing facilities where stationary power can be used (Chapter 4, this volume).

In vegetable and fruit production, major technical advances have been associated with drip irrigation, fertigation, plastic mulch and new varieties.

Drip irrigation

Drip irrigation is a water- and labour-saving way to irrigate plants. Hoses with regularly spaced drip holes are laid permanently at the centre of beds; when the water is turned on, the drip system delivers water at the root base of the plants. Water is not wasted between the rows or in evaporation as with flood, moving rig, or centre-pivot irrigation.

Fertigation

Fertigation uses the same drip irrigation hoses to deliver liquid fertilizer efficiently to the roots of plants. A farmer usually applies dry fertilizer before planting vegetables and then supplements during the growing season with fertigation. A positive externality of fertigation is reduced water pollution from leaching and runoff of agricultural chemicals.

Plastic mulch

In the production of vegetables and tomatoes, plastic mulch is frequently used with raised and rounded seedbeds. Long clear sheets of plastic are laid over the entire bed, pierced only where the young seedlings are planted. Plastic mulch reduces weeds, promotes growth (especially in hot-season plants such as tomatoes) and blocks micro-organisms from moving from the soil to the plant. The result is less need for hand weeding, herbicides, fungicides and other plant protection measures. Plastic raises the soil temperature, reduces water evaporation and increases the total photosynthetic activity of plants.

No-till farming

In dry-land farming, the gradual change from intensive seedbed preparation and cultivation to no-till farming has greatly reduced the demand for labour and some other inputs in major field crop production of the Midwest and South regions of the USA. The change in tillage practices started with the relatively high fuel prices of the mid-1970s and was speeded along by the soil conservation requirements of the US 1990 Farm Bill. The net impact on input demand with reduced intensity of tillage and number of field operations has been a reduced demand for labour, large horsepower tractors, mouldboard ploughs, heavy discs and fuel. These savings are partially offset by increased demand for chemical herbicides and specialized no-till planters.

Milk production

Historically, milk production has been relatively labour intensive, with year-round twice per day feeding and milking. The labour intensity of dairying has been reduced by mechanized milking, automated milk production and feeding records, automated feed distribution based on performance, and automated cleaning of dairy barns. Although totally automated milking systems exist that use electronic sensors, robotic milkers and video cameras, they have not been popular among US dairy farmers. Farms with large dairy herds have recently discovered that immigrant (largely Mexican) farm workers are more cost effective than totally automated milking systems.

Biotechnology

Significant advances in biological sciences associated with recombinant DNA make it possible for human ingenuity to create new living things that are not achieved by the works of nature. With strengthening of intellectual property rights (which started in 1970 with the Plant Variety Protection Act and continued in the 1980s with patent protection for living organisms, plants and non-human mammals), the private sector has found it profitable to undertake a larger share of US agricultural research. Although hybrid maize varieties have been developed largely by the private sector since 1940, the heavy focus of the private sector on development of new crop varieties in other areas has occurred only since 1986. For example, the private sector has become the major source in variety development for tomatoes, lettuce, soybean, other beans and peas (Huffman and Evenson, 1993; Fuglie *et al.*, 1996). Breeding for uniform ripening date has had major labour-saving advantages in melons, other fruits and vegetables because it reduces the number of 'field pickings', or facilitates mechanical harvesting of produce for processing.

Tomato production

Tomatoes are one of the large US fruit and vegetable crops and the technology has changed over time. The US tomato industry produces about 2 million short tons annually, divided between fresh market production, concentrated in Florida, and processed tomatoes, concentrated in California (Plummer, 1992; USDA, 1999b,c). Fresh market tomatoes are hand-picked and processed tomatoes are mechanically harvested (USDA, 1999b).

Fresh tomatoes

Fresh market tomato varieties have been developed that are medium sized, firm when purchased by the consumer, and flavourful when eaten. To reduce disease and insect-pest problems as the plants grow, these tomato plants are tied to individual wooden stakes or to lines strung between stakes, which is also a labour-intensive operation (USDA, 1999b). Calgene developed one of the first genetically modified tomatoes: the 'Flavr Savr'. These extended shelf-life tomatoes last about 2–3 weeks on the store shelf or about a week longer than mature green-harvested tomatoes. Unfortunately for the USA, extended shelf-life tomatoes performed better in Sinatra and Baja, Mexico, than in the USA (Plunkett, 1996). Also an uproar over genetically modified foods developed in Western Europe during the late 1990s; consumer demand was dramatically reduced, and these tomatoes were removed from the market.

Controlled-environment tomatoes (greenhouse and hydroponically grown) that are harvested 'vine ripe' have experienced rapid growth since 1999. These tomatoes have been largely imported from The Netherlands, Canada and Israel, but producers in the USA are entering the growing market for these high-quality tomatoes. The tomatoes have greater uniformity than open-air tomatoes and, it is claimed, improved taste. Many are being marketed 'on-vine' in clusters to convey an appearance of freshness to consumers (USDA, 1999b). The hand labour in these hothouses is somewhat different from that for traditional open-air staked tomatoes and can approach year-round work.

Processed tomatoes

Tomato varieties for processing have been bred for a pear or cylinder shape, high-solids content (5–9% compared with only 3–4% solids in old varieties), uniformity in ripening date, and generally tougher skins. With these attributes, they are less susceptible to pests while growing near the ground and can be harvested mechanically (Schmitz and Seckler, 1970). The mechanical tomato harvester was developed and adopted widely in California processed tomatoes in the late 1960s. It operates much like a small-grain combine, cutting the plants off near ground level and pulling them into a separator, where the tomatoes are shaken off the vines and sorted by gravity through a screen on to rolling conveyor belts. Until the early 1990s, four to six workers riding the machines undertook further hazardous hand-sorting of chunks of dirt and green tomatoes from the ripe tomatoes. During this era, payments to growers were frequently docked for excessive dirt and green tomatoes when loads were delivered to processing plants.

During the early 1990s electronic sorters were developed and attached to mechanical tomato harvesters. These electric-eye sorters were a major technical advance: they sense the colour of material on rolling conveyor belts and use air pressure to 'blow' green tomatoes and chunks of dirt off the belts. The remaining ripe tomatoes are then elevated into wagons or trucks. The electronic sorters have reduced the amount of hazardous hand-sorting and the number of workers riding on the tomato-harvesting machines, and have improved the quality of the product delivered to processors by largely eliminating the green tomatoes and dirt.

Fruit harvesters

Mechanical harvesters have also been developed and widely adopted in some areas for soft fruit (e.g. cherries, peaches, plums), and hard fruit (e.g. apples) for processing, and for nuts (USDA, 1999a). These harvesters are typically a two-part motorized machine with one part being driven on each side of a row of trees to be harvested. One of the two parts of the machine grips the tree and shakes it hard enough to make virtually all the nuts or fruit fall on to 'sloping to the middle' canvases, and they are then conveyed into boxes. After harvesting, the machine releases its grip and both parts move to the next tree. These machines greatly reduce the labour needed for harvesting and eliminate the hazardous work of harvesting trees from ladders.

Institutional innovations

Innovations in institutional arrangements are primarily of two types: increased use of farm labour contractors and coordination production processes for vegetables.

Farm labour contractors

Farm labour contractors (FLCs) are middlemen who, for a fee, will recruit, schedule, supervise and pay farm workers. Although FLCs have existed at least since the early 1960s, their intermediary services have grown rapidly since IRCA. They are frequently Hispanic and have few physical and financial assets. Under IRCA, employers can be fined for knowingly hiring undocumented workers. FLCs are less likely to be checked for undocumented workers by the US Immigration and Naturalization Service than are growers. In this environment, the number of FLCs in the West and Southeast USA has grown since 1986 (Taylor and Thilmany, 1993; Martin and Taylor, 1995). In the West (largely the states of Arizona, California, Oregon and Washington), there are currently about 50% more FLCs than in the late 1980s, and 100% more contractor labourers (Runyan, 1999).

Coordination

Contracts specifying a vertically coordinated production process are now typical for virtually all processed vegetables and most fresh-market lettuce, carrots and tomatoes. The contractor typically specifies which seeds are to be used, the varieties to be grown and which fertilizers and other chemical inputs are to be used, and may even specify that the contracting firm provide these inputs to growers. For fresh produce, coordination is important to year-round availability; and for processed vegetables, it more efficiently uses the processing capacity (Dimitri, 1999).

Growth and Changes in the Labour Intensity of Agriculture

The production of crops and livestock uses labour, capital services, materials and land as inputs. Changes in input ratios – for example, materials-to-labour and capital-to-labour – summarize structural change in agriculture. Without technical change, larger output requires larger input; but when there is technical change, added output may not require additional conventional inputs. Total factor productivity (TFP) is one widely used measure of technical change: it is the ratio of the quantity index of outputs produced by farmers divided by an index of the inputs used in production by farmers.

After 1986, growers in the USA switched from directly hiring seasonal workers to obtaining labour through farm labour contractors. Hence, to obtain an accurate picture of changing farm labour utilization over time, contract labour should be included with hired labour. The USDA, however, has traditionally included contract labour with intermediate inputs or materials. This practice is unfortunate.

The USDA Economic Research Service has been working over the past decade to construct a new set of state aggregate accounts for agriculture for all 50 states and to construct new TFP indexes for the USA (Ball *et al.*, 1997) and for all states. In this chapter the focus is on California and Florida with comparisons with the whole USA and to Iowa, which is a major producer of agricultural products and located in the middle of the country. The discussion focuses on the long period 1960–1996 and 10-year subperiods 1976–1986 and 1986–1996, over which to compare pre- and post-IRCA impacts on farm labour use.

Changes between 1960 and 1996

For US agriculture over 1960–1996, the average annual (compound) rate of growth of (real) total farm output was 1.8%, crop output 2.1% and livestock output 1.4% (Table 3.1 and Fig. 3.1). For California agriculture, total output grew at the rate of 2.2% per year, crop output at 2.4% and livestock output at 1.8% (Table 3.1 and Fig. 3.2). For Florida agriculture, output has been growing at a faster average annual rate than in the USA and in California: total output at 2.7%, crop output at 2.8% and livestock output at 2.2% (Table 3.1). The rate of output growth was faster over 1960–1980 than over 1980–1996 (Fig. 3.3). For Iowa, growth has been somewhat slower than the national average or for California and Florida: total output grew at 1.1% and crop output at 2.3%, while livestock output declined at an annual rate of −0.3% per year (Table 3.1). The abrupt dips in total farm output in 1974, 1983, 1987 and 1993 (Fig. 3.4) were due to adverse weather. Hence, over the period 1960–1996, there was rapid growth in farm output for the USA, California, Florida and Iowa, and crop output grew faster than livestock output. Hence, crop output was a growing share of total farm production.

The amazing story for US agriculture during the 20th century and during 1960–1996 was the lack of growth of farm inputs, given the rapid growth in farm output (Huffman and Evenson, 1993; Ahearn *et al.*, 1998). Over 1960–1996, the average annual rate of growth in US agriculture of total input, capital (i.e. machinery, equipment, buildings and breeding stock) and labour were negative (−0.3, −0.2 and −2.2%, respectively) but the growth rate for materials (annually purchased inputs) was positive (0.7%), as shown in Table 3.1. The farm labour

Table 3.1. Aggregate performance indicators (average annual percentage change) for California, Florida, Iowa and US agriculture, 1960–1996. (Source: California, Florida, Iowa and USA data from Eldon Ball, Economic Research Service, US Department of Agriculture.)

Real quantities	California	Florida	Iowa	USA[a]
Outputs				
Total (TQ)	2.20	2.69	1.07	1.84
Crop	2.42	2.84	2.31	2.13
Livestock	1.79	2.24	−0.28	1.40
Inputs				
Total (TX)	0.45	0.57	−0.60	−0.33
Materials	1.51	2.03	0.14	0.69
Capital	0.31	1.30	0.18	−0.24
Labour	−0.63	−0.35	−2.48	−2.23
Self-employed	−2.81	−1.38	−2.63	−2.48
Hired and contract	0.12	0.11	−1.14	−1.74
Ratios				
TFP = TQ/TX	1.75	2.12	1.67	2.17
Materials/labour	2.14	2.37	2.62	2.92
Capital/labour	0.94	1.65	2.66	1.99

[a]For the USA, contract labour is included with materials rather than hired labour.

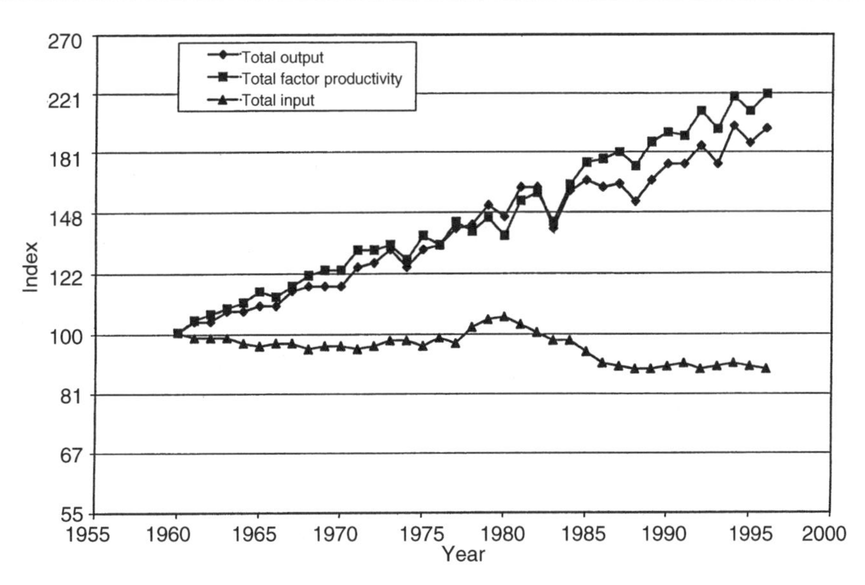

Fig. 3.1. Total output, total input and total factor productivity of US agriculture, 1960–1996 (1960 = 100).

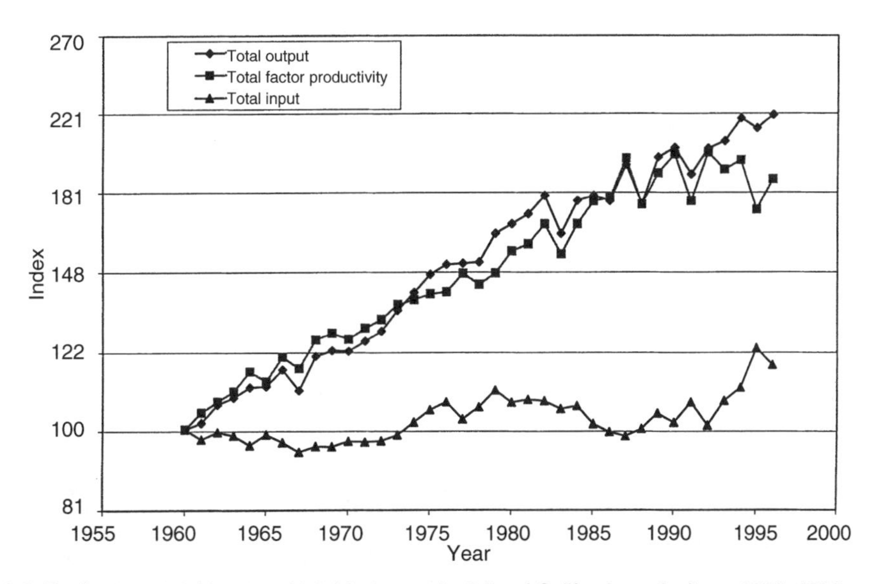

Fig. 3.2. Total output, total input and total factor productivity of California agriculture, 1960–1996 (1960 = 100).

input showed a rapid rate of decline over the whole period (Fig. 3.5). In California, total inputs, materials and capital input grew at 0.5, 1.5 and 0.3%, respectively, and labour declined at −0.6% per year. Although the farm labour input had a negative long-term trend, the growth rate was positive over selective subperiods, i.e. 1971–1975 and 1986–1996. The decline in the mid-1960s (Fig. 3.6) was due largely to the adoption of the mechanical tomato harvester. For Florida, the pattern of input growth was similar to California: total input grew at the average rate of 0.6%, materials at 2.0% and capital at 1.3%, but labour declined at −0.4% per year (Table 3.1 and Fig. 3.7). For Iowa, the pattern was

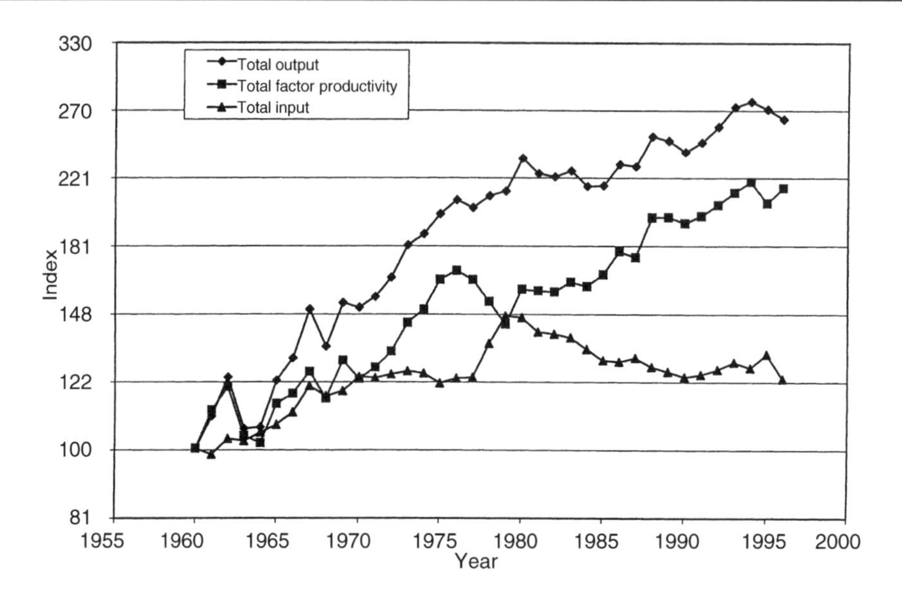

Fig. 3.3. Total ouput, total input and total factor productivity of Florida agriculture, 1960–1996 (1960 = 100).

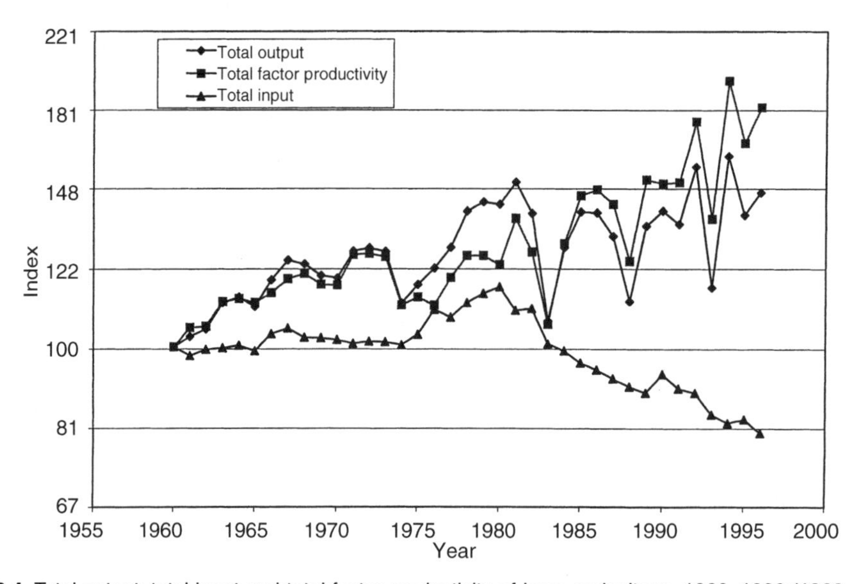

Fig. 3.4. Total output, total input and total factor productivity of Iowa agriculture, 1960–1996 (1960 = 100).

similar to the USA but not to California and Florida: the total input index and labour declined (−0.6% and −2.5%, respectively) and materials and capital increased at 0.1 and 0.2% per year, respectively. Although the long-term trend is negative for farm labour, positive growth occurred over 1974–1979 (Fig. 3.8), when both fuel and chemical input prices and farm field crop prices rose to high profit levels.

The changing composition of the labour input can be explored further by examining the performance of self-employed and hired-and-contract labour separately. For the per-

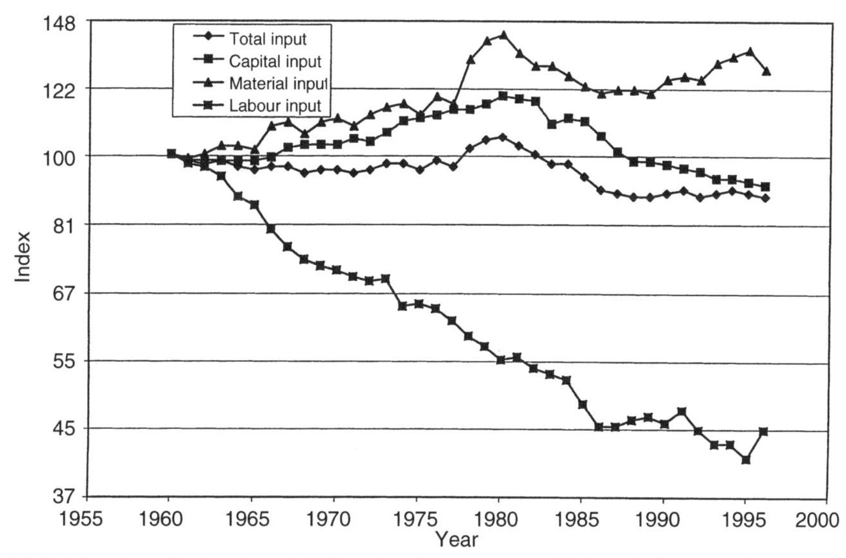

Fig. 3.5. Total input, capital input, material input and labour input of US agriculture, 1960–1996 (1960 = 100).

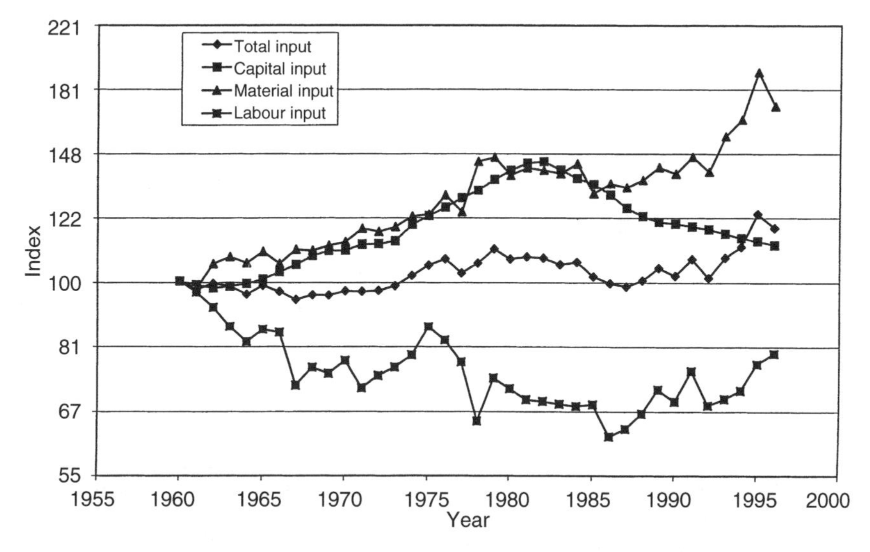

Fig. 3.6. Total input, capital input, material input and labour input of California agriculture, 1960–1996 (1960 = 100).

iod 1960–1996, the average rate of decline of self-employed labour was relatively large for the USA and the three states (Table 3.1), and it was much larger than the rate of decline of hired and contract labour. (Because contract labour is included with materials at the national level rather than with hired labour, the rate of growth of the labour input for US agriculture has a small negative bias and those of materials-to-labour and capital-to-labour ratios have a small positive bias.) In California and Florida, hired and contract labour actually increased a little. In California (Fig. 3.9), self-employed labour was relatively unchanging over 1974–1989 and at approximately half of its early 1960s value,

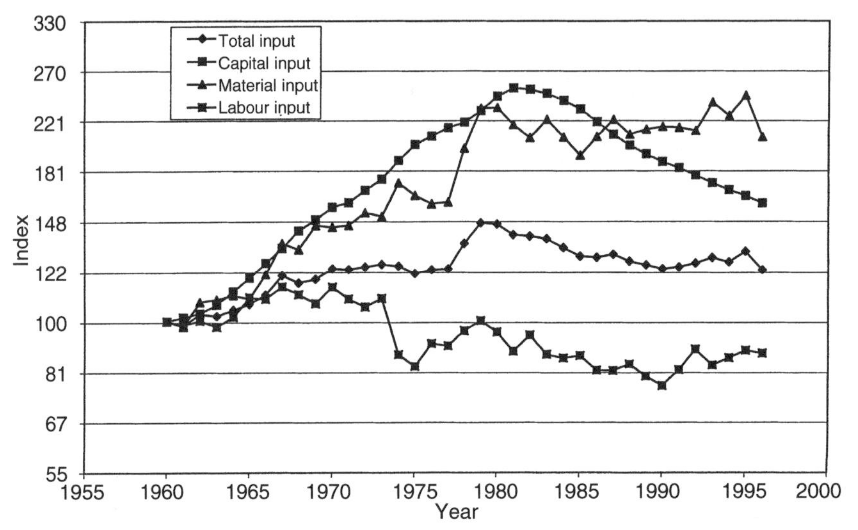

Fig. 3.7. Total input, capital input, material input and labour input of Florida agriculture, 1960–1996 (1960 = 100).

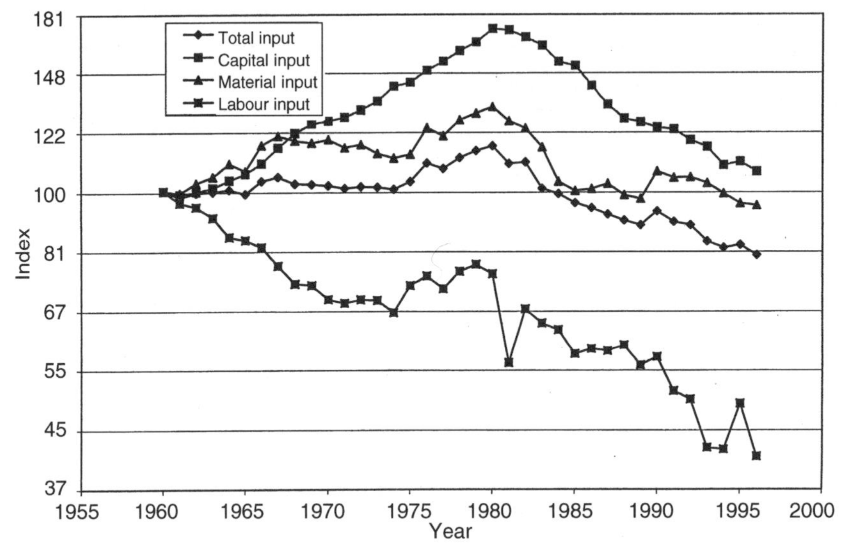

Fig. 3.8. Total input, capital input, material input and labour input of Iowa agriculture, 1960–1996 (1960 = 100).

then it decreased significantly. Hired and contract labour had a negative trend over 1960–1965, in part due to adoption of the mechanical tomato harvester. In Florida, although the long-term trend in self-employed labour is negative, Fig. 3.10 shows that the path is quite irregular. Over 1960–1971 hired and contract labour had positive growth, but then its path became cyclical. We do not have a good explanation; perhaps it was the weather.

In Iowa (Fig. 3.11), where a strong long-term negative trend exists in self-employed labour, little change occurred over 1970–1980. This was a period when Midwest cash grain farming was relatively profitable. Adverse weather events explain sudden dips in 1983 and 1993. The quantity of hired and contract

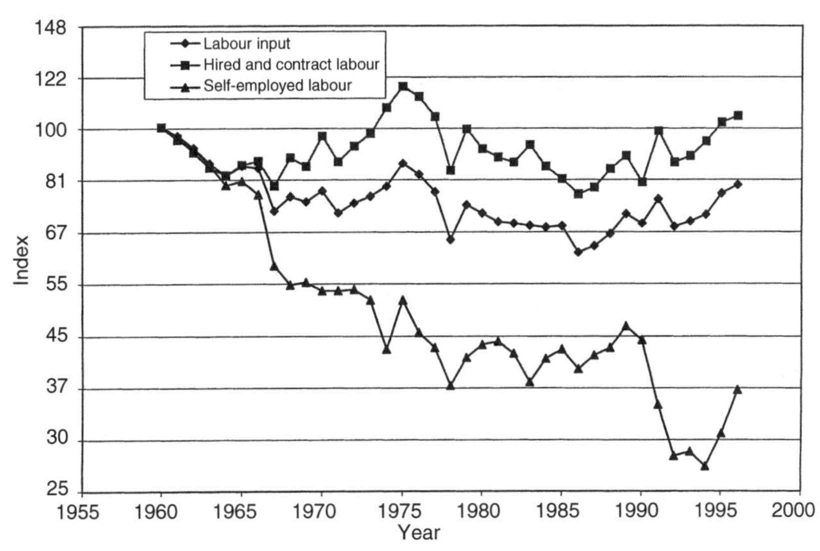

Fig. 3.9. Total farm labour, self-employed labour, and hired and contract labour of California agriculture, 1960–1996 (1960 = 100).

Fig. 3.10. Total farm labour, self-employed labour, and hired and contract labour of Florida agriculture, 1960–1996 (1960 = 100).

labour increased by about 75% over the 1970–1980 period, then the long-term negative trend re-emerged.

Thus, over 1960–1996 agriculture became more hired and contract-labour intensive relative to self-employed labour, especially in California and Florida (Table 3.1 and Figs 3.9 and 3.10). Furthermore, hired farm workers in the 1990s were more likely than all US wage and salary (and farm family) workers to be male, Hispanic, young, low schooled, never married and non-US citizens (Huffman, 1996; Runyan, 1998).

Turning to factor intensities in agriculture, the materials-to-labour input ratio and the capital-to-labour ratio are expected to

Fig. 3.11. Total farm labour, self-employed labour, and hired and contract labour of Iowa agriculture, 1960–1996 (1960 = 100).

respond to a persistent change in relative input prices and technical change that is biased in a particular way, e.g. labour saving. During 1960–1996, the materials-to-labour ratio was increasing at an average annual rate of 2.9% per year for the USA, 2.1% in California, 2.4% in Florida and 2.6% in Iowa agriculture (Table 3.1). For the same period, the capital-to-labour ratio was increasing at an average rate of 2.0% per year for the USA, 0.9% in California, 1.7% in Florida and 2.7% in Iowa. Thus, the labour intensity of agriculture in the USA, measured both as materials-to-labour and capital-to-labour, has declined over 1960–1996. In general these changes reflect substitution of materials and capital for labour and substitution of hired and contract labour for self-employed labour.

Total factor productivity over 1960–1996 grew at 2.2% per year for the USA, 1.8% for California, 2.1% in Florida and 1.7% in Iowa. Hence, real farm output was growing much faster than real farm inputs under the control of farmers. It is an amazing story that output growth came primarily from TFP growth and not from input growth. Over 1958–1996, only one of 37 sectors of the US economy had a higher rate of TFP growth than agriculture (Jorgenson and Stiroh, 2000).

Subperiod trends

In 1986, the Immigration Reform and Control Act (IRCA) was passed which gave amnesty to about 1.3 million undocumented individuals who could show prior work experience in US agriculture (Martin *et al.*, 1995). Since then, newly legalized individuals have brought their families and friends to the USA to settle in rural areas, especially in California and Texas, and to work in agriculture and other industries. In Table 3.2, average annual growths of outputs and inputs are reported for subperiods 1976–1986 and 1986–1996, chosen in order to focus on pre- and post-IRCA impacts.

The labour index in California and Florida declined during 1976–1986 but *increased* over the 1986–1996 period. In contrast, the labour index for Iowa declined rapidly in both subperiods, and for the USA the indexes showed a steep decline over 1976–1986 but no change over 1986–1996. For hired and contract labour, California and Florida showed a decline over 1976–1986 (of −3.8% per year for California) but the ratio increased after IRCA. In contrast, in Iowa, where fruit and vegetable production are unimportant, the index of hired and contract labour declined at

Table 3.2. Aggregate performance indicators (average annual percentage change) for California, Florida, Iowa and US agriculture, subperiods 1976–1986 and 1986–1996. (Source: California, Florida, Iowa and USA data from Eldon Ball, Economic Research Service, US Department of Agriculture.)

	California		Florida		Iowa		USA[a]	
Real quantities	1976–1986	1986–1996	1976–1986	1986–1996	1976–1986	1986–1996	1976–1986	1986–1996
Outputs								
Total (TQ)	1.59	2.18	2.67	1.35	1.38	0.50	1.86	2.00
Crop	1.84	2.22	0.98	1.60	3.56	0.41	2.19	2.01
Livestock	1.17	2.17	1.33	0.43	−1.49	0.15	1.29	1.99
Inputs								
Total (TX)	−0.75	1.70	1.06	−0.51	−1.50	−1.59	−0.78	−0.21
Materials	0.36	2.43	2.67	−0.01	−2.04	−0.56	0.12	0.68
Capital	0.38	−1.56	0.57	−3.24	−0.53	−2.88	0.63	−1.38
Labour	−2.98	2.57	−1.07	0.69	−2.45	−3.63	−3.60	0.00
Self-employed	−1.41	−0.81	−3.32	2.05	−2.40	−3.55	−3.31	−1.38
Hired and contract	−3.79	3.04	−0.30	0.25	−2.77	−4.06	−3.00	−0.51
Ratios								
TFP = TQ/TX	2.34	0.48	0.56	1.95	2.89	2.09	2.71	2.21
Materials/labour	3.34	−0.14	3.74	−0.70	0.41	3.09	3.72	0.68
Capital/labour	3.36	−4.14	1.64	−3.93	1.92	0.75	2.97	−1.38

[a]For the USA, contract labour is included with materials rather than hired labour.

−2.8% per year during the pre-IRCA period and at −4.1% during the post-IRCA period. The materials-to-labour ratio increased during 1976–1986 in California, Florida, Iowa and the USA; during 1986–1996, it declined in California and Florida but increased in Iowa and the USA. The capital-to-labour ratio rose rapidly during 1976–1986 in California, Florida, Iowa and the USA; during the post-IRCA decade, it declined in California and Florida at more than 4% per year, but in the USA the decline was much smaller (−1.4% per year), and in Iowa it grew rapidly (at 3% per year). The IRCA, which increased the supply of Hispanic immigrant low-wage seasonal agricultural workers to hand-harvested and labour-intensive agriculture of California and Florida, and the growing demand for fresh fruits and vegetables are major factors leading to different outcomes in California and Florida than in Iowa and the remainder of US agriculture. This is likely to continue as long as cheap and productive seasonal agricultural labour is available.

Conclusion

Agriculture in the USA has faced a continuous stream of new technologies from public and private research and development over the past four decades. Some have been labour saving. Also, new immigration legislation and Border Patrol policies have affected the ease with which farm workers enter the USA illegally from Mexico, and about 1.2 million previously illegal workers from all countries were legalized in the late 1980s through IRCA. Granting legal status to some family members but not to others has effectively increased the opportunities for work in the USA by all members of seasonal agricultural worker families and the supply of farm workers from what it would otherwise have been (Huffman, 1995; Martin *et al.*, 1995). The net result has been a sizeable increase in availability of labour for seasonal agricultural services during the 1990s.

Technical change in much of US agriculture can be described as labour-saving over the 1960–1996 period. However, hand-harvesting is required in fresh fruit and vegetable production, to obtain high quality products for the supermarkets. Breeding for dwarf trees has been an innovation that made hand harvesting easier for most fruit trees. Also, mechanization has been possible in non-harvest components of these crops, such as field preparation, planting and application of plastic, and in the production of other crops and livestock.

With relatively abundant immigrant farm workers in the western and southern states for working in labour-intensive fruit and vegetable production, it comes as no surprise that hired and contract labour has become an increasing share of US farm labour, especially in California and Florida. This change seems to have been facilitated by the growth of the institution of farm labour contracting, especially in the West and Southeast. Furthermore, relatively abundant and reliable immigrant workers have undoubtedly slowed the discovery, development and adoption of some mechanized technologies for agricultural production, especially in vegetable and fruit production.

Special problems have been associated with inflows of immigrant workers to agriculture and settling out into rural areas and small towns in the USA in the 1990s. Subsidized state and federal housing units are generally unavailable to farm worker families who cannot prove the legal status of all family members. Some states require proof of legal status before issuing drivers' licences, and this means that some farm workers cannot drive a car legally in the USA. When an immigrant family is split between the USA and another country (e.g. Mexico), obtaining useful health insurance coverage and covering hospital and medical expenses are also major problems.

Acknowledgement

Helpful comments were obtained from two reviewers and participants in the conference, *The Dynamics of Hired Farm Labour: Constraints and Community Response*. Matt Rousu and Jeng-feng Xu provided research assistance.

This is journal paper no. 2424 of the Iowa Agriculture and Home Economics Experiment Station, Ames, Iowa, Project 3077, and is supported by the Hatch Act and State of Iowa funds.

References

Ahearn, M., Yee, J., Ball, E. and Nehring, R. (1998) *Agricultural Productivity in the United States.* Information Bulletin No. 740. Economic Research Service, US Department of Agriculture, Washington, DC.

Ball, E., Bureau, J.C., Nehring, R. and Somwaru, A. (1997) Agricultural productivity revisited. *American Journal of Agricultural Economics* 79, 1045–1063.

Dimitri, C. (1999) Integration, coordination, and concentration in the fresh fruit and vegetable industry. In: *Fruit and Tree Nuts Situation and Outlook, FTS-285.* Market and Trade Economics Division, Economic Research Service, US Department of Agriculture, Washington, DC, pp. 23–31.

Fuglie, K., Ballenger, N., Day, K., Kotz, C., Ollinger, M., Reilly, J., Vasavada, U. and Yee, J. (1996) *Agricultural Research and Development: Public and Private Investments Under Alternative Markets and Institutions.* Agricultural Economics Report No. 735. Economic Research Service, US Department of Agriculture, Washington, DC.

Gale, H.F. (1996) Age cohort analysis of the 20th century decline in US farm numbers. *Journal of Rural Studies* 12, 15–25.

Gardner, B.L. (1992) Changing economic perspectives on the farm problem. *Journal of Economic Literature* 30, 62–101.

Huffman, W.E. (1995) Immigration and Agriculture in the 1990s. In: Martin, P.L., Huffman, W.E., Emerson, R., Taylor, J.E. and Rochin, R.I. (eds) *Immigration Reform and US Agriculture.* Publication No. 3353. University of California, Division of Agricultural and Natural Resources, Oakland, California, pp. 425–442.

Huffman, W.E. (1996) *Farm Labor: Key Concepts and Measurement Issues on the Route to Better Farm Cost and Return Estimates.* Staff Paper No. 280. Department of Economics, Iowa State University, Ames, Iowa.

Huffman, W.E. (2002) Human capital, education, and agriculture. In: Peters, G.H. and Pingali, P. (eds) *Tomorrow's Agriculture: Incentives, Institutions, and Innovations.* Ashgate Publishing Company, Aldershot, UK, pp. 207–222.

Huffman, W.E. and Evenson, R.E. (1993) *Science for Agriculture: a Long Term Perspective.* Iowa State University Press, Ames, Iowa.

Huffman, W.E. and Evenson, R.E. (2001) Structural adjustment and productivity change in US agriculture, 1950–82. *Agricultural Economics* 24, 127–147.

Huffman, W. and Moschini, G. (1999) The role of research and development and of new technologies. In: *Agriculture in the 21st Century – Surviving and Thriving.* Iowa Agriculture and Home Economics Experiment Station and University Extension, Ames, Iowa, pp. 17–22.

Jorgenson, D.W. and Stiroh, K.J. (2000) US economic growth at the industry level. *American Economic Review* 90, 161–167.

Martin, P.L. and Taylor, J.E. (1995) Introduction. In: Martin, P.L., Huffman, W.E., Emerson, R., Taylor, J.E. and Rochin, R.I. (eds) *Immigration Reform and US Agriculture.* Publication No. 3353. University of California, Division of Agricultural and Natural Resources, Oakland, California, pp. 1–18.

Martin, P.L., Huffman, W.E., Emerson, R., Taylor, J.E. and Rochin, R.I. (eds) (1995) *Immigration Reform and US Agriculture.* Publication No. 3353. University of California, Division of Agriculture and Natural Resources, Oakland, California.

Olmstead, A.L. and Rhode, P. (1993) Induced innovation in American agriculture: a reconsideration. *Journal of Political Economy* 101, 100–118.

Plummer, C. (1992) *US Tomato Statistics, 1960–90.* Statistical Bulletin No. 841. Economic Research Service, US Department of Agriculture, Washington, DC.

Plunkett, D. (1996) Mexican tomatoes – fruit of new technologies. In: *Vegetables and Specialties Situation and Outlook, VGS-268.* Economic Research Service, US Department of Agriculture, Washington, DC.

Runyan, J.L. (1998) *Profile of Hired Farmworkers, 1996 Annual Averages.* Economic Report No. 762. Economic Research Service, US Department of Agriculture, Washington, DC.

Runyan, J.L. (1999) *Ten Years of Farm Labour Contracting.* Economic Research Service, US Department of Agriculture, Washington, DC.

Schmitz, A. and Seckler, D. (1970) Mechanized agriculture and social welfare: the case of the tomato harvester. *American Journal of Agricultural Economics* 52, 569–577.

Taylor, J.E. and Thilmany, D. (1993) Worker turnover, farm labor contractors, and IRCA's impact on the California farm labor market.

American Journal of Agricultural Economics 75, 350–360.

USDA (1999a) *Fruits and Tree Nuts Situation and Outlook, FTS-285.* Economic Research Service, US Department of Agriculture, Washington, DC.

USDA (1999b) *Fresh Market Tomato Briefing Room.* Economic Research Service, US Department of Agriculture, Washington, DC. www.usda.gov/briefing/tomatoes

USDA (1999c) *Vegetables and Specialties Yearbook, 1970–97.* Economic Research Service, US Department of Agriculture, Washington, DC.

4 Family Settlement and Technological Change in Labour-intensive US Agriculture

Richard Mines
California Institute for Rural Studies, Davis, California, USA

Abstract

Labour-intensive agriculture in the USA faces a choice in the coming years. It can perpetuate a labour system based on a high-turnover, flow-through and low-wage labour market based on a stagnant technological model. This model depends on competing in world markets through low wages and implies a predominantly solo (unaccompanied by nuclear family) male workforce with the associated traits of instability and fewer days of work per year. Or the industry can evolve into a system based on technological change which enhances productivity, lowers costs and lightens tasks, expanding the supply of labour to older men and women. This model implies a lessening of the seasonal spikes, a smaller but more fully employed labour force and a stable family-based environment in farm areas. The chapter traces the roots of solo male migration and family settlement back to the post-war Bracero period and links the history of technological change with the nature and stability of the workforce. The chapter reviews specific alterations in practices and points the way to a transition to a stable labour force coupled with an appropriate technology. It underscores the possibility of reducing the social costs associated with an impoverished solo male labour force by introducing competition-enhancing change.

Introduction

The private and public sectors should cooperate in designing a system that takes into consideration the needs not only of the agricultural industry (i.e. farmers, bankers, input suppliers, retailers) and consumers, but also of the transnational communities supplying the labour that allows the industry to flourish. The traditional and current high-turnover, low-wage system in the USA cannot meet the needs of workers and may have outlived its ability to meet the needs of the industry groups. In fact, the current mix of technology and labour may leave the industry vulnerable to international competition. It is possible to eliminate or lessen many of the negative effects of the current system by transition to a system based on predominantly family rather than solo immigrant male labour.

The emphasis by national grower lobbying groups on a system of short-term all-male workers will postpone the necessary changes in technology and labour manage-

(eds J.L. Findeis, A.M. Vandeman, J.M. Larson and J.L. Runyan)

ment. Naturally, growers prefer to have a labour force that allows them to be flexible with respect to expensive investments. The grower magazine, *The Packer*, stated on 28 January 1989, 'with an abundant and reasonably priced labour force at hand ... why bother pushing for mechanization?' (P. Martin, October 1999, personal communication). This attitude is not shared by industry groups that support positive change.

The long-term goals of profitability for the industry and stability for the workforce should be sought in a joint manner. If properly conceptualized, they can be made compatible. To demonstrate the case for a transition to an orderly labour market, this chapter touches on two background issues and then discusses the social and technological alternatives facing us today. Firstly, the changing composition of the labour market in the USA over the last 10 years will be discussed. Then, the Bracero period and subsequent years will be reviewed for insights into the present situation. Finally, a gradual transition will be argued for, instead of a perpetuation of the *status quo*.

Ethnic Shift and High Turnover

Although the employment level in farm production in the USA has remained remarkably constant since the early 1970s at about a 1.3 million annual average (see Farm Labour Reports (various years), November), the composition of the labour force has shifted as one group of migrant workers has replaced another. These shifts have contributed to the maintenance of very low wages and poor working conditions, since the newly arriving groups have consistently been willing to accept low wages that more settled groups would reject. This undermining of labour standards occurs for basically two reasons: (i) the newcomers have lower costs, since their families live in a cheaper economy abroad; and (ii) they are unfamiliar with US labour protections.

Starting in California in the 1960s, the shifts in this workforce have meant an increasingly Mexican and Guatemalan labour force for the farm economy (Table 4.1). In California in 1965, a minority of farm workers were still of Mexican descent. By 1983, 71% were born in Mexico and by 1995 fully 94% spent their early years in Mexico.

The incorporation of the foreign-born has been characterized by taking over one task at a time. At first, Latin American immigrants began to dominate in harvests that included tasks such as picking, pulling and cutting. Then they took over the preharvest season tasks of hoeing, thinning, tying, propping and transplanting. Later, Latino immigrants became dominant in semi-skilled activities such as pruning, spraying, irrigation and grove maintenance. At approximately the same time they also took over the majority of

Table 4.1. Ethnic shifts in California agriculture (percentage of workers).

Group	California Assembly (1965)	UC-EDD survey[a] (1983)	California NAWS[b] (1994/95)
White, US born	43.9	4.5	1.1
Mexican	45.9	–	–
Hispanic, US born	–	16.5	4.0
Hispanic, Mexican born	–	71.3	93.5
Asian and Native American, US born	6.8	6.8	0
Black	3.3	0.9	0.1

[a]Survey conducted by the University of California (UC) in conjunction with the California Employment Development Department (EDD).
[b]National Agricultural Workers Survey (NAWS).

the packing, hauling and shipping tasks. Finally, Latino immigrants began working as supervisors doing recruiting, monitoring, checking and the transporting of crops.

According to National Agricultural Workers Survey (NAWS) data, the Latino immigrant farm labour force has increased from 60 to 80% nationwide in the period between 1989 and 1997. Although Latino immigrant farm workers have had a strong historical presence in the USA in regions of the western states and Florida, eastern states are also now becoming important destinations for many of these workers.

To demonstrate these changes over the last 10 years, the country can be divided into two regions: (i) the eastern USA, including all of the states east of the Mississippi River, except Florida; and (ii) the western USA, encompassing the rest of the country (i.e. all states west of the Mississippi River, plus Florida). These two regions roughly encompass the areas of traditional foreign incorporation into the farm labour market (the West) and the area of more recent incorporation (the East). Table 4.2 shows the percentage of foreign-born crop workers in the two regions for the period 1989–1997, based on data from the NAWS.

Table 4.2. Percentage of foreign-born crop workers in two regions of the USA, 1989–1997. (Source: National Agricultural Workers Survey for 1989–1997.)

Year	West and Florida	East
1989	83	40
1990	82	38
1991	78	26
1992	82	40
1993	81	57
1994	77	57
1995	79	59
1996	89	64
1997	83	77
Total	82	49
Number of workers	11,148	9,260

The data suggest that there has been little change in the region encompassing the West and Florida, where Latino immigrants already dominated at the beginning of the decade (Table 4.2). In the East outside of Florida, the proportion of foreign-born has doubled from 40% in 1989 to nearly 80% in 1997. The heavy dependence on immigrant farm workers has made the different regions of the USA more similar in the composition of their labour force.

There have not been major changes in the West over the period 1989–1997 in regards to the tasks performed by immigrant farm workers (Table 4.3). In the West and Florida all of the tasks were already dominated by the foreign-born at the beginning of the period. However, in the East, outside of the harvest task that also was already dominated by the foreign-born, there have been rapid shifts in the other tasks.

In the East, the preharvest and semi-skilled tasks became dominated by the new foreign-born groups and finally the shift occurred for the postharvest tasks as well. The proportion of the foreign-born in the preharvest tasks increased from 23% to 60%, in the postharvest tasks from 7% to 56%, and in the semi-skilled jobs from 13% to 61% over the last 10 years. At the beginning of the period, white and African-American US-born workers dominated the pruning, transplanting, packing and tobacco barn work in the East, but by the end of this period the participation of native-born workers was on a distinct decline.

What accounts for this rapid shift to an overwhelmingly Mexican (and Guatemalan) labour force? It is due in part to poor working conditions on farms that fail to keep veteran US and foreign-born workers in agriculture. Some part of this shift can also be explained by the availability of alternative low-wage jobs in the booming non-agricultural US economy. As the experienced domestic and foreign farm workers have left agriculture, most of the replacement workers have been recently arrived Mexican males. In fact, NAWS data indicate that among new (first-year) workers, 70% are recently arrived Mexican immigrants and another 8% are Guatemalans.

Low wages and a decline in earnings contribute to ethnic replacement in labour-intensive agriculture. Many farm workers are not able to work constantly through the year

Table 4.3. Place of birth of farm workers (by percentage of workers) by task, 1989–1997. (Source: NAWS, 1989–1997.)

	Preharvest		Harvest		Postharvest		Semi-skilled	
	US born	Foreign born	US born	Foreign born	US born	Foreign born	US born	Foreign born
West USA and Florida								
1989–1991	28	72	15	85	22	78	15	85
1992–1994	32	68	10	90	30	70	22	78
1995–1997	15	85	11	89	35	65	13	87
East USA, except Florida								
1989–1991	77	23	28	72	93	7	87	13
1992–1994	44	56	33	67	73	27	52	48
1995–1997	40	60	18	82	44	56	39	61

and there is evidence from the NAWS that there may be an actual earnings decline in real terms over time. The NAWS reports that the earnings of individual farm workers in the USA have stagnated at between US$5000 and US$7500 per year. Moreover, there are indications that farm workers are finding less farm work each year in recent years than previously. Table 4.4 shows that for foreign-born workers the percentage of time engaged in farm work in a given year declined from 58 to 48%. This trend affected the US-born farm workers even more severely, since the percentage of their time engaged in farm work in a given year declined from 50 to 35%. Two factors explain this shift. For the foreign-born, the shift occurred because the turnover rate has increased and a higher proportion of workers are first-year entrants from abroad. This explains why the percentage of time spent abroad is higher in the most recent period, as indicated in Table 4.4. For US-born farm workers, the drop in time spent in farm work occurred because the labour force increasingly comprises young summer workers rather than long-term US-born farm workers. This trend for both foreign and US-born farm workers indicates that yearly earnings in real terms may be declining.

Both trends indicate that the farm labour occupation cannot easily be used as a basis to build a career. In fact, only about 40% of farm workers have worked in US agriculture for a decade or more, and the turnover rates in recent years have been greater than 20% per year. The NAWS shows that in excess of 20% of the labour force in US agriculture has been in its first year in recent years. These trends

Table 4.4. Distribution of time spent (percentage allocated in last 12 months) engaged in various activities among crop workers in different time periods: three periods compared. (Source: NAWS, 1989–1997.)

Year	Farm work	Non-farm work	Not working in USA	Abroad
Foreign-born crop workers				
1989–1991	58.4	9.6	15.6	16.3
1992–1994	52.8	9.4	17.3	20.5
1995–1997	48.2	7.4	15.6	28.8
US-born crop workers				
1989–1991	50.2	20.3	26.3	3.1
1992–1994	42.5	21.5	32.7	3.2
1995–1997	35.3	19.1	37.2	3.3

mean that the current situation in US agriculture does not encourage settlement. Workers who bring their families to the USA would rather work in construction, manufacturing or services in an urban area and earn US$12,000 or US$15,000 a year so that they can survive. Most farm workers cannot earn a wage that will keep them in farm work.

The Bracero Period and Its Aftermath: Solo Males and Settlement

Agricultural guestworker programmes have been built and continue to be built on false premises. Firstly, the notion that workers come for short periods and then return has not proven accurate historically. Guestworker programmes are based on the underlying assumption that, after working a year or two in a foreign country, guestworkers will return to their countries of origin to be productive citizens. However, despite the deceptive name, many guestworkers remain in the receiving countries for their entire economic life. Secondly, the idea that these programmes can be prevented from adversely affecting domestic workers is contravened by existing evidence.

Based on the Bracero Program (a labour programme in which Mexicans went to the USA on short-term work contracts of 3–6 months) established in 1942 and continuing until 1964 between Mexico and the USA, historical evidence shows that many of the all-male *braceros* deserted their contracts or returned to the USA as undocumented workers. It is likely that most *braceros* stayed and worked in the USA for most of their economic lives (for a detailed historical overview, see Garcia y Griego, 1996; Calavita, 1992). Evidence from Congressional testimony and from ethnographic field work among ex-*braceros* demonstrates a strong tendency for *braceros* to have stayed in the USA (Mines, 1980; Calavita, 1992). However, although most stayed, many did not and those that did were able in many cases to bring their families. These family settlers have higher consumption costs and over time have become more familiar with US l[…] a result, they tend to demand […] than when they were unaccon[…] workers.

There is also historical evidence demonstrating that the presence of male-only guestworkers discouraged growers from raising wages. Taking, for example, four California crops that were largely worked by *braceros* and reviewing the wage changes from 1950 to 1974, a clear correlation between the presence of *braceros* and stagnant wages is seen. In the period 1961–1964, *braceros* harvested 79% of the tomatoes, 74% of the strawberries, 71% of the lettuce and 49% of the asparagus harvested in California (Holt, 1984). Table 4.5 shows that during the 1950–1964 period, wages were absolutely flat or declined in real terms. In all four crops, wages stagnated at about US$1.10 per hour in 1967 dollars. This occurred during a period of relative increases in real wages for production workers nationwide. However, starting in 1965 with the end of the Bracero Program, wages began to increase.

Wages for field workers reached a peak in 1978 and later dropped about 15% between 1974 and 1996. In 1998–1999, the US Department of Agriculture (USDA) reported another rise in hourly wages, but real wages for field workers still remain below the 1978 peak period. Under current circumstances, the institution of a large guestworker programme would likely have a similar effect to the 1942–1964 Bracero Program. It would tend to perpetuate a low-wage, high-turnover system and increase the stock of unauthorized workers in the USA.

Table 4.5. Real hourly wage (US$, 1967 = 100) of Bracero-affected crops in California, 1950–1974. (Source: California State Employment Service, Farm Labor Report No. 881A, 1950–1974.)

Crop	1950	1960	1965	1974
Asparagus	1.12	1.10	1.80	1.53
Lettuce	1.04	1.03	1.40	1.64
Tomatoes	1.11	1.00	1.37	1.58
Strawberries	1.18	1.13	1.38	1.51

Another flaw of the Bracero Program was that it failed to encourage the return of guestworkers to Mexico. Although employment on farms after the termination of the programme was replaced by a new system reliant on the influx of the unauthorized, initially many *braceros* settled with their families in many rural communities of the USA. In the period from approximately 1965 to the late 1970s, many *braceros* brought their families with them to the USA instead of returning to Mexico.

There are several reasons that explain this process. In 1965 there was a major revision of US immigration policy. Not only was the Bracero Program quashed by an alliance of unions, church and civil rights groups, but also in 1965 the Immigration and Nationality Act of 1952 was substantially amended in key provisions under the pressure of these same groups. The new act abolished the national origins quota system established in the 1920s, eliminating national origin, race or ancestry as a basis for immigration to the USA. This led to a more diversified pool of legal immigrants on the basis of family reunification and occupational qualifications (Portes and Rumbaut, 1996).

The Department of Labor's Labor Certification Program may have been particularly important in this process as former employers of *braceros* and unauthorized workers sought to legalize their workers at a time when employers felt their access to labour south of the border was uncertain. These ex-*bracero* employers even advertised in Mexico inviting their former workers to come north and frequently helped their ex-*braceros* to obtain legal status (Mines and Anzaldua, 1982). Permanent-settler core communities took root in rural California, including a substantial proportion of women from the various sending areas of Mexico (Palerm, 1991; Alarcon, 1995). In some cases, employers adapted their management practices to stimulate family settlement by instituting benefit programmes, including family housing for their workers (Mines and Anzaldua, 1982). In addition to all of these factors and relatively higher wages, the settlement of many families of ex-*braceros* was encouraged by the expansion of service programmes for farm workers and the upsurge of unionization among them (Zabin *et al.*, 1993).

The incorporation of Mexican-born women into these communities in the 1964–1976 period is corroborated by US Immigration and Naturalization Service (INS) statistics showing the relatively high proportion of females among Mexicans obtaining permanent resident status. In each year during that period, women constituted 48% or more of Mexicans becoming permanent residents. In the period from 1964 to 1972, over half were women.

Survey research provides additional corroboration of family reunification during the period between 1965 and the late 1970s. In the survey of California farm workers carried out in the summer of 1983 by the University of California in conjunction with the California Employment Development Department (the UC-EDD Survey), the cohort of workers who entered the country before 1965 was predominantly male. Those who entered during the late 1960s and early 1970s were about evenly divided between men and women. The age cohorts likely to have entered in the late 1970s and thereafter, again were predominantly male (Mines and Martin, 1986). The NAWS reinforces the view that this period was one of women joining men in the USA. From the NAWS, which has 10 years of data, it is possible to calculate the proportion of female foreign-born farm workers entering in each year since the 1950s. The overall average proportion of females among foreign-born crop workers is 15%. However, in the period from 1966 to 1969, 25% of entrants were women, and the proportion continued to be high into the late 1970s.

The experience of one group of immigrant farm workers illustrates this settlement process (Alarcon, 1995). There was a large concentration of migrants from Chavinda, Michoacan, in Madera County located in the San Joaquin Valley in California. Although there had been Chavindeños in that area since at least the 1940s, the arrival and settlement of families did not become important until the early 1970s. An analysis of the year of arrival of Chavindeño families shows that most of the families formed in Chavinda, arrived in Madera in the 1970s. In most cases the husband arrived first, and then the spouse and children followed after a period of time.

Coincident with this period of strong settlement was an upsurge in technological research to reduce the dependence on stoop labour. The introduction of mechanization for the harvest of processing tomatoes and cling peaches demonstrates the successes of mechanization at the end of the Bracero Program period. A good measure of this upsurge of interest in mechanization is that 25 agricultural engineers were working for the US Agricultural Research Service (ARS) on mechanical fruit aids in 1971, while today one half-time position remains. There exists a coincidence in the upsurge of solo, male immigrant farm workers and the decline in experimentation and adaptation of new technologies. A review of the agricultural engineering literature of the late 1960s also reveals that many knowledgeable observers were predicting a rapid technological change that has yet to occur. To cite one example, G.E. Rossmiller, the noted engineer, stated in introducing a 1969 volume on fruit and vegetable mechanization, 'The authors leave little doubt that the FV (fruit and vegetable) industry will be well on its way toward mechanical harvest of all but a very few specific commodities by 1975' (Cargill and Rossmiller, 1969).

It is critical for policy planning that family settlement coincides with improvements for farm workers, while periods of the heavy influx of solo males coincide with deteriorating conditions.

Transition or Perpetuation of the Current System

Since the Bracero Program, conditions faced by farm workers have been set largely by solo male newcomers who are willing and able to work for earnings below a sustainable wage. The social costs of this solo male system are borne by the transnational communities that provide the workers (Meillassoux, 1977; Bustamante, 1979). In particular, family separation means that a generation of Mexican children is being raised while their fathers are absent in the USA. Due to the absence of young males, rural Mexican production and development requires a population shift by which migrants to the USA are replaced by workers from smaller localities who do not have access to labour markets in the USA and are willing at least for a while to accept the prevalent wages in the area. However, frequently a local labour shortage occurs in Mexico and production is negatively affected. Lower production and income in Mexico renders it a weaker trading partner for the USA.

There are also social costs associated with having large numbers of solo males in rural communities in the USA. Their living and working conditions exacerbate problems of alcoholism and other forms of dependency (e.g. drug addition) (Kissam, 2000). In addition, social costs rise as farm workers, unable to make a living at farm work, leave the labour market and flow through to the urban sectors of the economy. This means that ex-welfare recipients and other low-wage workers are put in competition with and displaced by former farm workers in non-farm jobs.

The solo male model (exemplified by guestworker programmes) is not the only choice. In fact, a transition to a new type of agriculture based on a settled labour force coupled with appropriate technology and labour management practices would be compatible with a limited and controlled family-based immigration rather than a difficult-to-control solo male migration.

In fact, the current system of low-wage, high-turnover agriculture may be slowly losing its competitive edge with respect to international competitors. There is no doubt that the fruit, vegetable and horticultural (FVH) industry has expanded rapidly and is quite robust. In the past two decades, citrus production has gone up by over 60%, other tree fruit has been stable and vegetable production has increased by 75% (*Rural Migration Review,* 1999). The value of nursery production has also greatly expanded. Moreover, according to USDA statistics, those FVH firms with over $50,000 gross sales annually increased their net income markedly during the 1990s. However, a turning point may be approaching. Imports of fruits and vegetables are growing faster than and are larger than exports. Competition from abroad is growing rapidly.

Already, many low-wage countries such as Argentina, Mexico, Chile, Brazil and China and high-wage countries such as Italy, France, Holland, Australia and Israel may be able to outcompete FVH agriculture in the USA if it insists on using a low-wage strategy. Chinese apples, Mexican tomatoes, Chilean pears and Dutch peppers are examples of this competition.

The gradual adaptation to technological change and the creation of a stable labour force may be the best alternative over the medium to long term. It must be stressed that this change will require a concerted effort of all concerned to bring about success. The alternative of sticking to the *status quo* will only delay the inevitable realization that a low-cost labour policy is the wrong approach for the FVH industry. The USA has to compete with countries with less expensive inputs of water, land and labour and less stringent environmental and pest-control restrictions.

The first point that cannot be understated is the challenge of the task and the patience and cooperation among all groups it would require. At present, about 20 vegetable crops and 25 fruits still lack mechanical options. Problems of uneven maturation and of damage and short shelf life resulting from mechanical harvesting make it difficult to apply the available techniques. These commodities represent 1.4 million acres of vegetables and 2.2 million acres of fruit (Sarig *et al.*, 1999). It may take 10 or more years to develop technological options for seasonal workers for many of these crops. The tomato harvester took 12 years. Still, the change may be possible if undertaken with the proper resources and dedication.

Years of neglect and stagnation have left the USA behind in resources available to create technological alternatives for FVH agriculture. A commitment will be needed to build the human capital required to make the necessary changes. This will require government leadership, including a role for the cooperative extension service, and the cooperation of major firms who may benefit from the maintenance of the FVH in the USA. It should be remembered that past successful efforts at mechanization usually included a large government role. Also, since some of the sectors in need of change are small, coordination among them is necessary. The elements of the change, all of which encourage family settlement of farm workers, include: (i) technological innovations; (ii) employer and worker-advocate encouragement of settlement; and (iii) government encouragement and subsidization.

Technological innovations

1. Back-saving technological changes. These are often called task facilitators or labour aids (Table 4.6). They usually do not lower the cost for the employer and so they do not affect the demand for labour. However, by allowing older men and women to work, they expand the labour supply. These techniques reduce costs by reducing Workers' Compensation payments for musculo-skeletal diseases, one of the biggest problems for farm workers. They are much more common in Europe and Australia than in the USA (J. Thompson, Cooperative Agricultural Extension Service/

Table 4.6. Examples of labour aids and productivity enhancing technologies (already tested).

Productivity enhancing
- Mature green (bush) tomato harvester
- Tying of cauliflower leaves before harvest
- Mechanical harvester for fresh market onions
- Harvester for pickling cucumbers
- Cling peach harvester
- Tobacco harvester

Labour aids or task facilitators
- Conveyor belt for row crops to palletized areas
- Platforms, booms, derricks for harvesting or pruning trees
- Sleds for hauling strawberries, broccoli and celery
- Battery-run clippers for citrus harvest
- Deceleration tube for citrus harvest
- Three-legged ladder for citrus

University of California, Davis, 1999, personal communication). At times, they may increase productivity (e.g. material movers in a nursery environment) or they may slow down production. For example, derricks and booms in orchards may slow down the speed of the crew to the slowest worker.
2. Buck-saving productivity-enhancing changes. These technologies (e.g. mechanical harvesters and pruners) actually lower the total demand for labour. However, by lowering the demand at peak or spike seasons, these changes may actually even out the demand for a more year-round labour force through the year.
3. Introduction of new cultural practices, use of machine-appropriate cultivars and planting practices that anticipate mechanical techniques and lighten tasks in the short term. For example, the planting of dwarf trees, and the pruning and trellising of existing trees for easy access for pickers, will increase the off-season work and lighten tasks for current workers.
4. The staggering of crops and the encouragement of breeding of crops that come to fruition at different periods through the year. This will stimulate the settlement of a labour force that can make a living at farm work.
5. Possible gains in the packing sheds in the area of optical sizing and grading are very promising and should be stressed. This development favours complementary female employment for farm worker families.

Employer and worker-advocate encouragement of settlement

1. Labour-sharing management schemes should be encouraged across crops and areas. The Florida juice orange industry is actively promoting such ideas (G. Ing, Washington State Apple Board, Seattle, Washington, 1999, personal communication). The 9–10 month year may be sufficient to keep a settled labour force with low turnover, but one willing to accept seasonal tasks (G.K. Brown, Florida Citrus Mutual Project, Gaithersburg, 1999, personal communication).
2. The juice orange industry of Florida should be used as a model of how to organize a sector for technological change. Industry groups, as in Florida, should be set up in each sector to encourage technological change, competitiveness and the stabilization of the labour force. In Florida, the juice orange industry is well on its way to a full technological adjustment (G.K. Brown, Florida Citrus Mutual Project, Gaithersburg, 1999, personal communication).
3. Support of worker-advocate and union movements, who will need to reconceptualize the interest of a smaller but more skilled labour force, will be required. The high-turnover workforce of today cannot be adequately protected by these advocates in any case.

Government encouragement and subsidization

1. Specific subsidies for growers to experiment with new techniques should be granted to encourage the spreading of available technologies (e.g. raisin grapes).
2. Encouragement of the development of technologies for small growers and growers of organic and speciality crops should be a high priority.
3. In this context, it is crucial to emphasize that minimum labour standards enforcement will have to be greatly expanded. Collective bargaining laws should be expanded to agriculture and enforced as well. Improvements in productivity will not automatically be captured by workers.
4. Farm worker advocacy and service delivery groups should be used to help with housing, training, and delivery of social and health care services to the settling population. Budgets for these services need to be increased.
5. A large-scale but once-only regularization of the immigration status for the current labour force that does not have legal papers will be necessary if the industry is to have time to adjust. Many of the crops will need seasonal labour for years to come. Farmers and advocacy groups should be utilized, to minimize fraud in the programme and to

guarantee that the legalized workforce has experience in US agriculture.

Discussion of the Short- and Long-term Possibilities

Despite the fact that the remaining unmechanized sectors tend to be smaller and more difficult to mechanize, a transition is possible from the current (chaotic) labour market to one with stable characteristics. Social impact studies should be undertaken in each sector to determine the social changes required by the new technologies. Which workers will be displaced? What will be the skill needs and the numbers of those who will replace them in the altered industries?

Despite the long-term obstacles in many sectors, it should be recalled that technological change is not far away in many current crops. Some will be changes short of harvest mechanization; for example, the use of dwarf trees coupled with extensive trellising and pruning. As these initial relatively easy-to-obtain changes occur, and the industries see that a transition is possible and inevitable, resources and commitment may appear within industries in support of a more stable labour market.

According to agricultural engineers, the potential for change is particularly great at this time because of ongoing work in three areas: computer and sensor technology, which will be used mostly in packing houses; plant breeding, which may lead to uniform ripening, smaller trees and the spreading of harvests across more months; and, finally, the interest in fresh-cut presentations of vegetables and fruits, which favours mechanization (Sarig *et al.*, 1999). The harvesting of most processed vegetables, fruits and nuts and most plants grown below the ground have already been mechanized. Most fresh fruits and vegetables are a long way from a

Table 4.7. Technological availability for mechanization in US labour-intensive agriculture. (Sources: Sarig *et al.* (1999) and G.K. Brown, J. Thompson, G. Ing, D. Peterson and D. Linker, 1999, personal communications.)

Already mechanized	Nearly or already available (1–2 years)	A few years off (2–5 years)	Difficult (at least 5–10 years in future)
• Nuts, prunes, dates, figs pick • Tart cherries pick • Wine, juice grapes pick • Process tomatoes pick • Canned, dried or frozen vegetables pick • Beets, sugarbeets, potatoes, processed onions harvest • Beans, peas, cotton, grains	• Juice orange pick • Pickling cucumber pick • Pick of fresh cut or shredded lettuce, celery, spinach, etc. • Garlic, fresh market onion, green onion pick • Tying of cauliflower leaves before harvest • Specialty wines	• Raisin grape pick • Black olives, process apricots, peaches pick • Apples sauce, juice, slices • Tobacco harvest • Fresh hot peppers, snap beans • Fresh sweetcorn • Broccoli and cauliflower harvest[a] • Green bush tomatoes	• Strawberries, fresh blueberries pick, fresh caneberries (brambles) • Fresh stone fruits pick (e.g. pears, plums, peaches, apricots) • Table grapes pick, prune • Sugar prunes, sweet cherries • Fresh citrus pick • Fresh asparagus, melons, cucumbers, staked tomatoes, sweet peppers and squash pick • Box-packed lettuce pick

[a]Controversial: D. Linker reports possibilities if growers willing to improve packing sheds to handle product with waste material from the field. Sarig *et al.* (1999) report as many years off.

mechanical harvest. Still, there is a series of crops that are on the verge of technological change (Table 4.7). If no new low-cost labour programme or other means of holding down labour costs becomes available soon, growers in these sectors may opt for investing in change.

In the area of citrus and other tree fruit, changes may be not far off. The Florida orange juice industry has invested millions of dollars and considerable effort in developing several shake-and-catch methods of harvesting oranges suitable for its 600,000 acres of juice oranges. The prediction is that these will be implemented over the coming years. The change will substitute an altered and smaller labour force for the current 45,000 workers who climb ladders and pick the fruit by hand. There is discussion about picking fresh citrus as well. This may be made possible by extraordinary gains in the packing sheds. Optical sizing and grading equipment may be able to sort out immature, blemished, undersized and other undesirable fruit mechanically. The mechanical fresh citrus harvest, however, is quite a few years off.

The wine harvest, already 70% mechanized, will most likely continue its climb toward nearly full mechanization. The expanding new vineyards tend to be mechanized from the start. In the mechanized vineyards, a high proportion of year-round jobs are created because the cultural tasks of tying, trellising and caring for the grapes continue throughout the year and the peak harvest demand for labour is attenuated.

The pickling cucumber harvester was used more in the mid-1990s than it is today. Demand-reducing change will occur if the price of labour rises. As mentioned, the expansion of the market for valued-added lettuce, celery, spinach and other crops will allow for mechanized harvest of these crops. Mechanical garlic, fresh market onion and green onion pickers, and cauliflower tiers are close to mechanization. Vidalias (onions) in Georgia were machine harvested in 1999, and the cauliflower tier has extensive use; however, more experimentation is needed.

A few years off from a fundamental change is the 40,000-man pick of raisin grapes. The technology is there but the desire to commit to an expensive conversion with a long payback period is not present under current conditions. Acreage did increase by 50% to 1500 acres in 1999. In the black olive, processed peach and apricot industries, the need for hedge pruning, the breeding of more uniformly ripening crops and a faster sorter in the packing houses are holding back progress. Apparently, shake-and-catch for processed apples may be possible soon. The harvesting machines for fresh hot peppers, sweetcorn and snap beans are available. Mechanical harvesting of the cabbage family and green bush tomatoes is more controversial, but practical according to some engineers. Tobacco harvesters are available and are apparently in use in flat areas (Sarig *et al.*, 1999; D. Peterson, G.K. Brown, G. Ing, J. Thompson, D. Linker, 1999, personal communications).

All of these potential changes over the next few years together represent a significant reduction in the demand for labour, but, as can be seen by the list of difficult-to-mechanize crops, there will be a demand for a large seasonal labour force in the USA well into the future. These crops include fresh berries, stone fruit, table grapes, most fresh vegetables and boxed lettuce. Because of uneven ripening and extreme perishability, such crops may have to wait for some future revolution in robotics before mechanization is possible. However, it may be possible to manage the reduced demand for seasonal labour in these crops upon mechanization so that the current labour force (with some reliance on former farm workers) can meet the needs.

Conclusion

Labour-intensive agriculture in the United States does not need new guestworker programmes or the lessening of protections under the current programme. What it needs is a maintenance of the present workforce, and improvement in the conditions of work so that the workers will stay, bring their families and spend their entire careers on farms. The historical example of the period after the Bracero Program when relatively higher wages, the spreading of service programmes for farm workers, and the upsurge of unionization led to improved conditions and greater family reunification among agricultural labourers is important. Immigration and immigrant policies must actively support family rather than solo-male migration. Even under current conditions, a substantial portion of farm workers who were legalized under IRCA (about one in five of all farm workers) are still working in crop production. In fact, many are now in the process of becoming US citizens.

In general, farm workers who continue to perform farm tasks have incorporated skilled tasks in their portfolio of income-generating activities. The model of these successful farm workers can help in designing a work environment that will maintain the long-term commitment of a much larger share of those who do farm work to the industry. Higher wages and longer-term employment, better working and living conditions, access to home ownership, and access to education for children and employment opportunities for spouses are needed. New technologies and a stable farm labour force are not only desirable but also perhaps the most practical way for US labour-intensive agriculture to compete with other countries in the coming decades. At the same time, by reducing the magnet of thousands of new agricultural jobs, the social problems that accompany an excessive influx of low-wage workers will be lessened. The introduction of new technologies and the creation of a stable farm labour force will require subsidies for certain groups, but these may prove less expensive and more desirable than accommodating the perpetuation of the cheap labour market.

References and Further Reading

Alarcon, R. (1995) *Immigrants or Transnational Workers? The Settlement Process Among Mexicans in Rural California*. Report for the California Institute for Rural Studies, Davis, California.

Bustamante, J. (1979) Las mercancias migratorias. Indocumentados y capitalismo: un enfoque. *Nexos* 14.

Calavita, K. (1992) *Inside the State: the Bracero Program Immigration, and the I.N.S.* Routledge, New York.

California Assembly (1969) *Profile of California Farmworkers*. Sacramento, California.

California State Employment Service (1950–1974) *Farm Labor Report No. 881A*. California.

Cargill, B. and Rossmiller, G. (1969) *Fruit and Vegetable Mechanization*. RMC Report No. 17. Michigan State University, East Lansing, Michigan.

Farm Labor Reports (various years) *Quarterly Agricultural Labour Survey*. National Agricultural Statistical Survey, US Department of Agriculture, Washington, DC.

Garcia y Griego, M. (1996) The importation of Mexican contract labourers to the United States, 1942–1964. In: Gutiérrez, D. (ed.) *Between Two Worlds: Mexican Immigrants in the United States*. Scholarly Resources, Wilmington, Delaware.

Holt, R. (1984) Wages and conditions for California Braceros. PhD Dissertation, South Missouri University, Missouri.

Kissam, E. (2000) *No Longer Children*. Office of Policy, US Department of Agriculture, Washington, DC.

Meillassoux, C. (1977) *Mujeres, graneros y capitales: economia domestica y capitalismo*. Siglo XXI, Madrid, Spain.

Mines, R. (1980) Developing a community tradition of migration. *Monograph No. 3*. Program for US–Mexican Studies, University of California, San Diego, California.

Mines, R. and Anzaldua, R. (1982) *New Migrants vs. Old Migrants: Alternative Labor Market Structures in the California Citrus Industry*. Program for US–Mexican Studies, *Monograph No. 9*. University of California, San Diego, California.

Mines, R. and Martin, P. (1986) *Profile of California Farmworkers*. Giannini Foundation, Berkeley, California.

Palerm, J.V. (1991) *Farm Labour Needs and Farm Workers in California, 1970 to 1989*. Employment Development Department, Sacramento, California.

Portes, A. and Rumbaut, R. (1996) *Immigrant America: a Portrait*. University of California Press, Berkeley, California.

Rosenberg, H., Steirman, A., Gabbard, S. and Mines, R. (1998) *Who Works on California Farms? Demographic and Employment Findings from the National Agricultural Workers Survey.*

Agriculture and Natural Resources Publication No. 21583. Agricultural Personnel Management Program, University of California, Berkeley, California.

Rural Migration Review (1999) University of California, Davis, California, 5(4), October (online).

Sarig, Y., Thompson, J.F. and Brown, G.K. (1999) The status of fruit and vegetable harvest mechanization in the US. ASAE meeting presentation, Toronto, Canada, 22 July.

Zabin, C., Kearny, M., Garcia, A., Runsten, D. and Nagengast, C. (1993) *A New Cycle of Poverty: Mixtec Migrants in California Agriculture.* California Institute for Rural Studies, Davis, California.

Section II

Hired Farm Labour, Employers and Community Response

Section II focuses on the interrelationships between farm workers, their employers and the communities in which they live and work. As shown by the chapters included in this section, the hired workforce is changing in size and composition. Both of these dimensions of the hired labour force have implications for employers of farm labour, local communities, government policy and the countries or regions of origin, when immigration occurs. Important questions that arise include: What approaches can employers use to ensure an adequate supply of workers when needed by the farm operation? How can employers improve their work relationships with farm workers? What are the impacts of a changing workforce, and particularly an immigrant workforce, on local communities? To what extent are workers able to access public services, including those typically available to individuals and families in poverty? Finally, does the farm labour force have other 'spillover' effects or externalities – positive or negative – that should be considered and accounted for?

The first three chapters in Section II are case studies that assess changes in or affecting the hired farm workforce in a variety of circumstances. A comparison of the three chapters provides insights into the labour problem in farm production because the study areas have very different relationships with farm labour. Two of the states (Florida and Washington) have historically been heavily reliant upon migrant and seasonal farm workers for labour-intensive fruit and vegetable production. In contrast, dairy farmers in New York State have historically depended heavily on family labour and, to a lesser extent, on hired labour that is full-time. Dairy farms in New York State have not typically relied on immigrant labour – and labour that shows a certain degree of mobility – until now.

Specifically, Chapter 5 (Roka and Emerson) examines the number and economic conditions of farm workers in Florida, a state highly dependent upon farm workers in citrus and other fresh fruit and vegetable production. Chapter 7 (Thilmany and Miller) assesses the farm worker situation in Washington State, another state highly dependent on farm labour. In contrast, Chapter 6 (Maloney) looks at the dairy industry in New York State, which has turned to Hispanic workers to fill labour needs. Maloney's chapter, using a survey of employers in New York State, explores the question of how farmers and communities that have not been accustomed to an immigrant labour force can adjust positively to change. Chapters 5 and 6 seek to answer the question: what strategies can employers use? Chapters 5 and 7 also explore the critical policy issue of whether or not 'guestworkers' should be used, when shortages in the farm production sector occur. Or should real wages be allowed to rise?

The four chapters that follow take a more 'macro' view than the initial state-level studies. From a community perspective, one important issue is how communities are adjusting to changes in the farm workforce, when these changes occur. In the USA, for example, rural communities were affected in the 1990s by increasing populations of Hispanic farm and processing facility workers, very unlike the local populations. The issues are complex, and community responses have ranged from concern over lack of services (e.g. housing, health care, education and language-skill training) to hostility based on race and ethnicity.

Chapters 8 (Rosenbaum) and 9 (Denton) explore the impacts of farm labour on local communities or regions, using very different approaches. Rosenbaum quantifies the impacts of farm workers in southeastern Michigan, to answer the question: what are the impacts on the local economy if farm workers were not available to work in agriculture? Rosenbaum documents positive impacts on the local economy resulting from the additional economic activity that takes place linked to migrant and seasonal farm workers. The chapter argues that the hiring of workers generates benefits for local communities. What is problematic is that at least some local communities do not see it this way, particularly if hired farm workers are brought in from outside the community and if farm employers are not themselves 'local'. Taking a historical approach, Denton documents these problems, graphically and in detail, in a study of the introduction of Hispanic migrant workers into rural Kentucky.

The chapters that follow by Zahniser and Greenwood (Chapter 10) and Emerson and Roka (Chapter 11) assess: (i) whether the knowledge gained by farm workers in another country or region is transferable to the origin, resulting in long-term gains for those that migrate and return (Chapter 10); and (ii) whether changes in the distribution of income affect the incomes earned by farm workers by influencing the alternative employment opportunities open to them (Chapter 11). While the studies focus on the USA, the issues are applicable in a wide variety of countries that also employ immigrants in agriculture and that have witnessed a widening gap between the rich and the poor.

Section II concludes with three chapters that explore farm labour in different contexts (Australia, Canada and California), using approaches from sociology and anthropology. Chapter 12 (McAllister) describes the farm workforce in parts of Australia, asserting that it is the farm labourer who is at the very bottom of the agricultural production system. What is particularly interesting about this chapter is the last section, which describes (based on qualitative research) those who do farm work in Australia's fruit and sugarcane industries, ranging from retirees to backpackers to foreign workers. Chapters 13 (Mysyk) and 14 (Krissman) then describe the farm worker's condition from an anthropological perspective. Mysyk and Krissman argue that farm workers are the people left behind, exploited by large-scale employers and by policy that supports these employers. Unlike the earlier chapters in this section that focus on the economic outcomes of a system of employment that is largely competitive, these chapters argue that the true situation is one of concentrated economic power.

Section II demonstrates the difficulties and challenges of improving the condition of farm workers, despite a shared concern for this population – a population critical to the success of agriculture in much of the world.

5 Demographics, Income and Choices: Seasonal Farm Workers in Southwest Florida

Fritz Roka[1] and Robert D. Emerson[2]

[1]*Southwest Florida Research and Education Center, University of Florida, Immokalee, Florida, USA;* [2]*Food and Resource Economics Department, University of Florida, Gainesville, Florida, USA*

Abstract

Citrus and fresh vegetable growers rely on a large number of seasonal farm workers to plant, grow, harvest and pack the crops. Citrus acreage has risen dramatically, particularly in southwest Florida, with a corresponding increased concern about availability of harvesting labour. Due to the concerns over having an adequate number of farm workers, agricultural and community leaders in southwest Florida are interested in documenting the current size and demographic profile of farm workers in the region. This chapter summarizes survey results for the area and considers some of the options that agricultural employers and community leaders are considering that may attract a sufficient number of farm workers in the foreseeable future.

Introduction

Southwest Florida, an important citrus and fresh vegetable production region, employs a large number of seasonal agricultural workers. Beginning in 1995, seasonal labour demands increased significantly as large new acreages of young citrus groves began to mature and bear fruit. As in many such areas, southwest Florida has a limited local labour pool from which to draw during the peak harvest season. Consequently, the citrus and vegetable industries of southwest Florida depend upon a sizeable number of migrant farm workers to harvest crops and perform various field tasks during the growing season. The availability of an adequate number of seasonal farm workers who migrate in and out of the region has become an issue of increasing concern among agricultural producers. This chapter describes the employment picture of seasonal farm workers in southwest Florida. The picture includes an estimate of the workers employed by farming operations, farm worker demographics, and income-earning opportunities for workers on and off the farm. While growers in southwest Florida have not yet experienced any labour shortages, the expansion of the citrus industry in the local area and the expected increase in labour demand have caused agricultural employers to consider some alternatives that

 The Dynamics of Hired Farm Labour (eds J.L. Findeis, A.M. Vandeman, J.M. Larson and J.L. Runyan)

would attract an adequate number of seasonal farm workers.

This chapter is in four parts. Firstly, agricultural production and its economic importance to the southwest Florida economy are discussed. Secondly, the chapter summarizes labour requirements for the agricultural operations in the region and presents an estimate of farm workers available to meet the overall demand. Thirdly, demographic and income data collected through employer and farm worker surveys are presented. Finally, the chapter examines strategies currently being discussed among agricultural and community leaders to recruit and maintain an adequate number of seasonal farm workers.

Results presented here are based on surveys of farm workers and agricultural employers. The surveys were part of a local research effort initiated to examine the housing needs of farm workers. The study was funded by a coalition of interest groups, including county commissioners, state development agencies, farm worker advocates and agricultural employers. The University of Florida and the Southwest Florida Regional Planning Council coordinated the surveys, which were conducted during two agricultural seasons: 1997/8 and 1998/9.

In this chapter, a seasonal farm worker is defined as someone employed on a day-to-day basis regardless of their home base. Seasonal farm workers (for the purposes of this chapter) may be either permanent residents or migratory farm workers. By contrast, some labour regulations define seasonal farm workers on the basis of their permanent residency status, distinguishing them from migratory farm workers who leave an agricultural area once the production season has finished.

Agricultural Production in Southwest Florida

Southwest Florida is a five-county region including Charlotte, Collier, Glades, Hendry and Lee Counties. The southwest region is a major agricultural production area for Florida. More than 25% of the state's citrus acreage and 30% of the state's vegetable acreage are located in this region (Roka and Cook, 1998).

The acreage and approximate annual production value by commodity class are summarized in Table 5.1. Overall, agriculture in southwest Florida utilizes 40% of the land area and generates US$900 million of farm-gate sales annually. While citrus and vegetable acreages account for less than 20% of the farmed land, they generate more than 70% of the annual agricultural sales. More than 80% of the citrus acreage is devoted to round oranges (early, midseason, navels and late-season varieties), most of which are processed into juice products. Tomatoes account for more than a third of the vegetable acreage

Table 5.1. Acreage and value of agricultural production in southwest Florida. (Source: Roka and Cook, 1998.)

	Acreage		Value of production	
Commodity	Acres[a]	%	US$ ('000)	%
Citrus	180,000	13	250,000	28
Vegetable	45,000	3	400,000	44
Sugarcane	90,000	6	100,000	11
Pasture/cattle	1,100,000	77	50,000	6
Ornamental	5,000	< 1	100,000	11
Total	1,420,000	100	900,000	100

[a]1 acre = 0.4047 ha.

reported in Table 5.1 and all are sold in the fresh market.

As with most intensive fruit and vegetable production areas, the region produces and sells outside the region while channelling income and employment into the region. The indirect economic impacts from agricultural production in the region are estimated to be between 60 and 70 cents per dollar of farm gate sales (Mulkey *et al.*, 1997). Therefore, in addition to the US$900 million of farm sales, agricultural operations help to sustain another US$500 to US$700 million of economic activity in southwest Florida.

Southwest Florida has long been regarded as a winter vegetable production area. A favourable winter climate has allowed producers in southwest Florida to supply the northeastern US and Canadian terminal markets with fresh winter vegetables. It has only been in the last decade that southwest Florida has become a significant citrus production area. In 1986, southwest Florida accounted for less than 12% of the state's citrus acreage. Severe winter freezes during the early and mid-1980s destroyed significant citrus acreage in central Florida, resulting in extensive plantings of new groves in southwest Florida. Today, southwest Florida accounts for 25% of the state's citrus tree inventory and more than 20% of the state's annual citrus production (FASS, 1998b).

Table 5.2. Citrus and tomato acreage in southwest Florida.

Year	Citrus[a] (acres)[c]	Tomatoes[b] (acres)[c]
1986	72,480	15,400
1988	96,089	20,100
1990	126,252	19,800
1992	157,239	19,250
1993	166,940	21,600
1994	176,641	19,000
1995	177,867	18,000
1996	179,093	14,700
1997	179,948	15,650
% change 1986–1998	+148%	+1.6%

[a]FASS, 1998a. [b]FASS, 1999. [c]1 acre = 0.4047 ha.

Citrus acreage is almost 2.5 times the 1986 acreage levels (Table 5.2). Tomato acreage has fluctuated since 1986, rising by 40% in the peak production season of 1993/4, only to return to within 2% of the 1986/7 acreage during the 1997/8 season. Increased competition from Mexican growers and an expanding hot-house vegetable industry in Canada, The Netherlands and Spain has eroded southwest Florida's share of the fresh winter vegetable market (Van Sickle, 1996).

Seasonal Farm Worker Requirements

The production season for a particular vegetable planting is concentrated into a narrow window of time. For instance, an individual block of tomatoes requires only 80 to 90 days from planting until final harvest. However, sequential vegetable plantings in southwest Florida allow the production season to stretch from August until late May. Depending on the weather conditions, the first blocks of tomatoes are harvested in late October or early November. Harvesting of tomatoes, bell peppers and cucumbers continues until late April or early May. Watermelons are harvested between mid-April and late May. By early June, vegetable fields are cleared and left fallow due to unfavourable growing conditions until August, when the process starts over again.

A complicating characteristic of expanding citrus harvesting in southwest Florida is that it overlaps with vegetable production activities. Early grapefruit and tangerine varieties are harvested during early October. Harvesting of early and midseason-variety oranges begins in November and peaks during January. The late-season Valencia orange harvest starts in March and continues through May.

Citrus and vegetable growers depend on large numbers of seasonal farm workers to grow and harvest the crops. Citrus growers employ seasonal labour primarily for harvesting. Assuming an average yield for round oranges of 500 boxes per acre (about 1234 boxes ha^{-1}) (at 90 pounds or 40.8 kg per box) and an average harvest productivity rate of

10 boxes per hour, 50 labour-hours are required to harvest 1 acre (about 124 hours per ha) (Table 5.3). More than 55 million boxes of citrus were harvested in southwest Florida during the 1997/8 season, requiring between 5 and 6 million hours of harvest labour from seasonal farm workers (FASS, 1998a; Roka and Cook, 1998).

Vegetable growers utilize seasonal farm workers in a wider variety of tasks. In addition to harvesting crops, seasonal farm workers transplant, stake, tie, weed, prune and help during field clean-up procedures. Based on specific yield assumptions and estimates of average worker productivity rates, almost 200 labour-hours are required to grow and harvest 1 acre (0.4 ha) of tomatoes (Table 5.3). With 15,000 acres (6070 ha) of fresh market tomatoes in southwest Florida, more than 3 million h of seasonal labour are required each growing season. Payments to seasonal tomato farm workers are estimated to be approximately one-third of total production and harvest costs (Smith and Taylor, 1998).

It would seem to be a straightforward task to translate required labour-hours into worker counts by simply aggregating across employer payroll records. The major problem behind this method is the potential to double count a significant number of workers. Most seasonal workers are organized around crew leaders. While some crews have established relationships with single growers, others move from farm to farm depending on the type and availability of work. In addition, there is turnover within crews as individual farm workers shift among crews or even between production regions.

An alternative approach is to estimate the number of seasonal farm workers by quantifying three variables: daily acreage by field task, average worker productivity by field task, and average daily hours of employment per worker. All variables are tied to a specific reference week. Acreage and productivity rates combine to provide an estimate of total hours worked. Dividing by the number of hours per week contributed by an average worker translates total hours into a farm worker count relevant for the reference week. Acreage by field task and worker productivity rates are estimated from weekly crop reporting sheets and payroll records of employers. Hours of work performed are estimated from farm worker surveys reporting the hours worked during the reference period.

Using the above method, the seasonal farm workforce in southwest Florida was estimated to be between 16,000 and 17,000 field workers during the peak of the 1997/8 growing season (Roka and Cook, 1998). Peak employment of seasonal farm workers was assumed by between 4 and 11 January. By contrast, the Shimberg Center for Affordable Housing estimated that there were more than 25,000 migrant farm workers in southwest Florida during the 1995 season (Smith, 1997). Moreover, the Shimberg estimate did not

Table 5.3. Seasonal labour requirements per acre for citrus and tomatoes.

	Citrus	Tomatoes
Production level (per acre)	500 boxes	1400 cartons
Field tasks[a] (per acre)	–	114 hours
Harvest[b] (per acre)	50 hours	80 hours
Seasonal labour costs	27%[c]	33%[d]

[a]Includes transplanting, staking, tying, and plastic removal. Source: Smith and Taylor, 1998.
[b]Source: Roka, 1998.
[c]Assuming a delivered-in price of US$6.00 per box and a US$1.60 per box cost to harvest and transport fruit to the edge of a grove.
[d]Source: Smith and Taylor, 1998.

include the seasonal farm workers who were permanent residents. The worker counts from the above two studies converged, however, when family dependents were included. The Roka and Cook study estimated the population of seasonal (including migrant) farm workers and their family dependents to be between 35,000 and 48,000. The Shimberg Center's estimate for migrant farm workers and their dependents was 31,000. If migrant farm workers represent between 75 and 90% of the overall farm labour force, then the estimates developed by Roka and Cook (1998) are comparable to the estimate developed by the Shimberg Center. As an additional estimate of farm worker numbers in southwest Florida, the 1990 Atlas of State Profile estimated the total farm worker population in southwest Florida, including family dependents, to be almost 54,000 people (US Department of Health and Human Services, 1990).

The demand for seasonal farm labour in southwest Florida has dramatically increased over the past few years with the expansion of the citrus industry. More than 53,000 new citrus acres (more than 23,450 ha) have been planted since 1990. This represents an increase of nearly 2.7 million labour-hours to hand harvest the crop.

Demographic Characteristics and Income

Demographic profile

More than 1500 farm workers were interviewed in southwest Florida during the 1997 and 1998 production seasons. The objectives of the worker surveys were to develop demographic and income profiles of seasonal farm workers. Farm workers were asked about their agricultural experience, the type of work they performed, household composition, and whether or not they migrated for farm work. Standard demographic questions of age, gender and ethnic background were also included.

Personal interviews were conducted in the language of choice: Spanish, Creole, or English. The interviewer stressed during opening remarks that individual names and US Immigration and Naturalization Service (INS) documentation were not a part of the survey; consequently, the survey results could not address questions regarding residency and work authorization status.

The demographic characteristics of the farm workers interviewed between February and March of 1999 are summarized in Table 5.4. Responses were separated into three

Table 5.4. Comparison of seasonal farm worker demographics among citrus harvesters, vegetable field workers and vegetable packing-house workers. (Source: Roka, 1999.)

Characteristics	Citrus harvesters	Vegetable field workers	Vegetable packing house
Age (years)	31	31	32
Farm experience (years)	7	7	7
Demography (% of workers)			
Male	90	83	52
Female	10	17	48
Married/families	26	33	67
Single/separated	74	67	33
Ethnicity			
American white	0	0	0
American black	1	0	2
Mexican	91	79	51
Guatemalan	7	19	18
Haitian	0	1	26
Other	1	1	3
Migrant	75	75	54
Number of workers interviewed	85	460	65

employment categories: citrus harvesters, vegetable field workers, and vegetable packing-house workers. Field workers in both citrus and vegetable operations were predominantly male, Hispanic, and living apart from their immediate families. A slightly higher percentage of women were involved in vegetable production than in citrus harvesting. A far greater percentage of female workers and workers living with their immediate families were found in vegetable packing-houses. The percentage of migrant workers was higher among citrus and vegetable field workers than among packing-house workers. The greatest ethnic diversity was found among packing-house workers.

Table 5.5 illustrates the substantial changes that have occurred in the demographic profiles of citrus harvesters between 1970 and 1999. Over the 30-year span, four important demographic shifts have occurred: (i) farm workers are younger and have fewer years of farm experience; (ii) there has been an increase in the percentage of farm workers who are single men and a corresponding decrease in the percentage of farm workers who live with their immediate families; (iii) there has been an increase in the percentage of farm workers who migrate out of Florida once the agricultural season is finished; and (iv) there are fewer farm workers who claim to be US citizens. The majority of citrus harvesters in 1970 were US citizens. In 1997 the National Agricultural Workers Survey (NAWS) estimated the undocumented workforce within Florida agriculture to be at least 40% (Gabbard, 1997). By definition, an undocumented worker is not a US citizen.

Farm worker income

Agricultural employers are faced with competitive markets not only for the products they sell, but also for the seasonal labour they employ. The USA and southwest Florida, in particular, have enjoyed an extended period of economic prosperity. In southwest Florida the number of new jobs has increased at a faster rate than the growth in the region's population. Over the past 5 years, the unemployment rate has been declining. At the end of 1998, the unemployment rate was between 3 and 4% in the coastal counties of

Table 5.5. Comparison of citrus harvester demographics across time.

Characteristics	1970[a]	1990[b]	1999[c]
Age (years)	41	31	31
Farm experience (years)	18	7	7
Demography (% of workers)			
Male	86	95	90
Female	14	5	10
Married/families	58	na	26
Single/separated	42	na	74
Ethnicity			
American white	30	1	0
American black	60	11	1
Mexican	10	84	91
Guatemalan	0	0	7
Haitian	0	4	0
Other	0	0	1
Migrant	42	na	75
US citizenship	95	na	na

[a]Fairchild, 1975. [b]Polopolus *et al.*, 1996. [c]Roka, 1999.
na = not available.

Lee, Charlotte and Collier (Florida Department of Labor and Employment Security, 1998). Rural Glades and Hendry Counties had higher rates of unemployment: 7 and 8.5%, respectively, but these rates are between 3 and 6 percentage points lower than unemployment levels in 1995.

Starting wages for some of the entry-level non-agricultural positions in southwest Florida are presented in Table 5.6. These positions are low-skilled jobs and require minimal job qualifications. English ability is not essential so long as the position does not involve public interaction. All job categories listed in Table 5.6 pay above the US minimum wage (US$5.15 per hour). More importantly, these jobs offer a continuing employment period. In most cases, the job offers year-round employment with 40 hours per week. In some cases, hours per week are reduced to between 25 and 30 during off-season months (May through October). Although there is typically no extensive switching between farm and non-farm employment by the same worker, a pervasive empirical result in farm labour markets is that non-farm wage rates are a strong determinant of labour supply for agriculture (Chapter 11, this volume).

Table 5.7 summarizes the comparable income data collected from southwest Florida agricultural employers and farm workers. Much of the fruit and vegetable harvest is paid on the basis of piece rates. Assuming normal yields and average productivity, many workers can earn between US$7.00 and US$9.00 per hour while they are harvesting. Survey responses from farm workers tend to support employer data on hourly earnings during harvesting. The sample of citrus harvesters during 1999 earned more than US$7.00 per hour. The interview period, between early February and late March, coincided with the middle of the citrus harvesting period. During the same period, harvesting activities on vegetable operations were limited. The autumn/winter tomato season had passed and the spring tomato crop was just beginning. Consequently, jobs on vegetable farms were limited to field tasks paying minimum wage rates. While farm workers earn comparable hourly wages during harvest periods, they lack a sufficient quantity of harvesting hours to generate a comparable annual income. Farm workers report annual incomes from farm work of between US$7000 and US$9000, whereas a general labourer in a landscaping company can earn at least US$12,000 working 40 hours per week for 50 weeks at US$6.00 per hour.

Farm workers reported annual earmings of between US$7000 and US$9000. However, given the ambiguity of the questions concerning annual income and lack of verifying documents, it remains unclear whether an individual was reporting income from 12 months of employment or whether the reported annual earnings reflected less than 12 months and only farm wages earned in southwest Florida. At least 75% of the seasonal farm workers migrate out of southwest Florida after the harvest season. A third of the migrant farm workers interviewed between 1998 and 1999 reported returning home to

Table 5.6. Hourly wage rates and employment periods for entry-level positions at selected non-agricultural businesses in southwest Florida. (Source: B. Hyman, Senior Statistician, University of Florida, Southwest Florida Research and Education Center, phone survey, 20 October, 1999.)

Business	Hourly wage (US$)	Hours per week	Weeks per year
Fast-food restaurant	5.25–7.00	40	26
		30	26
Hotel housekeeping	5.25–6.40	40	26
		25	26
Golfcourse grounds crew	5.50–8.00	40	52
Landscape labourer	6.50–8.50	40	52

Table 5.7. Summary of 1998 and 1999 income data collected from employer and farm worker surveys. (Source: Roka, 1998.)

	Citrus harvesting	Tomato harvesting
Employer data	*Harvesting*	*Harvesting*
Piece rate	US$0.60–0.90 per box	$0.35–0.50 per bucket
Productivity (per hour)	10 boxes	20 buckets
Daily hours	8	6
Hourly earnings	US$6.00–8.00	$8.00–9.50
Daily earnings	US$50–60	$40–50
Farm worker data	*Harvesting*	*Non-specific field tasks*
Hourly earnings	US$7.40	US$5.20
Weekly hours	50	45
Weekly earnings	US$370	US$235
Good week	US$400	US$290
Fair week	US$320	US$190
Poor week	US$175	US$95
Annual income	US$7000–9500	US$6500–8500

Mexico. The remaining migrant farm workers travelled to other states for agricultural employment. Several community leaders reported that for those farm workers who remain in the southwest Florida region, unemployment insurance has been a supplemental source of income through the summer months.

Complementary seasonal employment in other southwest Florida industries is not a general option. The agricultural employment period coincides with the peak employment period for the other industries with larger proportions of the total employment in southwest Florida, namely tourism and construction. A few southwest Florida agricultural operations have farming operations in several states and are engaged in year-round production. For these companies, year-round work is offered to those farm workers who are willing to relocate to other production areas during the summer months.

Assuming that farm workers are rational economic agents, those workers who have the ability and knowledge of alternative employment options will move toward those jobs that improve their overall economic welfare. Comparing annual earnings between agricultural and non-agricultural jobs, agricultural employers are at a distinct disadvantage in recruiting and maintaining workers with roughly the same skill level if their workers seek year-round employment.

Options to Attract Seasonal Farm Workers

A stronger economy and fewer workers with ties to the southwest Florida community have increased concerns that there will be an insufficient level of farm labour during future production seasons. To alleviate this concern, agricultural growers will have to consider alternative means of attracting a seasonal farm labour workforce. From a grower's perspective, any options for seasonal farm workers must address two issues: (i) it must reduce the uncertainty factor with respect to future worker availability; and (ii) profitability must be either maintained or increased. Vegetable growers compete with producers from Mexico. Florida citrus growers compete with Brazil. If Florida citrus and vegetable growers are not able to maintain or increase profitability, some acreage currently in citrus and vegetables will move offshore.

Four strategies being considered by agricultural and community leaders in southwest Florida to recruit and maintain an adequate supply of seasonal farm workers are: housing; guestworker programmes; improving worker productivity; and increasing piece rates.

Housing

The bed capacity of permitted housing for migrant farm workers in southwest Florida is between 12,000 and 13,000. Clearly, the permitted housing stock is far below even the most conservative estimate of the migrant farm worker population. Several local news agencies have featured cases of overcrowded and decrepit living conditions endured by some farm workers. While there may be occasions when workers choose to save on housing expenses by sharing their living space with several companions, there is also emerging evidence that the relative shortage of adequate housing is forcing farm workers to make choices they would otherwise not make.

Citrus and vegetable growers who have invested in housing for their workers report a more reliable source of farm labour. Lower employee turnover rates reduce recruitment, training and administrative costs. Further, employers with housing have reported that their workers tend to be more productive, perhaps because the employer is able to utilize workers' time more efficiently. Providing housing is expensive and only larger growers with sufficient capital resources have considered building their own stock of housing. Developing farm worker housing typically involves a public interest. In the past, state and federal monies have been invested in southwest Florida to help to construct housing complexes. Recently, another US$30 million was awarded to the State of Florida with provisions to build affordable housing for single men working out of Immokalee (Collier County).

Guestworkers

Given that a large percentage of farm labour immigrates from Mexico and other Central American countries, either modifying the H-2A programme, or establishing a guestworker programme has received considerable attention. Growers need to recognize that an H-2A type programme will increase their direct recruitment costs. Furthermore, growers with experience in using the H-2A note that it takes a year or more to understand fully the procedures of the programme and tailor it to an individual operation.

Worker productivity

Improving worker productivity is an alternative that could increase the effective supply of seasonal labour. Not only would higher individual productivity rates lessen the demand for farm workers; it should also increase labour earnings for the remaining farm workers and, therefore, should increase their economic incentive to remain with agriculture. The mechanical citrus harvesters under development could potentially improve the productivity of a worker by tenfold. Providing workers with back supports or improving ergonomics to facilitate the physical tasks performed in the field could increase overall worker productivity by reducing fatigue and injury, thereby allowing a person more hours of employment.

Piece rates

Increasing the piece rate for citrus and vegetable harvesting has been advocated by a number of farm worker advocacy groups. So long as productivity and hours of work remain unchanged, increasing the piece rate will increase hourly and annual earnings of farm workers. Higher hourly and annual earnings should increase the number of people willing to do farm work. However, there may be situations when higher piece rates reduce overall farm worker income. Weather, disease and market uncertainties make productivity rates and employment hours highly variable among agricultural jobs. Higher piece rates directly increase the unit cost of production; during periods of low market prices, the corresponding higher unit cost may not warrant the full harvest of the crop. Consequently, fewer hours of employment may be available to workers.

Inherent in the argument to increase piece rates is the objective of improving the economic welfare of farm workers. Meeting that objective should not only encourage more people to do farm work, but also retain the workers currently engaged in agricultural employment. Increasing the effective earnings of a farm worker is one obvious approach to improve their economic welfare. However, if a farm worker gains access to better housing at a reduced cost, their economic welfare has improved, even without an increase in earnings. In addition to considering monetary incentives, it is important to consider the whole range of worker preferences. What are the worker's long-term (life) goals? How important is a stable family situation and the educational opportunities either for the worker or members of their family? Knowing what incentives are important to farm workers, the challenge for agricultural employers will be to design incentive packages that attract a sufficient number of quality farm workers while remaining consistent with the overall business goals of the farming operation.

If growers and agricultural employers in southwest Florida continue to rely on seasonal farm workers, they will have to make a greater investment in the seasonal labour market to secure and maintain the required number of workers. Any strategy will have limited potential if it reduces long-term profitability. A successful strategy should consider the preferences of farm workers and work to improve the economic welfare of those individuals who choose to remain with agricultural operations.

References

Fairchild, G. (1975) *Socioeconomic Dimensions of Florida Citrus Harvesting Labor.* ERD Report No. 75-2. Economic Research Department, Florida Department of Citrus, University of Florida, Gainesville, Florida.

FASS (1998a) *Citrus Summary, 1986/87–1997/98.* Florida Agricultural Statistic Service, Orlando, Florida.

FASS (1998b) *Commercial Citrus Inventory 1998.* Florida Agricultural Statistic Service, Orlando, Florida.

FASS (1999) *Vegetable Summary, 1986/87–1997/98.* Florida Agricultural Statistic Service, Orlando, Florida.

Florida Department of Labor and Employment Security (various years) *Local Area Unemployment Statistics.* Division of Jobs and Benefits, Labor Market and Performance Information, Tallahassee, Florida.

Gabbard, S. (1997) *Farm Workers in Florida – a Subset of NAWS.* Aguirre International, San Mateo, California.

Mulkey, D., Degner, R.L., Gran, S. and Clouser, R.L. (1997) *Agriculture in Southwest Florida: Overview and Economic Impact.* Staff Paper No. SP97-3. Food and Resource Economics Department, Institute of Food and Agricultural Sciences, University of Florida, Gainesville, Florida.

Polopolus, L., Emerson, R., Chunkasut, N. and Chung, R. (1996) *The Florida Citrus Harvest: Prevailing Wages, Labor Practices, and Implications.* Final Report to Florida Department of Labor and Employment Security. Division of Labor, Employment and Training, Tallahassee, Florida.

Roka, F. (1998) *1998 Employer Surveys.* Southwest Florida Research and Education Center, University of Florida, Immokalee, Florida.

Roka, F. (1999) *1999 Farm Worker Surveys – Summary Statistics.* Southwest Florida Research and Education Center, University of Florida, Immokalee, Florida.

Roka, F. and Cook, D. (1998) *Farm Workers in Southwest Florida.* Final Report to the Southwest Florida Regional Planning Council. Southwest Florida Regional Planning Council, Fort Myers, Florida.

Smith, M.T. (1997) *Migrant Farm Worker Housing Needs Assessment Methodology.* Shimberg Center for Affordable Housing, University of Florida, Gainesville, Florida.

Smith, S.A. and Taylor, T.G. (1998) *Production Costs of Selected Florida Vegetables 1996–97.* Circular No. 1202. Cooperative Extension Service, University of Florida, Gainesville, Florida.

US Department of Health and Human Services (1990) *An Atlas of State Profiles Which Estimate Number of Migrant and Seasonal Farm Workers and Members of Their Families.* Report to the Migrant Health Program, Bureau of Health Care Delivery and Assistance, Rockville, Maryland.

Van Sickle, J.J. (1996) Florida tomatoes in a global market. *1996 Proceedings of the Florida Tomato Institute.* University of Florida, Gainesville, Florida.

6 Management of Hispanic Employees on New York Dairy Farms: a Survey of Farm Managers

Thomas R. Maloney
Department of Applied Economics and Management, Cornell University, Ithaca, New York, USA

Abstract

In recent years there has been an increasing number of Hispanic employees working on New York State dairy farms. This survey, conducted in 1999, describes the experiences and current employment practices of New York dairy farm managers who employ Hispanic workers. The 20 employers in the study represented most of the dairies in New York that employed Hispanic workers. The purpose of the survey was to benchmark current employment practices on New York dairy farms employing Hispanic workers. The questionnaire examined a variety of employment-related issues, including language, recruiting patterns, wages, transportation, housing and cultural issues. Dairy employers have been very resourceful in recruiting and managing Hispanic workers. Despite culture and language differences, employers found positive ways to manage Hispanic employees.

While most of the working relationships with Hispanic employees have been positive, several challenges exist. Solving the language problem is the greatest initial challenge, since few Hispanic workers speak English. In addition, managers must understand cultural differences to avoid misunderstandings and interpersonal problems. Employers reported other challenges, including illegal immigration, community relations and employee turnover.

Introduction

The practice of hiring Hispanic employees on dairy farms in New York State is relatively new. Most New York dairy employers with Hispanic employees have employed them for 5 years or less. This contrasts with New York's fruit and vegetable industries, where hiring seasonal Hispanic employees has been common for over 40 years.

The motivation for conducting the research reported in this chapter results from the concerns of dairy employers who report difficulty in recruiting and retaining productive employees. This is in part due to the strong current economy and low levels of unemployment. New York dairy employers are seeking alternative labour pools. They are aware that other dairy farmers currently hire Hispanic employees

(eds J.L. Findeis, A.M. Vandeman, J.M. Larson and J.L. Runyan)

and many are asking if this is a viable option for them.

This chapter reports the current employment practices of dairy farm operators in New York who hire Hispanic employees. The objectives of the research reported here are to: (i) benchmark the current practices that dairy managers are using as they recruit, manage and compensate Hispanic employees; (ii) assess the wants and needs of Hispanic employees on dairy farms and how effectively those wants and needs are being met; (iii) develop a set of human resource management practices for dairy farm managers who employ Hispanic employees; and (iv) assess how dairy farm managers feel about their experience in managing Hispanic employees.

Methodology

A telephone survey questionnaire was developed to record the management experiences of dairy farm employers who hire Hispanic employees. The instrument was pretested on two New York State dairy farms. A list of dairy employers who hired Hispanic employees was compiled through the help of Cooperative Extension educators and contacts with New York dairy farmers. Farm employers on the survey list were located throughout upstate New York, with the exception of most of the northern counties. During the period of March through May of 1999, 20 dairy farm operators who employ Hispanic workers were surveyed. While those surveyed did not represent the entire population of dairy operators in New York State who employ Hispanic employees, they did represent the majority.

Nineteen employers were interviewed over the telephone. One employer chose to fill out the survey form and return it. Each telephone conversation lasted from 30 to 40 minutes. In each case, the individual interviewed was one of the people who supervised the Hispanic employees on a daily basis. In all cases the person interviewed was the farm owner or a partner in the business.

Results

The responses to the survey questions from the 20 dairy farm employers surveyed are presented here. Descriptive information provided by the survey participants has also been summarized. In some cases the responses to a question add up to more than 20 because more than one answer was chosen.

Farm profile

The majority of employers reported that the primary reason they began to hire Hispanic employees was that they believed they could not attract local employees willing to do the work required, which was primarily milking cows (Table 6.1). Several employers indicated that they believed the work ethic had changed and that American employees were no longer willing to do dairy farm work, particularly milking. Other employers indicated that unemployment was very low and there was strong competition for qualified employees.

The percentage of the dairy farm workforce made up of Hispanic employees in this study varied. In some cases Hispanic employees made up a substantial portion of the workforce. On seven of the farms studied, Hispanic employees made up 41% or more of the total dairy farm workforce. This can be an important factor if turnover among Hispanic employees on a given dairy farm is high compared with the rest of the workforce.

Employee profile

Mexican employees predominate in the western part of New York State and in the eastern part of the state. Employees from Guatemala are prevalent mostly in central New York, specifically in Cayuga County (Table 6.2).

Seven employers indicated that they employ Hispanic employees who live with their families on the farm. Employers also noted that employees who had their families with them were less likely to request exten-

Table 6.1. Farm profile, New York dairy survey.

Farm size	*Number of cows*
Range	125–3500
Average	869
Median	630
Years employed Hispanic employees	*Number of farms*
10	3
4–5	9
<4	8
Reasons for hiring Hispanic employees	*Number of farms*
Local employees not available or not willing to do work required	18
Hispanic employees came to farm seeking employment	2
Total workforce per farm	*Size of workforce*
Range	8–65
Average	18
Hispanic employees per farm	*Number of Hispanic employees*
Range	2–30
Average	6
Hispanic employees as percentage of total farm labour force	*Number of farms*
0–20	4
21–40	9
41–60	5
61–80	2
81–100	0

ded periods of time off to return to their home country. The majority of employers hired individuals, either young single men or men with families in Mexico or Guatemala. Those employees without their families are likely to return to their home country after a number of months. Another trend noted was that those individuals who left families in their home country were very likely to send all or most of their income home to their families.

The most common recruiting method was word of mouth. Employers, especially in eastern New York, would contact another dairy farm owner who employed Hispanic employees and request referrals from the employees themselves. Hispanic employees, especially in the eastern part of the state, appeared to be highly networked and capable of finding other Mexican employees who were willing to work on a dairy farm. Another trend that surfaced was that the longer an employer had employed Hispanic employees, the more likely the employer was to use word of mouth to recruit future employees rather than use a labour contractor. Half of the employers surveyed indicated that they used a labour contractor to recruit their first Hispanic employees. In most cases employers used a labour contractor who resided in New York State. The contractor receives a fee for each employee placed on the farm; the typical recruiting fee is US$500.

Hispanic employees work from 55 to 70 hours per week and usually a 6-day week. One employer had a rotation of 4 days on and 2 days off. On two farms employees worked 7 days per week. The most typical work schedule was 6 days of 10–12 hours each. The majority of employers reported that their employees came to the USA to work hard and to send their income back home. Therefore, they wanted to work many hours.

Table 6.2. Employee profile, New York dairy survey.

Countries of origin	*Number of farms*
Mexico	17
Guatemala	5
Cuba or USA	2
Family situation of Hispanic employees	
Individuals	10
Young single men	8
Families	5
All of the above	3
Recruitment method	
Word of mouth	14
Labour contractor	10
US Department of Labor, Rural Opportunities, H-2A programme	4
Employees came looking for work	2
Benefits provided	
Housing	20
Health insurance	3
Satellite television	3
Quality bonus	3
Retirement	1
How Hispanic employees meet their transportation needs	
Hispanic employees provide at least some of their own transportation (because they drive themselves or know another Hispanic employee who drives)	15
Farmer transports employees to necessary destinations	11
Farm provides all transportation	2

The majority of employers paid cash wages in the range of US$6.00–US$7.00 per hour, although wages could be as low as US$5.50 per hour and as high as US$9.50 per hour. Most of the jobs filled by Hispanic employees were milking positions. One employer provided a milk quality bonus of US$100 per month per employee in addition to cash wages. All of the employers interviewed provided some form of housing and most of the utilities. Several employers provided health insurance, satellite television, and a milk quality bonus. Other miscellaneous benefits included vacation, sick leave, meat, day care, uniforms and transportation. Some of the employers who provided satellite television did so at the request of the employees so that they could view soccer games or watch programmes in Spanish. In one case the employees paid for satellite television themselves.

The majority of employers reported that extended periods of time off were very important to their Hispanic employees; 13 of the 20 farms surveyed reported that their Hispanic employees had requested extended time off to return to their home country. The young single men and married men who came alone to work were usually not in the USA to stay. Their objective was to send their earnings home to their families and to return home after a period of time. In fact, it also appears that most employees did not have specific plans except that they were likely to return home within 2 or 3 years of taking a job. One employer reported, 'We will never have a one hundred percent Hispanic workforce because of high turnover.' Another employer said that the employees viewed work on a dairy farm as a temporary job. Employers also reported that their Hispanic employees were very good about finding a temporary or permanent replacement when they were leaving. Most employers viewed this very positively. However, some cautioned that not every new employee recruited was the same and it was the employer's role to see that the individual was qualified and willing to do the work. One employer reported that he had hired Hispanic employees for so long that when an employee wanted to leave he called previous employees residing in Mexico to see if any were willing to return to the farm. Based on experience, one employer reported that even though an employee said he was going home for a month or two and would return, there was no guarantee that the employee would return.

Employers reported that some employees arrived at the farm with only a few personal belongings. A few came with their own vehicles and some had a driver's licence. Hispanic employees who had a valid driver's licence were often allowed to drive farm vehicles to shop or go to doctor's appointments. Many Hispanic employees had their own network of friends and family. On days off or for recreation, friends or family members often provided transportation. Farm employers usually provided transportation to and from medical appointments, grocery shopping and laundry for those employees who could not transport themselves. Over half of employers reported that they provided some transportation for their Hispanic employees. When employers provided transportation, they usually did the transporting themselves. Two employers provided almost all of the transportation for their employees because there were no other options.

Language issues

Most employers acknowledged that language differences were at least a moderate problem (Table 6.3). Several reported that language was not a problem because there was someone on the farm who could translate. Most often this was a Hispanic employee who spoke English very well. On 14 of the 20 farms, managers spoke little or no Spanish. In these cases the employer either relied on the English-speaking ability of one or more Hispanic employees or in some cases used an outside translator to help to bridge the gap.

In most cases the employers relied on at least one employee speaking enough English to communicate with the employer as well as to be able to translate for the other employees and train the other employees. In several cases employers said that they hired interpreters especially for staff meetings, performance reviews and other important discussions. In one case the translator also provided English lessons for the Hispanic employees and Spanish lessons for the English-speaking employees. In eight cases employees were involved in taking English classes locally. However, some employers reported that after starting with English lessons, the employees began to lose interest or discontinued attending class. On five of the farms, the manager had taken Spanish classes.

Community issues

In only four cases employers reported that there were difficulties between Hispanic

Table 6.3. Language issues, New York dairy survey.

Extent that language differences are a problem	*Number of farms*
A great extent	4
A moderate extent	12
Not a problem	4
Do Hispanic employees speak English?	
Very little	6
A moderate amount	15
Very good command of English	1

employees and members of the community at large. For example, one employer reported that local landlords were unwilling to rent to employees from Mexico. The majority of employers indicated that they were not very involved in helping their employees adjust to the community. The nine employers who said that they helped their Hispanic employees to adjust to the community indicated that they offered rides to church, helped with shopping or directed employees to places to shop. Some employers also encouraged tutorial programmes and other services provided to migrant employees within the local community.

Quality of life

When asked how their Hispanic employees met their social needs, most of the employers responded that employees had a network of family and friends in other communities in New York State (Table 6.4). Some employees spent weekends or days off with family and friends up to 30 or 40 miles away from the farm. Six employers indicated that their employees attended church, an activity that allowed the employees to practise their religion and interact with others in the community.

When the employers were asked what their Hispanic employees did for recreation, some were not sure and others indicated that employees engaged in sports and other activities. Soccer was the most frequently mentioned sport. Many employers acknowledged that when the employee had time off, there may not be sufficient recreational or social activities to fill their time. Those employees that had satellite television particularly enjoyed Spanish programmes.

Other issues

Six of the 20 employers reported having difficulties with supervision that they attributed to cultural differences. Three of the employers interviewed were women and all of them reported that male Hispanic employees had difficulty accepting their supervision. In each case the women involved addressed the issue directly and resolved it.

Several employers also reported that Hispanic employees were greatly upset and insulted if someone shouted at them or reprimanded them in front of other people. Employers reported that because of their culture they are less tolerant of being shouted at than other employees might be. Several employers also indicated that the employees expected a strong chain of command and authority within the workplace, because this is what they were accustomed to in their home country.

Half of the employers surveyed indicated that there were instances where their Hispanic employees had broken laws, although most were minor infractions. Most law enforcement problems had to do with driving a motor vehicle, such as driving while intoxicated, speeding, driving without a

Table 6.4. Employee quality of life issues.

Means of meeting social needs	*Number of farms*
Family or friends in the area	14
Married with family living on farm	2
Employer not aware	4
How Hispanic employees meet their recreational needs	
A range of activities including basketball, soccer, television, Nintendo, gardening, shopping, going to town, movies	12
Employees don't have much to do in their spare time	2
Employer doesn't know or is not sure	6

licence or driving without registration. A few employers indicated that driving violations sometimes led to inspections by the US Immigration and Naturalization Service to determine if employees were in the country legally.

Seven employers indicated problems with alcohol. Several reported employees coming to work under the influence of alcohol and the need for discipline in those cases. Other employers established specific rules regarding alcohol. In one case there was a rule that unauthorized parties or social gatherings on farm property were not allowed. Employers also reported that there was no visible drug use by their employees.

Only four employers were aware that their Hispanic employees had specific concerns about feeling lonely or isolated. However, employers did acknowledge the potential for loneliness and isolation. Some employers indicated that when the employee lived with his immediate family on the farm, isolation was not a problem, but that for single individuals isolation could be a potential problem.

Only six out of 20 of the employers reported any friction or tension between non-Hispanic and Hispanic employees. The employers who did indicate difficulties pointed to problems of prejudice. Employers also reported that some existing employees felt that Hispanic employees threatened their job security. For those six employers who did report friction, most said that the people who had difficulties working with Hispanic employees had left and found employment elsewhere. In the cases where prejudice was evident, employers said it usually took the form of racial comments. The employers who reported prejudice also said that they took it upon themselves to make it clear to all employees that prejudice in the workplace would not be tolerated.

Advice to other employers

When employers were asked what advice they would give to someone who was just starting to hire and manage Hispanic employees, the answers were varied with some common themes. The most common theme was language. Advice included the following.

- Overcome the language barrier.
- Be calm and patient when dealing with Hispanic employees.
- Hire more than one Hispanic employee, to avoid problems of loneliness and isolation.
- Treat your Hispanic employees just like you would treat non-Hispanic employees.

Identified Issues

The New York State dairy employers interviewed for this study are convinced that Hispanic employees are a viable workforce option for the dairy industry. The practice of hiring Hispanic employees on dairies in New York State is relatively new and employers in this study have been very resourceful in recruiting and managing this new workforce. Despite language and cultural issues, dairy employers who have hired Hispanic employees generally report excellent work performance.

Language and cultural differences are the two immediate issues that managers face when introducing Hispanic employees to their business. Given the differences in language and culture, the employers in this study appear to have been successful in attracting Hispanic employees and showing them how to perform their jobs. As the employer–employee relationship developed in the months that followed, some employers reported that there were other issues to address. These included prejudice, turnover, isolation, alcohol abuse and immigration status. The following issues were identified during the survey.

Language

Bridging the language barrier is the first challenge that the employer faces when Hispanic employees are hired. In a few cases in the study, one of the farm managers spoke Spanish. In other cases a Hispanic employee who spoke English was asked to translate for the manager. The majority of managers in this study spoke little or no Spanish and most Hispanic employees spoke little or no English.

ISSUE

- Determine how to bridge the language gap when none of the managers or employees on the farm are bilingual.

MANAGEMENT STRATEGIES

- Dairy supervisors take Spanish classes at local colleges or other adult education sites.
- Dairy supervisors utilize a variety of language resources (including books and tapes) to learn key words and phrases important for day-to-day communication.
- Employers hire a translator to come to the farm during training, at staff meetings and when important employment policies are being discussed. A translator can also be used to tutor supervisors in Spanish and Hispanic employees in English.
- Hispanic employees take English classes at local schools and adult education sites.
- Employers provide encouragement and incentives for employees to learn English.
- Use portable electronic translating devices to help supervisors and employees to communicate effectively with one another.
- Purchase training tapes and other dairy management materials in Spanish and make them available to Hispanic employees.

Cultural understanding

All of the employers in this study reported that they had observed employee behaviours that were unfamiliar and probably related to culture. Employers also noted that to supervise effectively it was helpful to understand the culture.

ISSUE

- Employers need to understand the culture of their employees in order to be effective supervisors.

MANAGEMENT STRATEGIES

- Become familiar with employee culture by asking them to describe life in their home country.
- Read about the culture of employees and compare it with American culture.
- Become trained in how to supervise a multicultural workforce.
- Become trained in workplace diversity.

Prejudice

Most employers were impressed with many of the behavioural characteristics of their employees. They indicated that Hispanic employees were friendly and respectful and possessed a very strong work ethic. Most also reported that Hispanic employees got along well with their fellow employees. A small group of the employers indicated that there was occasional prejudice among non-Hispanic employees after Hispanic employees were hired. Several employers also stated that there were examples of prejudice within the community.

ISSUES

- Tension sometimes arises between current non-Hispanic employees and new Hispanic hires.
- Employers sometimes encounter prejudice in the community when Hispanic employees are new to the area.

MANAGEMENT STRATEGIES

- Address problems of prejudice in the workforce quickly and directly.
- Create opportunities for dairy supervisors to learn about the culture of Hispanic employees.
- Advocate for Hispanic employees within the community and help employees to adjust to the community.
- Take a proactive role in helping Hispanic

employees to adjust to the work environment on the farm.
- Keep lines of communication open among all employees and deal with problems quickly.

Employee turnover

Dairy employers sometimes remark that it is difficult to retain good local employees. Dairy farm owners who employ Hispanic employees also have concerns regarding turnover. Turnover of Hispanic employees can be high for several reasons. Even though employers check for proper immigration documentation, some employees enter the country illegally and risk being deported. Other employers report that if Hispanic employees get upset or offended, they may leave abruptly. Employers also reported that Hispanic employees want to go home for extended periods of time and, if not allowed to do so, they may leave anyway.

ISSUE
- There is potential for high turnover of Hispanic employees on dairy farms.

MANAGEMENT STRATEGIES
- Establish a flexible staffing system that allows Hispanic employees to leave and then return to farm employment.
- Ask employees when they start work to agree to give at least 2 weeks notice before leaving employment.
- Involve employees in finding their own replacements. This appears to work well, since Hispanic employees have strong networks of family and friends.
- Involve employees in training new employees and orienting them to the job before they leave.
- Encourage employees to return to the farm after their visit home.
- Encourage employees to come to the farm with their families.
- Check immigration documents carefully before hiring Hispanic employees.
- Treat employees with respect and dignity.

Isolation and loneliness

A few employers reported that loneliness and isolation affected their employees, especially when the farm was located in a very rural area. Some employers cautioned against hiring just one Hispanic employee, because of the potential for loneliness. Employers in this study indicated that they were relatively uninvolved with the social and recreational activities of their employees. They did, however, acknowledge these activities as important. A majority of the employers reported that their employees had family and friends within the region of the state where they lived, and that this network helped to alleviate feelings of loneliness and isolation.

ISSUE
- When Hispanic employees are thousands of miles away from home there is potential for them to feel lonely and isolated.

MANAGEMENT STRATEGIES
- Provide rides to church or other social functions.
- Encourage contact and socializing with family and friends.
- Provide opportunities for employees to make friends.
- Consider providing satellite television so that employees can view television programmes in Spanish.

Alcohol abuse

Alcohol abuse sometimes occurs with Hispanic employees, as it does with non-Hispanic employees. The majority of the employers in this study did not report alcohol-related problems. However, the employers who did, reported disruptions in the workplace and poor job performance as results. Four of the 20 employers reported that their Hispanic employees had been charged with driving while intoxicated. Three employers reported a problem with Hispanic employees showing up for work intoxicated and two reported work attendance problems as a result of excessive drinking on days off.

ISSUE

- Some employers report instances of alcohol abuse among Hispanic employees.

MANAGEMENT STRATEGIES

- Make employees aware of local laws regarding drinking and driving.
- Develop clear rules regarding alcohol and the workplace, and communicate them to all employees. State policies in an employee handbook.
- Encourage constructive recreational and social activities on days off so that drinking does not become the primary time-off activity.
- Encourage responsible use of alcohol.

Immigration issues

There was no survey question that specifically addressed the legal status of Hispanic employees but a number of employers raised the issue. Employers reported that even if they carefully checked the appropriate documents and it appeared that the Hispanic employees were legal, it was still possible that they were not and that they had entered the country illegally. Several employers expressed concern that they could face an immigration raid at any time and have some or all of their Hispanic employees deported.

ISSUE

- Hispanic employees may have entered the United States illegally despite possession of documents that appear to be legal, and the employer risks having some or all of the employees deported.

MANAGEMENT STRATEGIES

- Carefully check the immigration status of employees.
- Work through professional associations to shape immigration policies that will allow for a legal agricultural workforce.

Successful Human Resource Practices

The employers surveyed made many comments about how to manage Hispanic employees effectively and successfully. During the course of the survey, employers commented on practices that worked effectively to help them to meet their human resource goals. If employment of any workforce is to be effective, a number of human resource outcomes are needed. These include attracting quality employees, productivity, employee retention, work attendance and employee job satisfaction. It is as important to achieve these results with the Hispanic workforce as with any other workforce. During the course of the survey, employers described practices they used that worked effectively to help them meet their human resource goals. To achieve human resource objectives, dairy farm employers now and in the future will be challenged to adopt modern human resource management practices. The following list of successful human resource management practices by employers of Hispanic employees emerged from the survey interviews.

1. They work aggressively to overcome the language barrier, including learning to speak Spanish themselves.
2. They make a considerable effort to learn about the culture of their employees. This enables employers to better understand their employees as people, and to understand the supervisory techniques that are acceptable and unacceptable to employees from another culture.
3. They develop an organizational culture that accepts and appreciates the differences that individual employees bring to the workplace. They help all of their employees to recognize and appreciate the differences between cultures.
4. They establish employment policies and carefully communicate them so that all employees understand employer expectations for proper conduct on the job and on farm property (including housing). Once established, employment policies are uniformly enforced with all employees.

5. They make every effort to hire Hispanic employees who have legally entered the USA, thereby avoiding employee turnover due to deportation.
6. They acknowledge their employees' strong family ties and desire to return home periodically. Successful employers develop staffing systems that are flexible enough to allow for employees to return home for a period of several weeks or months and then return to the job.
7. They help to create and support social and recreational activities for Hispanic employees that will create a quality of life outside of the job.
8. They become involved in community relations to help community residents to accept and support Hispanic employees and to help employees to become oriented to the community. They become advocates for their Hispanic employees so that community residents will understand the importance of this workforce to the success of the agricultural community.

Recommendations for Further Study and Implications for Cooperative Extension

Several issues warranting further study emerged in the survey. They include: surveying Hispanic employees to assess their perspectives on dairy farm employment; studying community interactions with Hispanic employees; and studying immigration issues to determine how employers can accurately determine the immigration status of their employees. If there is a trend toward more Hispanic employees on Northeast dairies, Cooperative Extension may play several key educational roles:

1. Work with recruiters to help to ensure that information regarding language, culture and management is provided to employers at the time of hiring.
2. Work intensively with key employers who can provide an example for other employers to follow.
3. Conduct workshops and seminars to teach Spanish to dairy farm managers.
4. Conduct workshops and seminars to teach dairy managers to supervise and build a multicultural team.
5. Promote integration of Hispanic workers into communities.

It is important for communities to work with immigrants. It is also important for immigrants to learn to become part of the community.

Acknowledgements

The author extends his sincere appreciation to the dairy managers who took time to respond to the survey for sharing their knowledge and experiences. The author also thanks Dr Robert Milligan and Brian Henehan for their review of this manuscript and their helpful suggestions.

7 The Dynamics of the Washington Farm Labour Market

Dawn Thilmany and Michael D. Miller
Department of Agricultural and Resource Economics, Colorado State University, Fort Collins, Colorado, USA

Abstract

The Washington State farm labour market relies heavily on seasonal labour during peak harvest seasons. Employers argue that the seasonal labour market has tightened as a result of changes in immigration policy and economic conditions, yet they continue to increase the acreage of labour-intensive crops. This chapter presents agricultural labour market trends for Washington State, with specific attention to the degree of employment seasonality and how employers are managing seasonality given tight labour supplies. Empirical evidence suggests that more employers are changing the production enterprise mix to employ a stable number of workers for a greater period per year.

Introduction

Washington State's agricultural employment increased throughout the early 1990s to 1998, then stabilized through 2000 (Fig. 7.1). Prior to the 1990s, Washington followed the rest of the country with stable to declining employment. Recent growth was spurred by demand for labour-intensive crops (fruits, vegetables, and nursery crops) among both domestic consumers and export markets. Given the seasonality of Washington agriculture, a small share of the average annual employment level (91,610 workers in 2000, including owners, unpaid family workers and agricultural service workers) was employed year-round. For example, peak seasonal employment for apple growers alone was over 45,000 workers in 1998, an almost fourfold increase from the total demand for seasonal agricultural labour among all crops in January (Table 7.1).

The need for sufficient supplies of seasonal workers is a continuing concern for producers and employment officials. Agricultural producer organizations dispute the US General Accounting Office's finding that a sufficient farm workforce exists at the national level. Producer claims of labour shortages are the primary factor cited in legislation for granting amnesty to current workers through a new guestworker programme (Lipton and Thornton, 1997; Kiesling-Fox, 1998). There is little disagreement that agriculture requires some seasonal swings in employment levels, but previous

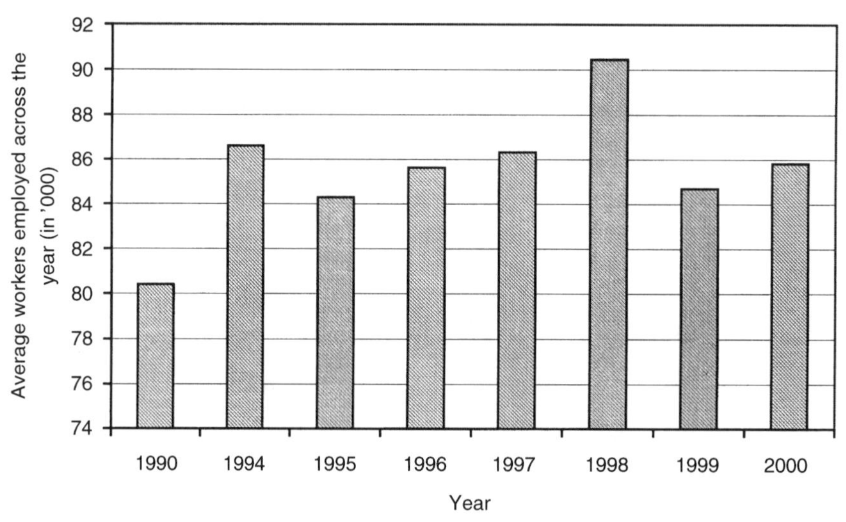

Fig. 7.1. Average annual agricultural employment: seasonal, year-round and unpaid family.

work by Thilmany (2001) suggests that there are labour management alternatives to smooth some of the seasonal peaks that contribute to labour shortages. Thus, it seems that the debate on a new amnesty or guestworker programme would be better informed by analysis of whether sustainable management practices may contribute to less extreme seasonal labour demand cycles. This chapter updates previous research to look at more recent Washington labour market trends and discusses findings from a survey of producers in the context of recent empirical evidence.

The first section of this chapter presents a brief overview of previous research on Washington State agriculture and its unique labour market characteristics, followed by an analysis of current agricultural wages and employment dynamics. To complement discussion of recent labour market trends, a brief review of an empirical model of labour management issues will be presented followed by a discussion of what these trends and findings contribute to the debate about seasonal farm labour.

Washington Farm Labour Market

Producers have rarely been able to manage production practices so that they sustain full-time local workers. Agricultural operations have lowered their demand for labour through continued mechanization, as evidenced by a continuous decrease in annual average farm employment nationally. Yet many enterprises continue to rely on large numbers of hired workers in peak seasons and debate the need for guestworkers to meet seasonal labour needs. Given the extreme seasonality of demand for labour among fruit, vegetable and horticulture producers, Washington State is an interesting market to examine.

Agriculture is a leading industry in Washington State in terms of both sales (3.46% of gross state product) and employment (3% of all jobs) (US Bureau of Economic Analysis, 1999). The value of agricultural production in the state totalled US$5.2 billion in 1998, making it the ninth largest agricultural crop state (by sales) in the USA. Still, Washington ranks fourth in the country with respect to hired farm labour expenses, demonstrating the significant role that labour issues play in the state's agricultural sector and employment base.

The dominance and growth of the apple and cherry industries in the state have contributed to the persistence of extreme seasonal swings in worker demand (Table 7.1). Unlike national trends, agricultural employment was growing in Washington up to a

Table 7.1. Employment of seasonal hired workers in Washington State by crop and month, 2000. (Source: Washington State Department of Agriculture.)

Crop	Jan	Feb	Mar	Apr	May	Jun	July	Aug	Sep	Oct	Nov	Dec
Apples	6,333	7,058	8,125	9,167	13,342	24,772	23,322	19,238	34,156	45,136	11,478	3,254
Cherries	324	401	498	432	162	11,564	14,570	83	18	43	22	404
Pears	301	460	530	662	403	493	591	5,092	3,782	386	627	692
Other tree fruit	52	221	552	578	518	334	809	1,581	1,780	0	141	183
Berries/grapes	1,176	1,094	1,930	1,825	1,720	3,950	8,000	5,487	2,902	1,730	1,568	1,428
Hops	4	18	842	850	1,722	845	145	90	1,722	24	49	52
Nursery/bulbs	755	883	2,487	2,826	2,414	1,966	1,777	1,877	1,266	1,020	1,287	1,474
Wheat/grain	56	30	115	303	444	406	616	1,485	247	137	94	56
Miscellaneous vegetables	662	630	576	706	1,465	1,762	2,578	2,544	2,711	2,685	1,393	748
Potatoes	291	207	935	1,242	1,095	752	1,883	2,146	2,543	2,907	1,079	410
Onions	875	904	854	570	297	1,440	1,450	1,403	1,381	1,043	835	859
Asparagus	10	8	570	5,338	7,113	6,096	615	219	90	126	0	0
Mint, cucumber and rhubarb	12	179	123	210	94	233	458	1,533	999	31	0	0
Other seasonal	357	330	637	1,645	1,141	1,658	2,418	2,334	2,191	1,684	1,211	362
Total	11,208	12,423	18,774	26,354	31,930	56,271	59,232	45,112	55,788	56,952	19,784	9,922

peak in 1998, after which numbers returned to mid-1990 levels (Fig. 7.1). Moreover, the seasonality of labour demand persists, as evidenced in Table 7.1. Historically, this seasonal labour demand pattern was of less concern, since surplus migrant labour was available from other states that experienced labour demand peaks at earlier stages of the production year (California, Texas and Arizona). There is anecdotal evidence that the flow of the western migrant stream has slowed, with tighter labour supplies and more year-round employment options for former migrants in the service sector.

Worker Migration and Turnover

Washington agricultural employers have an incentive to claim that labour shortages exist as it is the only method available to initiate special worker programmes, such as the H-2A programme or their preferred long-term solution of a guestworker programme (J. Jaksich, Washington, 1997, personal communication). The economic implications of continued dependence on seasonal and migrant farm labour are not simple or without long-term consequences. Bailey (1993) suggested that reliance on migrant labour has implications for the efficiency of local labour markets.

Anderson (1993) found support for the idea that increased adjustment costs or recruitment costs (such as those one would expect with a labour shortage) should gradually outweigh the incentive to treat labour as purely variable costs and lead to lower turnover. In the past, ample supplies of replacement workers kept such adjustment/recruitment costs low. Taylor and Thilmany (1993) argued that any labour shortage should be met by an employer response attempting to marginally decrease worker turnover. In the case of seasonal employment, employers may offer a longer duration of employment, higher wages (to offset fixed costs of finding new employment), or other incentives to attract return workers (such as a bonus to workers who return, or who refer workers). Worker turnover and reliance on migrants in Washington State throughout the 1990s may provide some interesting evidence about these incentives. As the labour market tightened (number of workers employed stabilized), Thilmany (2001) found that workers were able to secure more hours of labour throughout the year.

Cooper (1994) tested the hypothesis that market wages and wage variability are a primary source of information included in the migration decisions of agents. Those employers with higher and/or less variable wages fare better in their challenge to secure the return of previous workers, as well as a sufficient number of new recruits. Similarly, seasonal migrant workers may be more concerned about the duration of work available than the level of wages offered. In short, it is likely that several economic factors, including wage levels and variability, likelihood of employment and employment duration, all affect migrant behaviour. The latter part of the analysis in this chapter examines whether employers are implementing management practices that serve to lower turnover in an effort to avoid labour shortages in peak seasons.

Analysis of Washington Farm Labour Dynamics

The objectives of the analysis in this chapter focus on several questions posed by Washington Employment Department officials. Firstly, labour market trends suggest that employers have recently altered production enterprises or practices in response to labour shortages. Various graphs and tables are used to describe the recent history of the farm labour market, including workforce numbers, wages and earnings and regional labour demand. Trend analysis of employment data provides a baseline on labour supply and demand factors that influence employer perceptions on the availability of labour.

The second section of the analysis reviews the choices made by individual employers to manage labour through wage decisions and strategies to attract return seasonal workers from season to season. The objective of the farm employer survey was to differentiate the experiences of farm employers, who were placed in two categories based on responses: those who raised wages between 1994 and 1995, and those who attracted a higher share of return workers between those two years. It is argued that employers in each of these categories exhibit a propensity to use management strategies to handle the challenge of securing an adequate workforce.

Statewide employment trends

Aggregate farm employment numbers for Washington State from 1990 to 2000 are presented in Fig. 7.2. The clearest trend is that the demand for workers peaked in 1998, and continues to remain above pre-1990 levels. Yet average hours worked by farm labourers (a substitute for number of workers) have not risen as consistently. The year 1994 is an exception in both cases, since rumours of a bad apple crop kept migrants from travelling to Washington State, thereby forcing employers to utilize available workers more fully. This fact explains the marked increase in employed hours for the average worker, and the 1994 season was the beginning of recent debate on shortages. Still, more recent employment numbers (and labour demand) continue to escalate, albeit at a slower rate of increase (Payne, 2001).

Recent years represent a positive trend to employment officials. The number of workers required in agriculture remained steady after a peak in worker demand in 1998, while average hours worked increased significantly. This stabilization of the farm workforce may represent a market-clearing equilibrium as workers work more hours or are employed for a longer duration by employers hoping to recruit enough workers. This trend is confirmed by data showing that, on average, workers were employed by a smaller number of employers in 2000 compared with 1998 (when this information was first collected). This inference from the aggregate data is also consistent with the fact that Washington's agricultural production is moving away from the traditional, heavy reliance on apples to a more varied crop mix. Consequently, a more diverse crop mix would suggest a less pronounced labour demand cycle, and seasonal peaks are lower in magnitude than those reported in Thilmany (2001) (Table 7.2). There are employers who may find this stabilization in worker numbers unacceptable if they

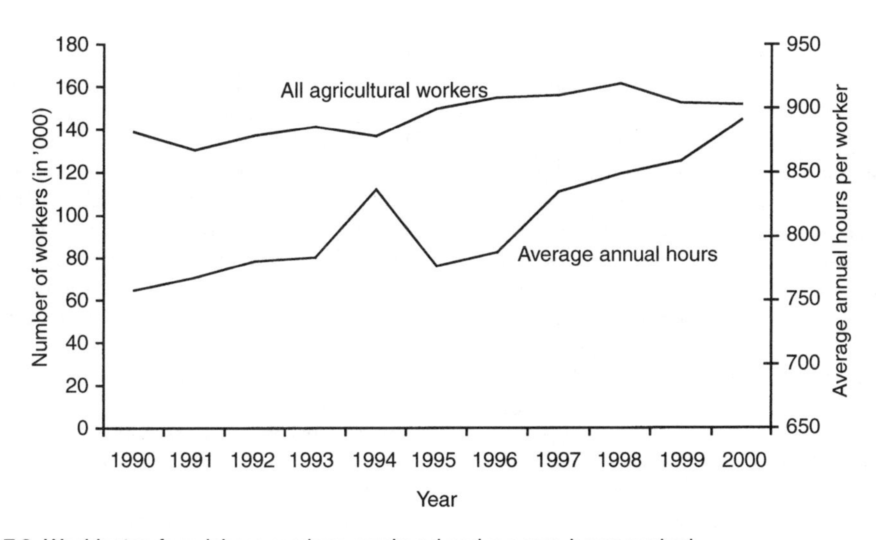

Fig. 7.2. Washington farm labour: workers employed and average hours worked.

Table 7.2. Average annual and hourly earnings of farm workers in Washington State, 2000.

Sector	Annual earnings (US$)	Percentage of agricultural employment (%)	Average hourly wage (US$) 1997	2000
Agricultural production: crops	2,502	80.6	–	–
Wheat	2,461	2.5	9.75	10.26
Irish potatoes	2,727	2.8	9.01	8.63
Field crops, except cash grains	2,693	6.2	7.72	8.32
Vegetables and melons	2,538	5.9	7.80	8.58
Berry crops	2,097	4.0	7.13	8.00
Grapes	2,849	4.0	7.12	9.03
Deciduous tree fruits	3,296	42.3	7.88	8.50
Ornamental floriculture/nursery products	4,093	5.5	8.57	8.75
Agricultural production: livestock	3,411	4.0	–	–
Beef feedlots	6,337	0.4	10.44	8.09
Beef, except feedlots	4,264	0.5	9.00	9.22
Dairy farms	9,727	2.9	10.70	10.38
Agricultural services	3,745	14.4	–	–
Crop preparation services	3,578	9.3	9.02	8.77
Farm management services	1,657	2.9	8.42	8.20
All agricultural workers	8,782	881 hours for 2.59 employers		9.76
Worked for agricultural and non-agricultural employers	12,194	1220 hours for 3.80 employers		10.00
Worked in agriculture only	7,272	745 hours for 2.05 employers		9.76

do not 'smooth' their labour demand, as lower seasonal supplies may hamper their ability to get crops harvested as quickly as they prefer.

Wages and earnings

Comparable to national statistics, earnings for Washington State's agricultural workers are low, but hourly and annual earnings vary substantially across agricultural sectors (Table 7.2). Hourly wages are competitive with alternative employment options, but the lack of year-round employment keeps annual earnings low. During the month of December 2000, the number of unemployment insurance (UI) claims from crop workers peaked at 9650 from a low of 2670 during the previous October, an increase of over 250% (Payne, 2001). Although annual average claims were up in 1999 from 1998, there was a significant decline in claims for the 2000 season. None the less, the seasonal demand for UI assistance persisted up until 2000, indicating that attempts at 'regularizing' the demand for farm labour were not successful until the recent trend toward employment stabilization. It is also important to note that 60% of farm workers in 2000 did not qualify for UI coverage, since they did not work 680 h during the year.

Not surprisingly, a high turnover rate exists among farm workers. Of the 152,474 workers employed in Washington agriculture at any time during 1999 (in contrast to the 91,160 employed on average), it is estimated that only 55.3% remained in agriculture in 2000. Although there was a higher share of workers returning to agricultural employment in 2000 than in previous years, there are still significant numbers leaving the industry. There are many workers who choose to work in both farm and non-agricultural jobs to improve annual earnings. Table 7.2 presents the average annual earnings of all agricultural workers, as well as those who work in only agriculture (about 69% of the total) and those with both types of employers (31%). Those who work for both agricultural and non-agricultural employers do not earn significantly more on an hourly basis, but had almost double the hours worked (1220 versus 745), number of employers (3.8 versus 2.05) and annual earnings (US$12,194 versus US$7,272) in 2000 (Payne, 2001).

This pattern represents a possible solution for supplying seasonal agriculture, but there are some drawbacks. There is an explicit (days without pay in transition) and implicit (employer-specific skills that do not fully transfer) cost to workers from switching employers twice as often during the year. This, together with the greater stability of non-agricultural employment (as evidenced by significantly lower unemployment claim seasonal cycles), may affect the poor retention rate of workers in agriculture. There is no clear motive for workers to return to farm operations if hourly earnings are similar and no long-term prospects are available. The earnings of workers who left farm jobs after 1999 averaged US$10,525, well above the US$7272 earned by farm workers and below the US$12,194 earned by those who combined farm and non-farm jobs. The complementary seasonality of some of these non-agricultural employment opportunities (i.e. food processing directly succeeding harvest) is the primary reason that some workers remain working in agriculture after securing non-farm work.

Regional patterns

The impact of farm labour issues is not uniform across the state of Washington. Almost 80% of the farm employment is located east of the Cascade Mountains, with Yakima, Chelan-Douglas, Benton-Franklin and Grant counties representing almost 60% of agricultural employment (Table 7.3). It is also

Table 7.3. Total employment, agricultural employment and share of agricultural employment in Washington State and selected areas, 2000[a]. (Source: Washington State Employment Security Department.)

Area	Total employment (no. of workers)	Agricultural employment (no. of workers)	Agriculture as percentage of total employment (%)
State total	2,887,500	85,820	3.0
Western Washington	2,293,250	17,420	0.8
Eastern Washington	594,250	68,400	11.5
State agricultural area			
Columbia Basin	38,260	9,580	25.5
Adams County	7,420	2,440	32.9
Grant County	33,380	7,970	23.9
North central counties	81,670	16,710	20.5
Chelan and Douglas	49,070	10,660	21.7
Kittitas	14,020	1,160	14.4
Okanogan	18,580	4,890	26.3
South central counties	105,000	21,800	20.8
Klickitat	7,800	1,120	14.4
Yakima	97,200	20,680	21.3
Southeastern counties	111,790	13,730	12.3
Benton and Franklin	87,700	10,700	12.2
Walla Walla	24,090	3,030	12.6
Eastern counties	255,040	5,740	2.3
Lincoln	4,330	1,060	24.4
Spokane	196,900	1,370	0.7
Whitman	19,050	1,560	8.2
Other eastern counties	34,760	1,750	5.0

[a]Total employment and agricultural employment have been adjusted to eliminate the effect of dual job-holding. Detailed averages may not add to total, because of rounding.

interesting to note the reliance on agriculture for each county's economy as measured by the percentage of total employment in agriculture. These patterns are not surprising, given the varying importance of agriculture throughout the state. The lack of alternative employment opportunities in these counties during agriculture's off-season dampens prospects for attracting return workers to agriculture from one year to the next. The resulting influx of migrant workers during peak seasons represents a challenge to the economic and social infrastructure of some areas (Thilmany and Carkner, 1996). Discussion of the community-wide social and economic implications is beyond the scope of this study, but is important in the debate on labour policy and programmes.

Migrant and seasonal workers

According to 2000 UI records, the average Washington farm worker is Hispanic (81.7%), male (72%) and under 40 years of age (52%). Approximately 21% of the agricultural workforce is migrant labour (defined as those who are not permanently settled in the area where employed beyond the term of employment). This share increases significantly in peak seasons and there is over 25% turnover annually out of agriculture from year to year (Wahlers, 1999). Farm worker earnings remain low, even though hourly earnings are similar to peer workers in non-agricultural industries, because of the extreme seasonality of agriculture.

Washington State employment officials estimate that 30–60% of the seasonal agricultural workforce is made up of illegal or undocumented workers. During the period when this study's survey was conducted (1995) about 13% of all Washington seasonal farm workers were migrants (8.5% were interstate and 4.4% were intrastate workers). The proportion of migrants is considerably higher during periods of peak activity, with migrants representing nearly one out of every five workers during the peak season. These data from state employment sources are consistent with the share of interstate migrants (20%) reported by growers in this study (Table 7.4).

Empirical Findings from the Employer Survey

Farm labour data and the employer survey

The primary source of data for the estimation analysis in this study is employer tax records. Nearly all farm employment since 1990 is covered by the Employer Security Act in Washington State. To augment these data with a more detailed account of labour activities by crop, Washington conducts a monthly In-Season Farm Labor Survey with voluntary responses from about 600 employers.

Although aggregate farm labour estimates exist, there is little information on the migratory nature of the workforce, job duration for individual workers, primary state of migrant residence, or the reliance of individual employers on workers of varying characteristics. To augment available data with more detailed information, this study also includes data collected from an appendix to the 1995 August Farm Labor Survey. Questions on employers' perceptions of the share of workers returning in recent years, origin and share of interstate migrants hired, and employers' decisions to raise wages were added to the standard questionnaire.

The response rate to the additional employer questions was almost 50% of the survey sample. To ensure sufficient representation from all sectors, especially those of greatest interest to a study of seasonal workers, the sample was compared with the general characteristics of the Washington farm labour market. The employers surveyed represent 10% of state farm employment (total and seasonal), and represent the crop and regional mix of the state fairly well.

Descriptive statistics from the survey are presented in Table 7.4. The means of these variables are in either number of workers, dollars per hour (for wages), or share of

Table 7.4. Variable definitions and descriptive statistics ($n = 242$).

Variable	Definition	Mean	Variance
TOT	Mean total employment (number)	69.7850	94.0450
SEAS	Mean seasonal employment	54.9790	69.2190
Tree fruits		26.232	
Vegetables		6.452	
Nursery		1.399	
Grains		1.287	
WAGE	Mean hourly wage	US$5.67	2.0090
Tree fruits		US$6.90	
Vegetables		US$6.69	
Nursery		US$7.49	
Grains		US$8.02	
SC	South central area (share of employers)	0.3100	0.4630
NC	North central area	0.2020	0.4030
CB	Columbia Basin area	0.1360	0.3440
SE	Southeastern area	0.1740	0.3800
VEG	Vegetable producer	0.1490	0.3680
NURS	Nursery producer	0.0620	0.2420
FRUIT	Tree fruit producer	0.4920	0.5010
WHEAT	Wheat producer	0.0580	0.2340
VARTOT	Total labour variability (= σ_{lab}/μ_{lab})	0.4329	2443.1000
VARWAG	Wage variability (= σ_{wage}/μ_{wage})	0.8144	262.1800
LENGTH	Length of employment (in months)	2.5860	2.2840
NOCROP	Number of crops produced	3.2930	1.9900
NOACT	Number of crop activities	7.6650	5.0410
BEST	Best workers retained for longer periods	66.1%	47.40
REF	Share referred by other workers	56.33%	44.0820
REC	Share recruited by employer	17.38%	32.3570
HARV	Employs primarily harvest labour	58.7%	49.30
TRANS	Employs transportation workers	3.3%	17.90
PACK	Employs packing labour	10.3%	30.50
GEN	Employs general farm workers	7.0%	25.60
INTER	Share of interstate workers	20.33%	30.4180
CA	Most interstate workers from California	36.40%	48.20
MX	Most interstate workers from Mexico	6.60%	24.90
OR	Most interstate workers from Oregon	7.0%	25.60
TX	Most interstate workers from Texas	12.80%	33.50
PSEA	Peak in seasonal employment (high/mean)	16.4990	22.0030
CON	Constant	1.0000	0
Dependent Variables			
RAISE	Wage rise in 1994 and/or 1995	47.1%	50.00
RETURN	Less than 50% return workers each year	74.4%	29.59

employers who fall into various categories solicited in the survey. A couple of details should be noted. The mean wage is well below the average wage, since those employers that were less specialized (including labour contractors) generally paid closer to minimum wages than those operations that specialized in specific commodities. The share of interstate workers does not add up to one, since not all employers responded to using migrant workers (while some that did, employ migrants employed from more than one area of the country). Also, the means for the two dependent variables included in the estimations are presented at the bottom of the table.

Although the means for each of the variables are interesting in themselves, the previous discussion implies that there are several interrelated factors among labour markets and employer choices. Thus, econometric models were used for empirical analysis of the survey. To assure the best econometric fit possible, correlation between variables was calculated. There are no variables highly correlated enough to justify specification problems, although some regional and commodity variables were in the 0.4–0.6 range for correlation coefficients.

The models examine the likelihood of securing at least 50% return workers from year to year and the likelihood of an employer offering wage increases in 1994 or 1995. In each case, a two-stage probit model was developed and estimated with the dependent, dichotomous variable defined by an employer's self-reported labour management experiences. The inclusion of migrant worker variables will test whether reliance on such workers affects upward-wage pressure and turnover.

1. Model 1 focused on an employer's decision to raise wages during 1994 and 1995. An increase in wages, rather than the absolute wage level, was analysed because the primary objective of the study was to determine whether employers had made any labour management decisions that would suggest an increase in factor demand or a perceived shortage of workers. This study assumes that increased wages may be a strategy that employers use to signal increased factor demand or to counter labour supply changes and assist in recruitment (or attract return workers).

2. Model 2 explored the differences in inter-year seasonal worker turnover. The dependent variable was equal to zero if the employer reported that less than 50% of their seasonal workers returned each year. (Optimally, the analysis could be performed on a continuous left-hand variable equal to the share of return workers. However, by convenience or design, most employers reported the return worker share in 10% increments. Thus, econometric estimation that assumed a continuous dependent variable would be biased. The 50% level is not arbitrary, but a clear dividing point for the sample.) Worker turnover among firms is a function of a variety of factors, including wages, total employment, regional and commodity-specific labour market conditions, and duration of employment (Taylor and Thilmany, 1993). The second model will test what employer-specific factors influence the share of return workers in a firm's labour supply.

Estimation results from the econometric models are presented in Table 7.5. For commodities, the reference group among the dummy categories was other field crops (including potatoes, hops and other seasonal crops) and for types of workers this reference group represents employers who hire mostly pre-harvest workers (pruners, weeders, planters, etc.). Estimated coefficients are reported and those variables with significant results are denoted. A measure of elasticity, the 'effect at means', was calculated for each variable to assist in interpretation. The effect at means shows the relative effect of a variable on the probability of an employer being included in one of the dependent variable groups. This statistic was calculated by estimating the difference in the probability due to a small change in each variable while holding all other variables constant at their means (Taylor and Thilmany, 1993).

Findings and discussion

Several different factors were significant explanatory variables in the employer's decision to offer wage rises during 1994 and/or 1995 (Table 7.5). Among commodities, nursery and vegetable producers were more likely to raise wages, a finding that may coincide with the increased production (and thus factor) for these agricultural products over this period. Between 1993 and 1995, vegetable and nursery production increased by 11.2% and 12.2%, respectively (National Agricultural Statistics Service, 2000). Others

Table 7.5. Employer's choice to raise wages (Model 1) and employers who report 50% or more return workers (Model 2).[a]

	Model 1		Model 2	
Variable	Coefficient	Effect at means (% change in probability)	Coefficient	Effect at means (% change in probability)
TOT	−0.0142**	−0.40	0.0102*	0.30
SEAS	0.0136	0.36	−0.0092	−0.24
WAGE	1.3678**	7.57	0.5346*	0.99
SC	0.3689	1.94	−0.2450	−1.21
NC	−0.5827	−1.89	−0.3780	−2.16
CB	0.2364	1.27	−0.0484	−0.22
SE	−0.2266	−0.87	0.0835	0.35
VEG	0.7165*	5.51	0.7230*	2.02
NURS	1.0630**	12.12	1.1298*	2.07
FRUIT	0.1656	0.74	0.8670**	4.22
WHEAT	−0.7235	−1.74	−0.3342	−2.04
VARTOT	−0.3121	−0.06	−0.5650**	−0.11
VARWAG	−0.1523	−0.05	0.9046	0.30
LENGTH	−0.0428	−0.05	0.0569	0.06
NOCROP	−0.2518**	−0.34	0.2831**	0.38
NOACT	0.1105**	−0.41	−0.0986**	−0.37
BEST	0.1023	0.44	−0.3137	−1.28
REF	−0.0013	−0.03	na	na
REC	−0.0007	−0.01	na	na
HARV	0.1558	0.04	0.2974	0.08
TRANS	−1.1632*	−0.02	0.8014	0.01
PACK	0.9384**	0.04	−0.0027	−0.00
GEN	1.1750**	0.04	0.2347	0.01
INTER	0.0016	0.01	0.0002	0.00
CA	0.7271**	0.12	0.2791	0.04
MX	−0.7191*	−0.02	−1.1901**	−0.04
OR	−0.5454	−0.02	−0.2578	−0.01
TX	−0.1984	−0.01	−0.6159*	−0.04
RAISE	na	0.00	−0.2602	−1.19
PSEA	−0.1085**	−0.66	0.0369	0.26
CON	−7.7630**		−2.8261**	
Right predictions	71.7%		79.3%	
Log-likelihood function	−132.43		−112.80	
Mean of dependent variable	0.471[b]		0.744[c]	

*,**Significant at 10% and 5% levels, respectively.
[a]Model 1 refers to the employer's choice to raise wages in 1994 and/or 1995; Model 2 refers to employers who report 50% or more return workers, year-to-year.
[b]47.1% of respondents raised wages.
[c]74.4% of respondents had a majority of workers return.
na = not applicable.

would suggest that, since fruit producers still provide the majority of jobs, those producers with employment earlier in the season may need to offer higher wages to encourage workers to return earlier than the peak orchard harvest season.

Among types of workers, employers of packers and general labourers were more likely to raise wages, while transportation employers were less likely to do so. There were no significant regional results. Finally, those firms that hired California migrant workers were more likely to give rises, with the opposite true for Mexican workers. These findings are likely related to the differential skills and other job opportunities for those respective groups of workers.

Producers with a larger workforce were less likely to raise wages (possibly due to its large effect on cost structure). Rises were more likely to be offered by those employers with relatively high current wages. Although predicted levels were used in the two-stage modelling, an upward bias for employers who increased wages is expected since the reporting period is after the rises were given. Employers with more diverse cropping operations were less likely to give a rise; however, those firms with the highest number of distinct crop activities (i.e. harvest, thinning) were more likely to give a rise.

Employers who reported high retention and return rates among seasonal workers exhibit several interesting characteristics related to their operations, size and ability to smooth employment demand peaks. Among the producers of specific commodities, fruit, vegetable and nursery employers realized a higher share of return workers. No significant relationships were found among the regions (as was the case in the other model). Larger employers were more likely to retain workers, most likely due to their visibility as an employer, but it may also be based on worker perceptions that employment will more likely be available from larger employers during peak periods. Higher wages had the expected effect, increasing the probability of a worker returning each year.

The stability of a firm's employment record played a significant role in their ability to attract return workers. Increased variability of employment increased interyear turnover. Similar to the previous model's findings, enterprise diversity measures had differing results. A more diverse crop mix increased the likelihood of a firm attracting a large share of return workers, and vice versa for the number of crop activities. Finally, a relatively high use of Mexican and Texan workers increased turnover, most likely due to distance and worker-specific characteristics.

Conclusion

Thilmany (2001) noted that earlier Washington State agricultural employer trends suggested that employers could make managerial choices that could curtail any further increase in demand for labour. The labour market information provided in this chapter confirms those conclusions. Although some seasonal peaks in labour demand are likely to persist with the seasonal nature of agricultural production, both the change in farm production enterprise mix and increased average annual hours worked by workers suggest a 'smoothing' of labour demand cycles. This is an encouraging trend in the eyes of Washington State employment officials and those communities that would prefer a more year-round, fully employed workforce to marked influxes of migrant workers with little connection to the community.

Depending on future immigration policy initiatives, employment officials view dependency on seasonal and/or migrant workers as a potential downfall for many producers if legal or regulatory issues make new supplies of migrant workers riskier to secure and hire. This study's findings, together with previous research, would suggest that there are some workable alternatives to guestworker programmes, or at least that some labour management strategies should be encouraged in addition to programmes meant to supplement farm labour supplies. It is likely that more fully utilizing current workers is the best option for the employers and their communities, but it will require progressive managerial strategies and production practice changes by employers in addition to increased information and job-matching efforts from employment officials.

Acknowledgements

The authors wish to thank anonymous reviewers and participants in *The Dynamics of Hired Farm Labour: Constraints and Community Responses* conference for suggested comments and improvements to this research. The authors are indebted to Jeff Jaksich and the Washington Employment Security Department for allowing access to the data and employer surveys. Funding for this study was provided, in part, by the Colorado State University Agricultural Experiment Station.

References

Anderson, P.M. (1993) Linear adjustment costs and seasonal labour demand: evidence from retail trade firms. *Quarterly Journal of Economics* 108, 1015–1042.

Bailey, A.J. (1993) Migration history, migration behavior and selectivity. *The Annals of Regional Science* 2, 315–326.

Cooper, J.M. (1994) Migration and market wage risk. *Journal of Regional Science* 34, 563–582.

Kiesling-Fox, S. (1998) Don't tell farmers there's no labor shortage. *Farm Bureau Focus on Agriculture,* February. http://www.fb.com/views/focus/fo98/fo0223.html

Lipton, D. and Thornton, M. (1997) *FB:* agriculture labor solution needed now. *Farm Bureau Focus on Agriculture,* September. http://www.fb.com/news/nr/nr97/nr0924.html

National Agricultural Statistics Service (NASS) (2000) Agriculture in Washington. *Agricultural Production Summary,* Washington State Office. http:// www.nass.usda.gov / wa / annual100/ general.pdf

Payne, L. (2001) *Agricultural Workforce in Washington State 2000.* Washington State Employment Security, Olympia, Washington.

Taylor, J.E. and Thilmany, D. (1993) Worker turnover, farm labor contractors and IRCA's impact on the California farm labor market. *American Journal of Agricultural Economics* 75, 350–360.

Thilmany, D. (2001) Farm labor trends and management in Washington State. *Journal of Agribusiness* 19, 1–15.

Thilmany, D. and Carkner, R. (1996) Immigration issues in rural Washington. Paper presented at the conference *Immigration and the Changing Face of Rural America,* Ames, Iowa, July.

US Bureau of Economic Analysis (1999) *GSP 1997–1999 Report for Washington Agriculture.* From GDP by Industry, 1947–1999.

Wahlers, R. (1999) *Agricultural Workforce in Washington State 1998.* Washington State Employment Security Department, Labor Market and Economic Analysis Branch, Olympia, Washington.

8 The Economic Impact of Migrant Farm Workers on Southeastern Michigan

Rene P. Rosenbaum
Julian Samora Research Institute and Department of Resource Development, Michigan State University, East Lansing, Michigan, USA

Abstract

The views of farm workers and their employers continue to dominate what constitutes the farm labour problem, but a community perspective is emerging that emphasizes the impact of agricultural labour on rural areas. This chapter develops a model of migrant and seasonal farm worker (MSFW) labour as an economic development event to examine the economic impact of the MSFW labour market on the rural area of Branch, Hillsdale, Jackson, Lenawee, Monroe and Washtenaw Counties in southeastern Michigan. The study shows that the farm worker population represents a significant economic development event to the region and the state. It is estimated that the 1257 farm workers in the region contributed over US$23 million to the regional economy, or US$18,000 per worker. Because a diverse and large number of community interests seem to benefit from the presence of farm workers in the area, it is suggested that communities should consider taking positive steps to maximize the benefits to the local economy from the presence of migrant and seasonal farm workers. The findings also show that the economic contributions of farm labour to the local economy depend on whether the farm labour force is composed of domestic migrants, seasonal or foreign workers. The study results imply that the use of legal foreign workers through the H-2A programme or some other programme would reduce the positive impacts of this event to the community.

Introduction

The views of farm workers and their employers continue to dominate what constitutes the farm labour problem, but a community perspective is emerging that emphasizes the impacts of agricultural labour on rural areas. Although studies linking the farm labour population to local communities are more common today, few studies have described the role that migrant and seasonal farm workers (MSFWs) play in the local economy of a receiving community. In the rural areas where they work, seldom has migrant and seasonal farm labour been treated as a community economic development event, a form of economic change that contributes to the local economy.

This chapter makes a contribution to the emerging community view of the farm labour problem by examining the economic impact of the MSFW labour market on the rural area

 The Dynamics of Hired Farm Labour (eds J.L. Findeis, A.M. Vandeman, J.M. Larson and J.L. Runyan)

of Branch, Hillsdale, Jackson, Lenawee, Monroe and Washtenaw Counties in southeastern Michigan. The next section provides a synopsis of the three different perspectives of the farm labour problem. This discussion is followed by development of a model of farm labour as an economic development event to measure the economic impacts on rural areas from the presence of MSFWs. MSFW-dependent agriculture and the MSFW population and labour market in southeastern Michigan are then described and the seasonal farm labour economic impact model is applied to measure the economic contributions of the farm worker population in Michigan's southeastern region for 1997. The findings are used to gauge the potential economic impact on the local economy of the H-2A national policy initiative in the USA, a guestworker programme that would sharply increase the solo male versus family proportion of the migrant workforce. The chapter concludes with a summary of the research findings and a statement about the significance of viewing the farm labour problem as an economic development event.

Perspectives of the Farm Labour Problem

The traditional labour market approach to understanding the farm labour problem is not inconsistent with the politics of the USA in the last century and a half. According to Paehike (1999), US politics has centred on the struggles between economic values such as capital accumulation, enhanced trade, efficiency and economic growth, and equity values such as wages, working conditions, social welfare, public health and public education. Analogously, the current dynamics of the hired farm labour market can be seen in terms of the struggle between the economic values of farm worker employers and agribusiness (and consumers), and the equity values of farm workers and their advocates. In large measure it is the effective expression of the values of these two groups that has dominated the policy and politics of hired farm labour in the USA. Their voice remains at the centre of many of the issues that contribute to the current dynamic of the US hired farm labour market.

Even though the interests of farm workers and farmers currently lay claim to what constitutes the farm labour problem, historically the interests of the latter group have dominated the politics and economics of the situation. The low farm income problem in various forms has always been the central concern of agricultural economists. Historically, policies to improve the economic status of farmers have been based on the view that farmers are caught in the 'farm problem' created by price and income-inelastic product demand in agriculture. Prior to the 1950s, parity prices, production adjustments and marketing efficiency were the acceptable ways to deal with the farm problem. Another line of thought, popular since the Second World War, was the view that farm prosperity depended on the elimination of 'the redundant claimants against aggregate farm income'. Johnson (1959, p. 47) wrote: 'Stated simply, the farm problem is the result of the employment of more labour in agriculture than can earn as large a real income as the same labour could earn elsewhere in the economy.' Hence, the link was made between farm prosperity and the need to accelerate technological advances and off-farm mobility (Fuller, 1984).

In the case of MSFW-dependent agriculture, solutions to the farm problem have focused on securing an adequate supply of low-wage workers to care for and harvest fruits and vegetables. This approach remains central to the guestworker legislative policy being debated in the US Congress. Agricultural interests are calling for an alternative to the H-2A non-immigrant guestworker programme that allows the US agricultural sector to bring in seasonal foreign workers on a temporary basis when domestic workers are unavailable. The main objective is to put in place a foreign worker programme that does not require employers to demonstrate to the US Department of Labor that guestworkers are needed before they are admitted. A recent legislative proposal aims at encouraging guestworkers to return to Mexico and other countries of origin by calling for a 25% deduction of the worker's wages to be paid only in the country of origin if the worker appears in person (Martin and Taylor, 1998). The lobbying efforts by major agribusiness organizations for changes in agricultural guestworker policy to ease the importation of foreign workers have yet to succeed. Contrary to the growers' views that a labour shortage exists, recent reviews of the H-2A programme conclude that there does not appear to be a national agricultural labour shortage, except in some specific crops or geographic areas (GAO, 1998).

While the economics and prosperity of the farm are the central concern in the economic values perspective of the farm worker problem, in the equity values perspective the long-standing concern remains improving the persistently relatively low returns for labour services in farming. The historical plight and position of the migrant worker near the bottom of the US labour market prompted farm worker equity concerns and a belated call for action in the 1960s. As a result, the federal government began programmes in the mid-1960s to help migrant workers and their families. These federal programmes multiplied during the 1970s and 1980s and, by 1992, 12 different programmes spent over US$600 million annually to assist migrant and seasonal farm workers and their families (Martin and Martin, 1994).

But in spite of all the federal efforts, MSFWs remain one of the most economically-disadvantaged occupational groups. Fuller (1984) noted that many of the problems faced by migrant farm workers in 1980 were the same ones they had faced 30–40 years before. His key policy concern was 'the long persisting adverse conditions experienced by the "forgotten people" who harvest our crops'. Although a broad array of policy proposals were offered and reviewed at the 1980 national conference on seasonal agriculture labour markets in the USA, R. Emerson emphasized the importance of considering farm worker policy within the broader spectrum of poverty, regardless of the occupation. This view of the farm labour problem recognized the dynamic nature of the labour markets. According to Emerson (1984, p. 504), viewing farm worker policy within this broader spectrum of poverty was important because 'a focus solely in the context of the existing farm labour market tends to point one in the direction of maintaining current participants within the market rather than considering the overall welfare within the economy'. Within this broader perspective, income maintenance programmes were thought to offer considerable appeal over various programmes that targeted specific occupational groups (Emerson, 1984).

Clearly, the sentiment at the time was to give attention to moving workers out of farm labour. Aiding the transition of displaced and current farm workers out of the unskilled labour pool into more remunerative and stable employment remains a preferred policy approach. It can be contrasted with the approach to protecting and aiding those remaining in agriculture through legislation, regulation and direct services. However, what Fuller (1984) said two decades ago still rings true: migrant workers remain one of the most economically disadvantaged and impoverished occupational groups in the USA (Martin, 1988; Oliveira, 1992; Martin and Martin, 1994; Griffith and Kissam, 1995).

The unsatisfactory conditions of US migrant farm labour for most of the 20th century have failed to change the agricultural sector and its dependence on a migrant labour system that relies on cheap and ethnic minority labour. Although the federal assistance programmes established to help migrant workers and their families have enabled many individuals and families to 'escape' agriculture and the farm labour market, these programmes do nothing to raise the income level of migrants still in the farm labour market (Martin and Martin, 1994). In other words, when these programmes are successful and farm workers are able to leave the migrant stream, other migrant workers simply replace them; the migrant labour system is not changed – it is only that the faces of the workers who need help are different. Existing legislative protection and the regulation of wages, hours and working conditions do not seem to protect those remaining in agriculture.

The community perspective of the farm worker problem is emerging amid concerns over an increasingly global economy. There is a growing interest in locally-based community economic development strategies that emphasize locally-determined community objectives. Scholars and practitioners of community development are increasingly valuing locality-based community development and citizen participation approaches over other approaches to address social problems (Aigner *et al.*, 1999). With their gain in relative prominence, these new and emerging values representing community interests complicate and challenge the old debate between economy and equity that has historically dominated the politic in the USA.

The community view that is emerging does not discount the interests of farmers and farm workers. The approach recognizes these interests but expands the analysis of the labour market situation to include the interests of and impacts on a broader range of community stakeholders. One distinguishable line of thought within the community perspective is the emphasis on community attitudes toward farm workers. The growing demand for predominantly immigrant agricultural labour across the USA has raised resident reactions and concerns in rural areas. Farm workers continue to face resentment from white residents, open acts of discrimination, and occasionally acts of violence. In the 1990s, communities witnessed an increasing number of conflicts between immigrants and natives. All kinds of social problems have been documented: police harassment, housing, increasing poverty, educational access for migrant children, health services, increasing needs for bilingual services in hospitals and courts, and concerns over the integration of farm workers staying in the USA year-round (Hedges and Hawkins, 1996; Martin *et al.*, 1996). Community leaders remain ill-equipped to address the resulting tensions.

Another line of thought in the community view of the farm labour problem considers the economic impacts of seasonal agricultural labour markets on rural areas. Some labour market research from California in the 1990s, for example, identified negative impacts of workers on national and local economies, arguing that immigrant workers take over jobs from domestic workers and freeze low wages into place (Martin and Taylor, 1998). Other economic impact research, however, suggests that the presence of a migrant and seasonal agricultural labour market has a positive impact on the local economy of receiving communities (Sills *et al.*, 1993; Trupo *et al.*, 1998).

Farm Labour as an Economic Development Event

Specifying the relationships

The typical way to conduct an economic impact study is to focus on the contributions to the economy that result from an economic development event. An alternative way is to estimate the cost to the local economy from the elimination of such an event. Although this approach captures a worst-case scenario, it has been selected because it is useful to highlight the economic contributions of farm workers. The approach permits analysis of the different impacts associated with the different scenarios and changes that could occur in the absence of the event.

The economic impacts from the loss of a migrant and seasonal farm labour market in a local economy are felt throughout all its sectors. In the case of the private sector, the absence of the MSFW population directly influences the MSFW-dependent agricultural sector, the agricultural sector with links to food manufacturing industries in addition to several other industries in the local economy. In Michigan agriculture, MSFWs perform a variety of tasks associated with at least 46 different fruits, vegetables, Christmas trees and floriculture crops grown across the state (Office of Migrant Services, 1998b). One immediate short-term impact from the disappearance of farm workers would be a farm labour shortage. Under such labour market conditions, wages paid to farm workers would have to increase to induce entry of new workers into the farm labour market. If high school students and other people not in the labour market are recruited into the farm labour market, it is unlikely that wages would increase enough to meet the demand for labour at current levels of agricultural production. Over the long term, the increased claims against farm income and the resultant decrease in farm earning potential is a disincentive for growers to remain in labour-intensive agriculture.

Rather than substitute migrant labour with local labour, an alternative option available to growers is to substitute migrant workers with mechanical harvesters. However, this option is not viable either. Not only is mechanization of agriculture less than complete, but growers also prefer workers to machinery because farm wages continue to fall relative to the price of machinery. If crops are harvested mechanically rather than by hand, the return would be lower because of the ability to hand-harvest crops multiple times. In those cases where growers are able and willing to substitute machinery for labour, machine manufacturers and service providers would be substituted for labour as claimants against aggregate farm income. Even if economically feasible, growers may not be willing to substitute machines for migrant labour, because the adaptation of machinery would lock growers into particular crop production and marketing options over multiple growing seasons (Martin and Martin, 1994). The negative effects on the environment from the use of machinery would also constitute a social cost.

In addition to the options of replacing migrant labour with local workers or machinery, farm operators have the option to switch to alternative agricultural production that is not labour intensive. They also have the option to sell or lease their farm land and leave agriculture altogether. The option of switching to alternative agriculture could have serious consequences on the earning potential of individual farm operators as well as on the structure of agriculture. Under this scenario, the most significant impact to growers would be the lost revenue from producing less profitable crops. This decrease in earning potential could negatively affect the overall operation of farms because in many instances labour-intensive agriculture is the only profitable agriculture; it is used to subsidize the rest of the farming operation (Trupo *et al.*, 1998). For agriculture, the loss of farm workers would undoubtedly lead to some changes in farm land use into pasture and mechanized field crop production. It would also mean less diverse farming operations and, hence, higher risks for farm operators. A related potential impact for farmers choosing to remain in agriculture despite the loss of MSFWs is the negative effect on land prices. In the absence of MSFWs, the earning potential of farm operators is assumed to decrease. As farm income decreases, the future value associated with the productivity of the farm land would also decrease. This decease in land prices could also influence local tax revenues (Trupo *et al.*, 1998).

Rather than remain in agriculture and produce less profitable crops, evidence suggests that farm operators are more likely to leave agriculture and sell or lease their farm land. That is the case in the state of Virginia, for example, where 80% of surveyed farm worker employers reported that they would retire from farming and sell their farms rather than engage in alternative crop or livestock production (Trupo *et al.*, 1998).

The decision by farm operators to leave farming influences agriculture in a variety of ways, depending on whether the land is sold or leased to other farm operators or sold to developers. Assuming that farm land is sold or leased to other farm operators, the impact would be fewer farmers, absentee farm land ownership or farm land ownership consolidation, larger farms and a less diverse agricultural industry. If, on the other hand, growers sold their land to developers, less farm land and fewer farmers and farms would result. Thus, the loss of the migrant and seasonal farm labour market could also potentially affect the availability and ownership of farm land in addition to the profitability, number, size and diversity of farms in a local area.

The economic impacts from the loss of migrant and seasonal labour markets extend beyond the change in the economic performance of MSFW-dependent farm operations and the resultant changes in the structure of agriculture (e.g. the size of farms, number of farms, product diversity, rural land ownership and land use changes) in a local area. For example, sectors with firms that sell inputs or value-added services to MSFW-dependent agricultural producers will also be affected if these producers choose to sell their operations altogether or switch to less labour-intensive crops or livestock production. The loss of this market to the local economy is important in light of the fact that the costs of producing and marketing labour-intensive agriculture are greater than the costs of producing, harvesting and selling traditional grain crops (Trupo *et al.*, 1998). Irrigation and grading and packing equipment as well as costs associated with migrant housing purchase, construction, maintenance and utilities directly influence the local economy. Although growers bear many of the higher costs associated with more profitable labour-intensive agriculture, from the community perspective, these costs represent an increased income flow to the local economy compared with income flows to the local economy from traditional grain crops or livestock production.

Another industry affected by the loss of the MSFW labour market is food manufacturing. Food manufacturing is influenced in two ways, because it relies on the output of migrant-dependent agriculture as well as hiring migrant labour directly in its operations. Given that some farm workers are employed in the processing sector, their absence could present labour shortages in that sector and possibly higher wages for existing workers. Over the longer term, the loss of the farm labour market would likely mean a decline in the number of growers engaged in fruit and vegetable production as well as a reduction in land for fruit and vegetable production. This, in turn, could cause food processors difficulty in acquiring the necessary production volume to achieve economies of scale in processing operations. If an insufficient volume of local production is available, the processing sector may find it too costly to continue operations.

Thus, in addition to the primary and multiplier impacts to the local economy associated with the immediate loss of economic activity in the food processing sector, local economies could see a reduction in processing, canning and freezing production altogether. The loss of the processing sector that could result has implications for many other business operators in a local economy, in addition to farm operators that rely on MSFWs. These include farm operators that do not rely on MSFWs but sell to the processing sector, suppliers of other inputs to the food manufacturing industry, and the local public sector that would incur a reduction in tax revenue associated with the loss of economic activity in the food manufacturing sector.

The impacts on agriculture and the food manufacturing industries will vary according to whether migrant-dependent agricultural employers remain in agriculture and switch to other crops or livestock production, or get out of agriculture altogether and sell or lease their farm land. The agriculture and food processing sectors are not the only private industry sectors affected by the absence of MSFWs: also affected are the retail and service-providing sectors directly utilized by the MSFW population. The benefits from farm worker expenditures extend to the grocery, consumer goods, clothing, gasoline retailer and other service sectors of the local community. These farm worker household expenditures also produce a fiscal impact associated with the tax revenue collected by the public sector. This injection of spending will have further expansion effects as local residents spend and respend their dollars. The loss of the MSFW population would cause the direct and indirect effects of the spending of farm worker income in the local area to vanish.

The non-profit and public sectors that service the migrant and seasonal population will also be influenced by the loss of the migrant and seasonal farm labour market. The general perception is that migrant farm workers are 'strangers in the fields' at their temporary workplaces. Their special needs and problems are not met by their employers or by local assistance programmes in the areas where they temporarily reside. This explains why federal and state initiatives now exist to address the needs of migrants. Since the 1960s, the federal departments of education, health and human services, labour, agriculture, and other federal and state agencies have put in place migrant labour programmes to service their needs. The various childcare, health benefits, food stamps, job training, legal assistance, elementary, high school and college assistance and housing programmes now available to migrants are typically operated by private non-profit and public entities that receive transfer payments to provide these services. In the case of non-profit corporations, these transfer payments help to pay salaries and for supplies, emergency assistance, medical care and food stamps. They also provide a variety of programmes and services to benefit the migrant and seasonal farm worker community.

Reference has been made to the tax revenue reductions associated with the loss of economic activity in the local economy. The public sector also benefits from the direct administration of migrant farm worker programmes available to the migrant and seasonal farm worker population. Salary payments and other expenditures associated with the administration of these programmes represent a direct infusion of funds to the various sectors of the local economy. As in the case of private and non-profit sector expenditures, these transfer payments also will have a multiplier-like impact on the local economy.

The faith-based organizations, in particular, provide a variety of programmes and services to benefit the migrant and seasonal farm worker community. Over the years these non-profit sector entities have devoted a fraction of their budgets to providing services related to the presence of MSFWs. In the absence of farm workers in the region, however, these expenditures would likely be diverted to other local needs.

Methods and limitations

There are many types of community impacts associated with an economic event such as the temporary use of migrant workers in agriculture. Demographic impacts, for example, reflect changes in the size, location and composition of the population that can result from this development event. Fiscal impacts result from changes in local government revenues due to expenditures of the farm workers and employers that affect the local tax base. Economic or monetary impacts are associated with the changes in the level and distribution of local employment, income, sales and wealth. The presence of farm workers also has longer-term impacts on land use and farm structure, including farm size, number of farms, and the diversity of agricultural production.

This study is largely limited to the economic or monetary impacts of the farm worker population on the local economy of a six-county region in southeastern Michigan. Providing a detailed assessment of *all* effects on the rural area from the presence of the migrant and seasonal agricultural workers requires an extensive analysis of primary and secondary data and a serious commitment of time and financial resources. Such an endeavour is beyond the scope of this chapter. Even when the multitude of potential changes in a community's social, demographic, environmental, land use, industrial diversity and concentration and fiscal dimensions are excluded from the analysis, estimating the monetary impacts of the MSFW population in rural areas remains a challenging task. Chief among the reasons for the difficulty is the availability of even the most basic information. This problem is made more complicated when a county or regional focus, rather than a state focus, is adopted for analysis.

To estimate the economic loss to the local economy of southeastern Michigan from the absence of the farm worker population, the following monetary losses will be analysed: the loss in value of production from the foregone production of MSFW-dependent crops; the decline in community revenue from the loss of farm worker expenditures in the local economy; the loss in community income from the decline in transfer payments to MSFWs; and the loss of community income from the reduction in farm worker housing costs. Areas not factored into the analysis included: (i) the direct impact on the economy from the expenditures of faith-based organizations on the MSFW population; (ii) the value of the foregone factor input costs associated with labour-intensive agricultural production; and (iii) the indirect forward linkages associated with food manufacturing. Although the impact of the absence of farm workers in the area on the processing sector is deemed important, it was difficult to identify the relationships that exist between the agricultural production in a local area and its linkages to the local processing sector. Also, no attempt was made to measure the indirect and induced effects caused by additional rounds of spending by directly and indirectly affected firms and sectors in the local economy. A variety of input–output programs is available to estimate these indirect and induced effects, but it is believed that these models would only have a distorting effect on the calculations. The unique consuming patterns of the MSFW population and the exporting nature of agricultural production, which complicate its relationship with the processing sectors, make the use of these models hazardous in estimating these effects. Given these complications, only the direct impact estimates are provided. These are based on what is deemed reliable data based on surveys, interviews and the Agricultural Census.

Table 8.1 is a simplified model for conceptualizing the net monetary gain to the community from the farm labour economic development event. The steps undertaken in this study to quantify the economic contributions of the MSFWs to the region, as well as the methods and data used in the analysis, are described as follows:

1. Private sector monetary changes:

- Estimates were made of the total farm labour-related expenditures by growers used to maintain the farm labour population during their working months. These expenditures mainly include direct housing and utilities costs.
- Farm worker earnings and the amount spent locally were estimated from a farm survey conducted in the summer of 1997.
- Changes in cash receipts from land allocated to high-value crops were used to estimate the value of lost production from the disappearance of MSFWs. The dollar value of production from crops utilizing farm labour was compared with the dollar value of production from the next best alternative crop, assumed to be traditional grain crops. The crops utilizing MSFWs were estimated using a survey conducted by the Office of Migrant Affairs in cooperation with Michigan State University Extension. Estimates of the dollar value of production were obtained from the Michigan Agricultural Statistics Service.
- The number of migrant and seasonal farm workers in the six-county region, their earnings and how they spent their earnings were estimated. Data on the number of MSFWs were obtained from an interview with an agricultural employment specialist from Michigan Works, a private placement company.

2. Non-profit and public sector monetary change:

- Non-profit organizations servicing the MSFW population were identified and transfer payments to this population were estimated. A survey of all the agencies providing farm worker services was conducted to get estimates on these expenditures. These were then divided between the public and non-profit sectors.

Even though a comprehensive analysis of all the impacts associated with the presence of MSFWs in a receiving rural area is not provided, the monetary impact analysis is still useful. Identifying the impacts on the private, public and non-profit sectors of the local economy will enable community stakeholders and farm worker advocates and employers to point to the contributions of the MSFW population. The findings represent one type of information to help communities better understand the links between migrant labour and the local economy. It helps to dispel community concerns that the MSFW population is a drain on the local economy. With an improved understanding of the contributions of the migrant population, the community will be better poised to take steps to maximize the benefits to the local economy from the presence of this population.

Table 8.1. Simplified model for measuring net monetary gain to the community from a farm labour economic development event. (Adapted from Shaffer, 1988, p. 242.)

Private sector monetary changes
  Net change in farm labour-related expenses to the employer
+ Net change in business sales
+ Net changes in cash receipts from high-value crop acreage compared with next best land use

= Net private sector monetary gain

Non-profit sector monetary changes
  Net change in child care non-profit sector revenue
+ Net change in health care non-profit sector revenue

= Net non-profit sector monetary gain

Public sector monetary changes
  Local school revenue net gain
+ Public services revenue net gain
+ Local government revenue net gain
+ Other public sector revenue net gain

= Net public sector monetary gain

Community net monetary gain
  Net private sector monetary gain
+ Net non-profit sector monetary gain
+ Net public sector monetary gain

= Community Net Monetary Change

MSFW-Dependent Agriculture and MSFWs in Southeastern Michigan

The study area, southeastern Michigan, saw the arrival in the 1920s of the first Mexicans from south Texas to migrate to work in agriculture. The need for Mexican labour is inextricably linked to the sugarbeet companies that started operations in the region at the turn of the century. Local workers could not be relied upon to do backbreaking fieldwork, and farmers would not even agree to plant sugarbeet unless the refining mills could guarantee an adequate supply of labour. Before 1920, sugarbeet companies recruited large numbers of Belgians, Hungarians, Moravians and Bohemians from nearby cities to work in the beet fields. After 1920, the sugarbeet companies found it necessary to recruit from outside the immediate area of southeastern Michigan and northern Ohio. They began recruiting 'Mexicans' from the Laredo area of south Texas, transporting them to Michigan by train at the start of the season and returning them at the end of the season by the same means. By the 1930s some workers began to come in their own vehicles. After the 1935 farm labour strikes in the region, 'Texas Migrants' became the main source of labour. Blissfield, in Lenawee County, became the main drop-off point for Mexicans seeking agricultural work throughout Michigan (Rosenbaum, 1996).

Today, MSFW-dependent agriculture in southeastern Michigan is no longer just sugarbeet production; it is quite diverse and the demand for migrant and seasonal farm workers cuts across a large variety of field operations. Of the 46 labour-intensive crops grown in the state, 39 are grown in Michigan's southeastern region. As Table 8.2 shows, MSFWs are rarely hired specifically for field crops like sugarbeets any more. Workers predominate in the planting and harvesting of fruits and vegetables; field crop employment still exists but serves to fill unemployment gaps as employers attempt to retain workers until fruits and vegetables are ready for harvest. Increasingly, MSFWs are being hired to work in floricultural and Christmas tree operations.

The various field crop, vegetable, fruit, berry and floricultural MSFW-dependent commodities produced in each county in the region in 1997 are shown in Table 8.2. Monroe, with 31 different labour-intensive crops, had the most diverse labour-intensive production in the region. Washtenaw and Lenawee followed with 27 and 26 labour-intensive crops, respectively. Jackson registered only three labour-intensive commodities. All six counties relied on MSFWs in some aspect of the hay harvest operation. Five of the counties were MSFW-dependent in pumpkins, potatoes, apples, strawberries, soybeans and nursery and bedding plants. The number of labour-intensive commodities averaged over 19 per county in the region.

As Table 8.2 shows, it is a mistake to think of MSFWs as only employed in fruits and vegetables. Field crops, floriculture and nurseries also are migrant-labour dependent. Likewise, it is a mistake to think that MSFWs are only used in harvesting operations. Although the harvesting task is performed in most crops, the majority of the crops rely on migrant labour for multiple tasks. Asparagus is the only commodity where MSFWs have been used solely in harvesting operations. The large number of crops dependent on migrant labour also contribute to the diverse number of tasks performed by migrant labour in the region. Table 8.3 identifies the types of work performed by MSFWs on a crop-by-crop basis for the agricultural commodities grown in the region.

Not much information is available on MSFW-dependent agriculture in southeastern Michigan beyond the information shown in Tables 8.2 and 8.3, from the Office of Migrant Services for the state. To get a better picture of MSFW-dependent agriculture in the region, Table 8.4 combines several data sources to estimate the number of acres of MSFW-dependent agriculture and the value of production for each county. Information was not available on all the MSFW-dependent crops: in some instances, data were withheld for some counties to avoid disclosing data for individual farms; nor were comparable data on floriculture crops and Christmas tree production available. Because of these limitations, information is recorded in Table 8.4 for only 7 of 17, 3 of 11, 19 of 26, 17 of 32 and 10 of the 28 labour-intensive crops grown in Branch, Hillsdale, Lenawee, Monroe and Washtenaw counties, respectively.

It should be noted that Jackson County was excluded from Table 8.4 despite the presence of a migrant labour camp housing 25 farm workers and 10 seasonal workers residing elsewhere. The Census data show that Jackson County produced such labour-dependent crops as asparagus, cantaloupes, squash, pumpkins, tomatoes, sweetcorn, apples, watermelon and blueberries. It was excluded from this section of the analysis because the Office of Migrant Services survey used to identify the labour-dependent crops did not register any of the fruit and vegetable crops grown in the county as dependent on MSFWs. Presumably the 35 workers registered for Jackson County worked in hay, potatoes or soybeans (Table 8.2). In the case of potatoes, the Census withheld the data to avoid disclosure.

Soybeans and hay were excluded from the analysis in Table 8.4 because they were included in the category of field crops that were not considered MSFW-dependent for purposes of this study. The large acreage associated with these crops would have distorted the acreage figures for the more labour-dependent crops in the table. The tasks of soybean weeding and hoeing and hay baling and moving performed by MSFWs, although sequentially essential to the production process of these commodities, represent a relatively minor addition to the value of production of these crops.

It was also assumed that all farms growing the crops in Table 8.4 relied on migrant and seasonal labour, which may not be the case if self-pick or family-only operations exist. Because the value of production figures on a county-by-county basis were not available, state average value-of-production figures for labour-intensive crops were multiplied by the crop acreage in each county to

Table 8.2. Labour-intensive crops in Michigan grown and not grown in southeastern Michigan, 1997. (Source: Office of Migrant Services,1998b.)

	County						
Agricultural commodity	Branch	Hillsdale	Jackson	Lenawee	Monroe	Washtenaw	Total number of counties
Vegetables							
Asparagus		+		+	+	+	4
Beans (snap, pole, green)	+				+	+	3
Broccoli					+	+	2
Cabbage				+	+		2
Cantaloupes				+	+		2
Carrots				+		+	2
Cauliflower					+		1
Celery							0
Cucumbers	+			+	+	+	4
Greens (mustard, turnip)					+	+	2
Lettuce							0
Onions	+			+		+	3
Peppers, bell (sweet)				+	+	+	3
Pumpkins	+	+		+	+	+	5
Radishes						+	1
Sweetcorn	+	+		+	+	+	5
Sugarsnap peas							0
Sugarbeets				+	+		2
Tomatoes	+			+	+	+	4
Mushrooms						+	1
Potatoes	+		+	+	+	+	5
Squash (summer, winter)				+	+	+	3
Zucchini					+	+	2
Fruits and nuts							
Apples	+	+		+	+	+	5
Cherry, sweet							0
Cherry, tart						+	1
Grapes							0
Peaches				+	+	+	3
Pears				+	+		2
Plums				+	+		2
Berries							
Blackberries				+			1
Blueberries				+	+	+	3
Raspberries	+			+	+	+	4
Strawberries	+	+		+	+	+	5
Other crops							
Maize, seed							0
Hay harvest	+	+	+	+	+	+	6
Sod					+	+	2
Beans, dry edible	+				+		2
Soybeans	+	+	+	+	+		5
Christmas trees		+		+	+	+	4

Continued

Table 8.2. (*Continued.*)

Agricultural commodity	County						Total number of counties
	Branch	Hillsdale	Jackson	Lenawee	Monroe	Washtenaw	
Plants							
Bulbs	+	+			+		3
Nursery plants	+	+		+	+	+	5
Bedding plants	+	+		+	+	+	5
Flowers	+						1
Apricots/herbs							0
Barley/maize, field							0
Total	17	11	3	26	31	27	115

estimate the total value of production for the crops in each county in southeastern Michigan. Given that all labour-intensive crops in the floriculture sector and most crops in the field crop sector are excluded, it is believed that the estimated total value of production for labour-intensive crops in southeastern Michigan represents a lower limit of the total value of production involved.

The limitations of the information in Table 8.4 not withstanding, a number of salient features about MSFW-dependent agriculture in the region are nevertheless discernible. The data suggest a wide range in the significance of MSFW-dependent agriculture across the counties in the region. The counties can be organized into three tiers in terms of labour-intensive agricultural activity and dependence on MSFWs. Lenawee and Monroe are at the high end of MSFW dependence, followed by the counties of Washtenaw and Branch in the middle, and Hillsdale and Jackson at the low end. Lenawee County ranked first in land use and total value of production from labour-intensive crops, accounting for 47 and 46% of the total land use and value of production, respectively. Monroe County ranked second with 26 and 36% of the total land use and value of production, respectively. These counties combined for a total of 73 and 82% of the total land use and value of production in the region. The counties of Branch and Washtenaw combined for 13 and 16% of the total acreage and value of production in the region, respectively. Hillsdale County contributed around 1% of the acreage and 1.5% of the region's total value of production from labour-intensive agriculture. Jackson County relied on MSFWs for only three crops and data were not available on these crops to measure the county's contribution to total land use and value of production from labour-intensive agriculture in the region.

The three-tier classification of counties identified on the basis of their contribution to land in farm production and value of production is supported by information in Table 8.5 on where workers reside and where employers are located. The migrant and seasonal farm workforce in southeastern Michigan in 1997 was estimated at 1257 workers. Lenawee and Monroe Counties together accounted for 67% of the workforce and 74% of farm worker employers. Branch and Washtenaw Counties contributed nearly 29% of the workforce and 16% of farm worker employers. Jackson and Hillsdale accounted for less than 4% of the workforce and 11% of the employers.

Table 8.5 shows that the majority of workers were in the migrant stream; only about 10% were considered seasonal workers. Given the predominantly migrant character of the workforce, the overwhelming majority (83%) resided in labour camps (Table 8.6). The number of living units in each of the camps in the region also supports the three-tier county categorization of where labour-intensive agricultural production is concentrated in the region (Table 8.7).

Table 8.3. MSFW-dependent agricultural commodities and type of work done by MSFWs. (Source: Office of Migrant Services, 1998b.)

Agricultural commodity	Type of work
Apples	Processing, training, thinning, harvesting, packaging, loading, pruning
Asparagus	Harvesting
Beans (dry edible)	Hoeing/weeding
Beans (snap, pole and green)	Weeding, harvesting, grading, packing
Bedding plants	Potting, planting, shipping
Blackberries	Cleaning, hoeing, harvesting, packaging, shipping
Blueberries	Harvesting, packaging, shipping
Broccoli	Transplanting, weeding, harvesting, packaging
Bulbs	Planting, weeding, harvesting, shipping
Cabbage	Transplanting, weeding, harvesting, packaging, shipping
Cantaloupe	Transplanting, weeding, harvesting, shipping
Carrots	Thinning, hoeing, weeding, sorting, packaging, shipping
Cauliflower	Transplanting, hoeing, weeding, harvesting
Cherries, tart	Harvesting, process line, pruning
Corn, sweet	Weeding, harvesting, grading, packing
Cucumbers	Hoeing, weeding, thinning, training vines, harvesting
Flowers	Not available
Greens	Harvesting, packaging
Hay harvest	Baling, moving hay
Mushrooms	Planting, harvesting, packaging
Nursery plants	Potting, planting, transporting, shipping
Onions	Transplanting, weeding, harvesting, sorting, bagging
Peaches	Pruning, thinning, harvesting
Pears	Pruning, harvesting
Peppers, bell	Transplanting, hoeing, weeding, harvesting, sorting, packaging
Plums	Pruning, harvesting
Potatoes	Weeding, grading, packing
Pumpkins	Weeding, harvesting, loading
Radishes	Weeding, harvesting, loading
Raspberries	Cleaning, hoeing, harvesting, packaging, shipping
Sod	Tractor cut/roll, hand load/unload, delivering, unrolling
Soybeans	Weeding, hoeing
Squash, summer	Weeding, harvesting, packing
Squash, winter	Weeding, harvesting, packing
Strawberries	Planting, cleaning, hoeing
Sugarbeets	Thinning, hoeing, weeding
Tomatoes	Transplanting, weeding, hoeing, harvesting, packaging, shipping
Christmas trees	Planting, shearing, pruning, painting, harvesting
Zucchini	Harvesting, packaging

The migrant population is usually housed in labour camps, although about 17% of the workers do not avail themselves of labour camp housing provided by their employer and instead rent their own housing in the towns and villages near the place of work (Table 8.6). Although growers are not obligated by law to provide migrant housing, housing is an essential element in securing an adequate supply of seasonal agricultural workers. Sometimes there is a rental fee, but usually the employer absorbs the housing expense. It is suspected that employers without labour camps relied on growers with labour camps to house their workers, but the extent of this practice is not known. Seasonal workers usually commute from their residence to the place of work.

State and federal regulations ensure that migrant labour camps meet certain minimum

Table 8.4. Acreage and value of production for labour-intensive agriculture in southeastern Michigan, 1997. (Sources: 1997 Census of Agriculture (AC97-A-22), 1999; Michigan Agricultural Statistics, 1998–99, Michigan Department of Agriculture 1998 Annual Report; and calculations by B. Gardner from 1997 Census of Agriculture and State Enterprise Budgets.)

Crops by county	Acres	State average value per acre (US$)	Total value of production (US$)
Branch			
Beans, snap	422	1,262	532,564
Pumpkins	51	1,200	61,200
Sweetcorn	81	1,393	112,833
Tomatoes	516	2,667	1,376,172
Apples	164	1,819	298,316
Raspberries	2	3,400	6,800
Strawberries	16	4,632	74,112
Total	1,252	–	2,461,997
Hillsdale			
Pumpkins	21	1,200	25,200
Sweetcorn	34	1,393	47,362
Apples	200	1,819	363,800
Total	255	–	436,362
Lenawee			
Asparagus	10	988	9,880
Cabbage	{25}[a]	2,028	50,700
Cantaloupes	{20}	2,831	56,620
Carrots	245	3,011	737,695
Cucumbers	714	1,152	822,528
Peppers, bell	120	4,343	521,160
Pumpkins	131	1,200	157,200
Sweetcorn	97	1,393	135,121
Tomatoes	2,731	2,667	7,283,577
Squash	2	1,700	3,400
Potatoes	97	1,915	185,755
Sugarbeets	1,833	718	1,316,094
Apples	683	1,819	1,242,377
Peaches	45	2,890	130,050
Pears	2	1,111	2,222
Plums	7	1,209	8,463
Blueberries	3	3,033	9,099
Raspberries	5	3,400	17,000
Strawberries	24	4,632	111,168
Total	6,794	–	12,800,109
Monroe			
Asparagus	7	988	6,916
Beans, snap	96	1,262	121,152
Sugarbeets	278	718	199,604
Cabbage	173	2,028	350,844
Cantaloupes	119	2,831	336,889
Cucumbers	193	1,152	222,336
Peppers, bell	148	4,343	642,764

Table 8.4. (*Continued.*)

Crops by county	Acres	State average value per acre (US$)	Total value of production (US$)
Pumpkins	280	1,200	336,000
Sweetcorn	490	1,393	682,570
Tomatoes	1,228	2,667	3,275,076
Potatoes	1,579	1,915	3,023,785
Squash	36	1,700	61,200
Apples	284	1,819	516,596
Peaches	16	2,890	46,240
Pears	1	1,111	1,111
Raspberries	10	3,400	34,000
Strawberries	28	4,632	129,696
Total	4,966	–	9,986,779
Washtenaw			
Asparagus	46	988	45,448
Beans, snap	23	1,262	29,026
Cucumbers	13	1,152	14,976
Pumpkins	205	1,200	246,000
Sweetcorn	485	1,393	675,605
Tomatoes	33	2,667	88,011
Squash	19	1,700	32,300
Apples	291	1,819	529,329
Raspberries	24	3,400	81,600
Strawberries	48	4,632	222,336
Total	1,187	–	1,964,631
Overall total	14,454	–	27,649,878

[a]Determined by the average-size farm for the crop in the county.

Table 8.5. Migrant and seasonal farm worker workforce and employers by county, 1997. (Source: 'Farmworkers and farmworker employers', unpublished document, Michigan Jobs Commission, Adrian Office, 1997.)

County	Number of workers			Number of farm worker employers
	Migrant	Seasonal	Total	
Monroe	300	80	380	16
Jackson	30	5	35	2
Hillsdale	12	3	15	2
Washtenaw	181	12	189	3
Branch	168	5	173	3
Lenawee	432	33	465	12
Total	1123	138	1257	38

Table 8.6. Number of migrant and seasonal farm workers by housing arrangement by county, 1997. (Source: Michigan Jobs Commission, Adrian Office, 1997, unpublished.)

County	Workforce in labour camps	Workforce not in labour camps
Monroe	274	106
Jackson	25	10
Hillsdale	12	3
Washtenaw	181	8
Branch	168	5
Lenawee	388	77
Total	1048	209

standards. When five or more migrant workers occupy a site, it is required that the labour camp be inspected and licensed by the Michigan Department of Agriculture. According to the Department, there were 38 licensed labour camps in the region. The number of labour camps housing fewer than five workers is not known. The licensed camps met minimal standards such as roofs free from leakage and structurally sound, screened doors and windows, structurally sound floors, and supplies of electricity and water. If a camp is found to have serious deficiencies, the issuing of a licence is not recommended. M.G. Johansen, Environmental Manager for the department in charge of inspections, estimated in 2000 that growers spent approximately US$75 monthly in utilities and labour camp maintenance per living unit during the period of occupancy, which averaged 5.5 months over the six-county region. It is estimated that total housing and utility expenditures in the region for the season are about US$79,534 (Table 8.7).

The migrant and seasonal workforce in southeastern Michigan consists of both workers in family units and solo male workers (Table 8.8). The 156 family units accounted for 733 workers, or nearly 54% of the total workforce. About 11% of these family working units were seasonal; the rest were migrant. Owing to the fact that an overwhelming number of families were in the migrant stream, most (81%) lived in labour camps. As in the case of migrant families, the overwhelming proportion (83%) of migrant solo male workers resided in labour camps.

The family unit composition of the workforce accounts for the difference that exists between the size of the farm worker population and the farm worker workforce in the area. The migrant and seasonal population in southeastern Michigan is larger than the migrant and seasonal workforce because not all the family members are farm workers. In an average-size family unit of 6.04 persons, only 47% worked. Another 38% of the family members are non-workers below the age of 12 years and the remaining 15% consists of non-working adult members. When the non-working family members are added to the 446 solo male workers and 811 family workers in the region, it is estimated that the

Table 8.7. Labour camps, length of occupancy and number of units by county, 1997. (Source: 1997 Licensed Agricultural Labor Camp List, Michigan Department of Agriculture.)

County	Licensed labour camps (number)	Average length of occupancy (months)	Living units (number)	Estimated housing costs per camp per season (US$)
Branch	9	6.5	28	13,650
Hillsdale	1	5	1	375
Jackson	2	5	14	5,250
Lenawee	9	5.1	67	25,628
Monroe	13	6.4	50	24,000
Washtenaw	4	5.25	27	10,631
Total	38	–	187	79,534

Table 8.8. Families, family workers and solo male workers in labour camps and not in labour camps, by migrant and seasonal categories, 1997. (Source: Michigan Jobs Commission, Adrian Office, 1997, unpublished.)

Selected characteristics	Migrant	Seasonal	Total
Number of families	139	17	156
In labour camps	127	0	127
Not in labour camps	12	17	29
Number of family workers			
In labour camps	678	0	678
Not in labour camps	55	78	133
Number of solo male workers			
In labour camps	370	0	370
Not in labour camps	20	56	76

migrant and seasonal population in the region consists of 2168 people (R.P. Rosenbaum, 1998, Lenawee County Farm Worker Survey, unpublished).

Survey data for 1996 from Lenawee County farm workers show differences in income depending on whether the worker is a member of a migrant family, a resident seasonal family, or a solo male migrant farm worker. While migrant workers earned an average income of US$2228, seasonal workers earned an average of US$5057 and solo male workers earned an average income of US$2000. If these estimates are extended across the region, the 733 migrant family workers earned a gross income of US$1,633,124; the 134 resident seasonal workers earned a total of US$677,638 and the 390 solo male migrant workers earned a combined income of US$780,000. The combined total income earned by these three groups is estimated at US$3,090,762 (Lenawee County Farm Worker Survey, 1997).

The survey data also show that not all farm workers had the same propensity to consume in Michigan and locally. Different expenditure patterns were observed for each of the three groups of farm workers. Migrant and seasonal farm workers spent 47% of their income earned in the region in Michigan, of which 80% was spent locally. Resident seasonal farm workers, on the other hand, spent most of their income earned (86%) in Michigan, of which 68% was spent locally. By contrast, solo male workers spent just 30% of their income in Michigan, most of which (86%) was spent locally. The expenditures of the three groups aggregated at the state level totalled US$1,584,337, or 51% of earned wages. When aggregated at the local level, these expenditures equalled US$1,211,578, or 39% of total income earned.

Seven public and non-profit organizations were identified as providing services to migrant and seasonal farm workers in the region in 1997 (Table 8.9). These agencies administered at least US$1,238,391 in revenue to service the farm labour population. Over 72% of the funds were administered through state departments, although the funds represent federal as well as state funds. These state agencies administered the Migrant Education Program, Food Stamps, Day Care, Aid to Dependent Children, Family Medicaid, Migrant Hospitalization programmes and other medical and emergency service programmes. The remaining non-profit agencies administered approximately 28% of the revenue. These agencies administered health, childcare and employment programme services. As can be seen, the variety of sectors affected by the presence of farm workers is significant. In the case of the public sector operated programmes, the figures show only the cost of services provided. The figures would be larger if administrative costs were factored into the analysis.

The proportional representation of family units and/or solo male workers in the

Table 8.9. Migrant and seasonal farm worker service network expenditures in southeastern Michigan, 1997. (Sources: Interviews with agency administrators, November 1998; Office of Migrant Services, 1998c.)

Agency	Agency expenditures (US$)	
Community Action Agency		20,501
Migrant Education Program		
Jackson	114,578	
Washtenaw	79,680	
Lenawee	122,957	
Total		317,215
The Family Medical Center		19,000
HDI Washtenaw Mobile Unit		49,000
Family Independence Agency		
Monroe	276,979	
Washtenaw	3,000	
Lenawee	293,019	
Jackson	6,000	
Total		578,998
Telamon		
MHS	161,171	
402	55,006	
Total		216,177
Jobs Commission		37,500
Overall total		1,238,391

farm worker workforce makes a significant difference in the amount of income from transfer payments that flows into the region from federal and state sources. Children-specific programmes such as the Migrant Education Program, Telamon's Migrant Head Start or the Family Independence Agency's Day Care or Aid to Dependent Children programmes would be eliminated if the workforce consisted of only solo male workers. When these programmes are eliminated, the transfer payment flow into the region is reduced by over 66%. The reduction in the transfer payment flow is actually greater, since children also benefit from the other Family Independence Agency programmes that are not specific to children. In addition, solo male workers are less likely to use migrant services compared with family workers, for a variety of reasons. For one, fewer may qualify since some of these programmes require that the beneficiary to be a legal resident; and there is a higher likelihood that the proportion of undocumented workers is greater for solo male workers than for family workers. Solo male workers are also younger on average and therefore less likely to utilize health services.

Contributions of MSFWs in Southeastern Michigan

Measuring the economic contributions of the MSFW population to the local economy of southeastern Michigan relies on the information from the previous two sections. The straightforward discussion follows the model presented in Table 8.1. The findings are then used to gauge the potential economic impact on the local economy of the H-2A national policy initiative currently being debated by the US Congress.

It was indicated in Table 8.4 that MSFW-dependent agriculture in the six-county area of southeastern Michigan generated a production value of US$27,649,878 from the use of 14,454 acres (5849.5 ha). This represents an average of US$1913 per acre (US$4724 per ha). Assuming MSFWs are not available and growers remain in agriculture and switch to growing traditional field crops, which earned an average value of production in Michigan of US$281 per acre (US$694 per ha), the production value from this acreage would have been US$4,061,574. Thus, the use of farm labour would have contributed an extra US$23,588,304 (Table 8.10). Lenawee and Monroe counties accounted for over 82% of the increased value of production.

The proportion of wages earned over the season that are spent locally can be used to estimate changes in business sales in the region. In the previous section it was demonstrated that farm worker expenditures contributed US$1,211,577 to the local business community, but employment compensation needs first to be subtracted from value added to avoid double-counting the income contributions to the local economy. This is because almost all farm workers are employed rather than contracted, and when estimating value added to the US economy, the Economic Research Service subtracts contract labour but not hired labour expenses from farm cash receipts (Michigan Department of Agriculture, 1999). When earned income (US$3,090,762) is subtracted from the added value of production from MSFW-dependent agriculture in the region, value added is US$20,497,542. With this adjustment, the contribution to the private sector from increases in value added and farm worker expenditures is US$21,709,119.

Housing maintenance and utilities are expenses associated with MSFW-dependent agriculture that are factored in when estimating value added. Because these expenses are deducted from cash receipts to estimate value added, and since these costs are not expenses associated with traditional field crop production, they represent an additional contribution to the local economy despite being an expense to agricultural employers. In 1997, these expenditures totalled US$79,534.

When these expenditures are combined with the income to growers and the business community, the total private sector impact from MSFW-dependent agriculture is estimated at US$21,788,653. It is worth pointing out that 94% of the contribution accrues to agricultural employers. An additional 5.5% accrues to the business community, primarily the retail sector. Although, relatively speaking, business expenses are a small amount, the amount of business income generated from farm worker customers can be substantial for particular individual firms near the labour camps and the surrounding area.

An additional source of income to the private sector comes from state and federal transfer payments that are administered through the public and non-profit sectors

Table 8.10. Added value of production of major crops harvested by farm labour in six-county area above that of field crops, 1997. (Source: Michigan Agricultural Statistics Service, 2000.)

		Value of production		Added value of production from MSFW-dependent agriculture	
County	Acres[a]	Traditional grain crops (US$)	MSFW-dependent agriculture (US$)	(US$)	%
Branch	1,252	351,812	2,461,997	2,110,185	8.95
Hillsdale	255	71,655	436,362	364,707	1.55
Lenawee	6,794	1,909,114	12,800,109	10,890,995	46.17
Monroe	4,966	1,395,446	9,986,779	8,591,333	36.42
Washtenaw	1,187	333,547	1,964,631	1,631,084	6.91
Total	14,454	4,061,574	27,649,878	23,588,304	–

[a]1 acre = 0.4047 ha.

Table 8.11. Direct economic impact (US$) of the migrant and seasonal farm worker population, 1997.

Type of monetary change	Region	State
Value added	20,497,542	20,497,542
Housing maintenance and utilities costs	79,534	79,534
Farm worker expenditures	1,211,577	1,584,337
Public and non-profit sector expenditures	1,238,391	1,238,391
Total direct impact from MSFWs	23,027,044	23,399,804

in the region. The service sectors, especially the health, education and childcare industries, are major beneficiaries of the estimated US$1,238,391 in transfer payments that annually flow into the region. It was not possible to estimate for this study how this revenue was divided between employee compensation expenses that accrued to the public and private sectors, and service expenses that accrued to the private sector.

Based on calculations of direct effects, the farm worker population contributed to a net monetary gain of US$23,399,804 to the state economy, of which US$23,027,044 was spent on the regional economy of southeastern Michigan (Table 8.11). These estimates are considered conservative because they exclude: (i) the majority of the field crops that rely on migrant and seasonal farm workers for hoeing and weeding; and (ii) all floriculture and nursery commodities. Even so, given that 1257 farm workers were employed in the region, the contribution per farm worker to the local economy is estimated at over US$18,000.

As was anticipated, the injection of income into the local economy from solo male labour is less than that of family workers. They not only spend less of their earnings, but they also receive less income from federal and state income. Agricultural employers could potentially pay less in housing because less housing would be needed for the same number of workers.

The state's agricultural employers have a long tradition of relying on domestic family units to meet their labour needs. Another deterrent is that the cost of using H-2A rather than domestic labour is considered more costly because of the transportation, income and employment guarantees that employers have to provide to these workers. However, the interest in foreign workers is growing and the H-2A programme is slowly beginning to take hold in Michigan. Although data are not available to make quantitative estimates of their impact, it is reasonable to conclude that their net positive impact on the local economy would be less than that of domestic workers. In addition to injecting less money into the local economy because of their lower propensity to consume locally and their lower need for services, they would also constitute a larger leakage because of the higher costs that agricultural employers would have to incur for their services.

Conclusion

This chapter has demonstrated that the farm worker population represents a significant economic development event to the region and the state. Labour-intensive agriculture represents a net benefit to both the producers and the community. The monetary impact of the farm worker population would be greater if the current migrant solo male workers were replaced with migrant and seasonal family working units. The use of legal foreign workers through the H-2A programme or some other programme would reduce the positive impact of this event to the community.

Communities should consider taking positive steps to maximize the benefits to the local economy from the presence of the migrant and seasonal farm worker population. A diverse and large number of community interests seem to benefit from the presence of the farm worker population. These stakeholders should be made aware of the impact that the farm labour population has on their industries. Issues related to the farm workers should not be left to only economic or equity-value advocates. Consideration should be given to regional and community income as well as farm income when making arguments on behalf of farm workers. Improving community attitudes towards the migrant and seasonal farm worker population should also be an integral part of any strategy to maximize the positive economic impact that the farm worker population has on the local economy.

References and Further Reading

Aigner, S.M., Flora, C.B., Tirmizi, S.N. and Wildox, C. (1999) Dynamics to sustain community development in persistently poor rural areas. *Community Development Journal* 34, 13–27.

Emerson, R.D. (1984) Summary. In: Emerson, R.D. (ed.) *Seasonal Agricultural Labor Markets in the United States.* Iowa State University Press, Ames, Iowa, pp. 482–523.

Fuller, V. (1984) Foreword. In: Emerson, R.D. (ed.) *Seasonal Agricultural Labor Markets in the United States.* Iowa State University Press, Ames, Iowa, pp. vii–xvi.

GAO (1998) H-2A agricultural guestworker program: changes could improve services to employers and better protect workers. (Abstract in *Month in Review, December: 1997 Reports, Testimony, Correspondence, and Other Publications.* (GAO/OOA-98-3) US General Accounting Office, Washington, DC.

Griffith, D. and Kissam, E. (1995) *Working Poor: Farmworkers in the United States.* Temple University Press, Philadelphia.

Hedges, S.J. and Hawkins, D. (1996) Illegal in Iowa: the new jungle. *US News and World Report, Special Investigation,* Vol. 121 (12), Sept. 23, 1996.

Johansen, M. (1997) *Licensed Agricultural Labor Camp List, 1997.* Michigan Department of Agriculture, Environmental Stewardship Division, Lansing, Michigan.

Johnson, D.G. (1959) The dimensions of the farm problem. In: Heady, E.O. (ed.) *Problems and Policies of American Agriculture.* Iowa State University Press, Ames, Iowa, pp. 47–62.

Martin, P. (1988) *Harvest of Confusion: Migrant Workers in US Agriculture.* Westview Press, Boulder, Colorado.

Martin, P.L. and Martin, D.A. (1994) *The Endless Quest: Helping America's Farm Workers.* Westview Press, Boulder, Colorado.

Martin, P.L. and Taylor, J.E. (1998) Poverty amid prosperity: farm employment, immigration, and poverty in California. *American Journal of Agricultural Economics* 80, 1008–1014.

Martin, P.L., Taylor, J.E. and Fix, M. (1996) Rural communities: the Latinization of rural America. *Rural Migration News* 2, 2.

Michigan Department of Agriculture (1999) *Michigan Agricultural Statistics 1998–1999.* Annual Report. Michigan Department of Agriculture and US Department of Agriculture, National Agricultural Statistics Service, Lansing, Michigan, August 1999, p. 8.

Office of Migrant Services (1998a) *Labor Intensive Crops by County Where Grown.* Family Independence Agency, State of Michigan, Lansing, Michigan.

Office of Migrant Services (1998b) *List of Crops on Which Migrants Work in Michigan.* Family Independence Agency, State of Michigan, Lansing, Michigan.

Office of Migrant Services (1998c) *Family Independence Agency Migrant Expenditures, 1997 Program Year.* Family Independence Agency, State of Michigan, Lansing, Michigan.

Oliveira, V.J. (1992) *A Profile of Hired Farmworkers, 1990 Annual Averages.* AER No. 658. Economic Research Service, Washington, DC.

Paehike, R.C. (1999) Environmental values and public policy. In: Vig, N.J. and Kraft, M.E. (eds) *Environmental Policy,* 4th edn. CQ Press, Inc., Washington, DC.

Rosenbaum, R. (1996) *Migration and Integration of Latinos into Rural Midwestern Communities: the Case of 'Mexicanos' in Adrian, Michigan.* JSRI Research Report 19. The Julian Samora Research Institute, Michigan State University, East Lansing, Michigan.

Shaffer, R. (1989) *Community Economics: Economic Structure and Change in Small Communities.* Iowa State University Press, Ames, Iowa.

Sills, E., Alwang, J. and Driscoll, P. (1993) *The Economic Impact of Migrant Farmworkers in Virginia's Eastern Shore.* Virginia Cooperative Extension Publication No. 448-214/REAP

R016. Virginia Polytechnic Institute and State University, Blacksburg, Virginia.

Trupo, P., Alwang, J. and Lamie, D. (1998) *The Economic Impact of Migrant, Seasonal, and H-2A Farmworkers in the Virginia Economy.* Virginia Cooperative Extension, Virginia's Rural Economic Analysis Program, Department of Agriculture and Applied Economics, Virginia Polytechnic Institute and State University, Blacksburg, Virginia.

US Department of Agriculture (1999) *1997 Census of Agriculture, Michigan State and County Data,* Vol. 1, *Geographic Area Series, Part* 22. US Department of Commerce, Bureau of the Census, Washington, DC.

9 Community Response to the Introduction of Hispanic Migrant Agricultural Workers into Central Kentucky

Beckie Mullin Denton
Department of History, Eastern Kentucky University, Richmond, Kentucky, USA

Abstract

In 1987, Georgia Vegetable Company (GVC) announced its intention to bring in Mexican migrant labour to work fields in Clark County, Kentucky. The entire episode provides a case study for better understanding of local response to changes in farm labour practices and to alterations within a community. Many residents responded to the announcement emotionally and in terms of prejudice and anti-immigrant attitudes. Citizens voiced their concern through more than 100 letters to the editor of the local newspaper. Public rallies and meetings were organized by a citizens' group that arose in the turmoil. When local government officials failed to respond as the opposition force insisted, a referendum campaign was carried out to change the form of government to a more responsive one. Yet the cause of the controversy was not entirely the result of prejudice, because the immigrants also became a symbol of a bitter struggle against GVC. Many citizens resented the control of local resources, such as water and land rights, by an outside agricultural firm. They were enraged at the potential damage to both the environment and human life that actions such as aerial spraying would produce. There was also an undercurrent of struggle between large and small landowners and perceived attacks on traditional farming practices. This case study of the Georgia Vegetable Company's introduction of Hispanic migrant labourers into Clark County, Kentucky, provides a lesson for, firstly, white and predominately white communities in Kentucky and elsewhere that are encountering an influx of Hispanics for the first time and, secondly, those firms contemplating the use of migrant labourers in areas where they have not been previously introduced.

Introduction

From the strawberry rows in California to the onion fields in Georgia, Hispanic migrant workers have followed the harvests across the USA. Kentucky, especially western Kentucky, has long used migrant agricultural workers, some of whom have been Hispanic, though accurate numbers cannot be found (Raupp, 1987g). Verification of the insignificant number of Hispanic migrant workers in Kentucky before 1987, and information on the tremendous influx via the tobacco industry in the late 1980s and early 1990s can be found in US Department of Agriculture (1994), Rosenberg and Coughenour (1990) and Rosenberg (1992).

The geographical region of central Kentucky (rather than the entire state) was chosen for this study and here it was not until 1987 that significant numbers of Hispanic migrant agricultural labourers began to be used, as reported in *The Winchester Sun* (4 May 1987). In that year, when Georgia Vegetable Company announced its intention to bring in Mexican migrant labour to work fields in Clark County, Kentucky, a predominately white local population confronted for the first time the likelihood of an influx of Hispanic workers. In 1980 the Clark County population was 93.55% white, 6.15% black and only 0.57% of Spanish origin. The total population of this rural county was 28,322 (Bureau of the Census, 1983). Winchester is its county seat.

Many residents responded to the company's announcement emotionally, declaring their resistance to the introduction of Mexican migrant workers who, they proclaimed, would increase the crime rate and spread disease. Their resistance was not merely a racial or xenophobic issue: the actions of Georgia Vegetable Company were also seen as a move by an outside agricultural corporation that threatened a way of life, a change in the community over which the local residents had little control. The entire episode, which is examined in this chapter, provides a case study for better understanding of local response to changes in farm labour practices and to alterations within a community.

The Announcement and Immediate Response

On 15 April 1987, at a meeting of the Board of Adjustments and Appeals, Georgia Vegetable Company (GVC) first publicly announced its plans to build a labour camp in Clark County for migrant workers (Raupp, 1987a). These labourers, mostly Mexican or Mexican-American, would harvest vegetables from June to October on the 1000 acres (405 ha) of farmland (or 1200 acres, according to Raupp, 1987e) that the company had leased mainly in Clark and also in the surrounding counties of Bourbon and Fayette (see also Nesbitt, 1987). According to R. Grist, the sole owner of GVC, the migrant workers would be housed in two Clark County locations. The first would be four rejuvenated tenant farmhouses in the Becknerville area, the second a labour camp to be built in the Clintonville area; both locations were in the west end of the county. The labour camp would consist of 20–25 trailers, housing 200–250 migrant workers (Ballard, 1987a; Raupp, 1987a,b).

It is uncertain if the owner of GVC knew what a historic action he was initiating with this announcement; however, aware or not, he was proposing 'the first attempt in Kentucky history to bring in and house migrant [agricultural] workers on such a massive scale' (as quoted in Raupp, 1987d). The owner of GVC had experienced success with the use of migrant workers in his operation based in Tifton, Georgia, and had received very few complaints from the community there; thus, he may have been unprepared for the ensuing reaction of Clark County residents to the proposal for migrant labour and a labour camp in their county (Raupp, 1987h).

Reaction was immediate and highly emotional. After receiving phone calls from citizens concerned about the use of migrants, County Judge Executive James B. Allen Jr explained at a hearing of the Clark County Fiscal Court that there was no legal method of halting the introduction and use of migrant workers (Ballard, 1987a). 'Nobody's really happy with it but it's going to happen,' Allen said (as quoted in Raupp, 1987d). 'So we've got to deal with it ... You don't put up a sign on the county line saying "No migrants allowed".'

According to Allen (J.B. Allen Jr, Kentucky, 1997, personal communication), the phone calls that he and other local officials received were quite numerous at first and some were even hysterical. When asked of the specific nature of the complaints, Allen said that they included everything. 'There was one woman ... who said that everybody here's going to get AIDS from the Mexicans,' Allen explained and continued, 'There were a lot of things that were just ridiculous and it was a very emotional issue' (J.B. Allen Jr, Kentucky, 1997, personal communication).

Public reaction mounted with continuing phone calls to local officials, inquiries and letters to the local newspaper, petition drives in opposition to migrants, and demands for public hearings. GVC announced a public informational hearing to be held on 3 May to give government officials the opportunity to question company representatives. Although the public would be allowed to attend, citizens were asked to refrain from addressing questions to GVC officials at the meeting and were restricted to the submission of written questions, which would be answered at a second meeting to be held at an undisclosed time (Raupp, 1987c).

An advertisement in the local paper on 1 May called for citizens concerned about the migrant workers and aerial spraying, which GVC had also announced, to attend a rally the following day. Approximately 200 people participated in the 90-min rally, during which protesters 'voiced their opposition to the pending presence of the migrants and raised questions concerning housing and sanitation' (Duke, 1987; Raupp, 1987d).

The 2.5-hour public informational hearing on the following evening was 'emotionally charged from the outset' (Raupp, 1987f). It was attended by 200 Clark residents, many of whom shouted their objections to GVC officials and waved signs in front of the television camera. Local government officials and GVC representatives discussed issues including acreage, housing, transportation, water use, payroll taxes, wages and waste disposal. Disregarding the proposed written question format, citizens 'began to dominate the discussion, at times with angry emotional outbursts' (Raupp, 1987f).

As opposition in Clark County mounted, it became organized under a group calling itself Power of the People Committee (POP). This group of predominantly rural residents called for the continuation of pressure on local, state and federal officials via a telephone and letter-writing campaign. Additionally, POP held meetings to organize and plan future opposition strategy (Ballard, 1987b; Raupp, 1987i,j).

Letters to the Editor

Partially in response to POP rallying cries, many Clark residents vented individual emotions in a flood of letters to the editor of the local newspaper, *The Winchester Sun*. In April, prior to the rise of sentiment over the GVC issue, 26 letters to the editor appeared in the newspaper; only one of these (27 April) was on migrants. In May, as the GVC controversy reached its peak, 103 letters to the editor were published. Of these, 51 were on GVC-related issues. B. Blakeman, the editor of *The Winchester Sun*, stated in the *Courier Journal* (Lexington, Kentucky, 8 June 1987), 'I don't think anything has generated more letters than the migrant-workers issue in the 26 years I've been here.'

The letters covered a wide range of topics but most can be grouped within two broad categories: those that feared changes the Mexican migrant workers would bring, such as an increased crime rate, and those that resented alterations in the community by GVC itself, such as the detrimental effect of aerial spraying both to humans and to the environment. As citizens failed to stop migrant use or to win a major battle against the agricultural firm, a third category of complaint was that local government officials had failed to respond to the concerns of local residents. Of the three categories, the perceived changes expected with the introduction of Mexican migrant labour received the most attention.

Mexican migrant worker-induced changes

The letters to the editor offer a unique insight into public response. The author of the first letter on migrants (27 April) expressed her concern for the migrant workers who might suffer abuse under GVC. This seems a valid point when one considers the obviously crowded conditions that would be imposed under GVC's plan to house 200–250 workers in only 20–25 trailers. However, the same individual who voiced her anxiety about the potential abuse of migrant workers did so only after a very lengthy and stereotypical description of the negative impacts of migrants on the community. She complained that migrants would endanger the community, increase the crime rate, overburden an already overcrowded public school system, spread infectious diseases, take jobs that should be given to local labour (the Lexington *Courier Journal* of 8 June 1987 said that the Clark County unemployment rate was nearly 9%) and cause an influx of illegal aliens. Toward the end of her letter she admonished her readers: 'Get on the phone, get out in your community and talk it over, mobilize now or learn to live with it.'

After this call for action, many Clark County residents wrote letters to the editor with similar complaints, such as the letter on 7 May that stated:

> What will ordinary citizens of Clark County gain by having the migrant workers here? Nothing! What do we have to lose? Plenty! One of the things we have to lose is our tax money, because we will most likely be expected to provide them food stamps, schooling, police protection and medical care. Other things we have to lose is a loss in our property values, a greater risk of being burglarized, raped, mugged or run over by a vegetable truck hauling the produce to market.

An editorial in the 6 May edition of *The Winchester Sun* further emphasized the prejudice in the rapidly mounting hysteria when it stated:

> Unfortunately, some Clark Countians, with stereotyped images of migrants dancing in their heads, have reacted with alarm to [the GVC announcement that migrants would be used] ... The net result of the entire situation is that long before the facts were in, objectivity was tossed out the window and the migrants found guilty of unimaginable horrors ranging from drunkenness, thievery and rape to spreading AIDS among our citizens.

Whether or not the concern of local citizens in regard to Mexican migrant workers was based on real or perceived fears, the results were the same. Yet the emotional and hostile response toward Mexican workers was not caused entirely by prejudice, but also by resentment toward GVC and the changes it would make in the community. The Mexican labourers had become a visible symbol and continuing reminder of the power of the outside agricultural firm.

Other changes in the community

GVC not only introduced Mexican migrant labour, but also proposed other actions that would directly affect the community. Of these actions, citizens responded emotionally to the company's announcement that it would utilize aerial spraying in its farming operations. Letters from local residents about aerial spraying frequently mentioned the close proximity of Becknerville School to the GVC spraying operations. Many citizens stated their concern about the health risks that the children might face from pesticide exposure. Also mentioned was the potential introduction of pesticides on to neighbouring properties and the possible contamination of stock ponds in the area.

The Clark County Fiscal Court was unable to ban aerial spraying. Restrictions were ordered by the court; however, these put only minor limits on an action that Clark County residents, expressing their views in letters to *The Winchester Sun* (April to August 1987) adamantly opposed (Ballard, 1987b). Local farmers had used aerial spraying on crops in the past, prior to the entry of GVC, and Clark residents had never publicly objected (J.B. Allen Jr, Kentucky, 1997, personal communication). However, the increased acreage (relative to that of small tobacco farmers) planted by GVC would require a wider expanse of spraying activity, increasing the danger of pesticide application to humans and to the environment. Additionally, pesticide use by tobacco farmers who owned property in the county and had an interest in protecting its resources may have been perceived as less dangerous than the use of aerial spraying by the outside agricultural firm that merely rented land and had no real interest in the well-being of the county. While the newspaper coverage and the letters to the editor were far fewer on the issue of aerial spraying than on migrant usage and the labour camp, the opposition to aerial spraying added more bitterness to an already acrid controversy.

Other issues regarding feared GVC-produced change surfaced in letters to the editors. Citizens complained of potential problems of water usage, possible traffic-related difficulties, and control of Clark

County land by an outside company. Finally, the majority of citizens who wrote letters were rural residents, and many of these were small tobacco farmers. It is probable that some of these local farmers resented what they perceived as attempts to change traditional tobacco farming to what was suggested as the more viable option of vegetable production. In a county with a distinct geographical division between smaller, owner-operated farming in the more mountainous east and the larger, 'agribusiness' farms in the west where GVC-rented lands were located, elements of old animosity between the two regions and two ways of life may have been present. All of these potential difficulties, real or perceived, produced fearful and often angry citizens who turned to local officials for help.

The 'Failures' of Local Government

Although many residents sought assistance from local government officials, help was not always forthcoming, as local politicians at times did not respond in the preferred manner. The first example of this lack of 'appropriate' response was on 23 April when the Fiscal Court announced that it had no legal means to halt GVC's use of migrant labour.

A second example was the failure of the Winchester Board of Commissioners to nullify a grant agreement with Kentucky Agricultural Marketing Cooperative (KAMC). The cooperative had been initiated to help local farmers to market vegetables as a supplement to tobacco farming and had been supported by a grant from the city. In turn, KAMC had brought in GVC to help the struggling cooperative (Nesbitt, 1987; Raupp, 1987d). It is probable that cooperative officials hoped that if GVC produced vegetables for KAMC to market, the cooperative would gain strength and local farmers, who had not previously supported the KAMC, would be won over. However, anger toward GVC's introduction of migrants now turned toward the cooperative, especially A. Graves, a local citizen who was an employee of the KAMC and one of the farmers leasing land to GVC. A local attorney, and POP member, brought an action to nullify the loan, claiming that the dual role of Graves constituted a conflict of interest. In addition, he stated that 'at least one coop official had "duped" the city into approving that loan [to the cooperative] by purposely withholding information that migrant labourers would be used in GVC's local harvesting efforts' (as quoted in Raupp, 1987j). An investigation requested by the Winchester Board of Commissioners and conducted by the Kentucky Commerce Cabinet, however, found no evidence to support POP's claims of misconduct (Raupp, 1987k,t). Again, Clark County officials were unable to act in the manner desired by the opposition group.

Finally, the local government failed POP members and other concerned Clark County residents in their demand to prevent the housing of migrant labour in the county. With the Fiscal Court's inability to stop GVC from employing migrant labour in the county, the issue quickly turned to the company's proposal for a labour camp. A zoning ordinance, already in effect in Clark County, prevented the placement of 20–25 trailers in Clintonville as proposed; however, GVC made it clear that it would seek a zoning change (Raupp, 1987l,m,n). Aware of the company's intention, Clark residents maintained their opposition efforts.

At a meeting on 26 May, the Planning and Zoning Commission awarded four zoning permits to GVC 'allowing the company to distribute at least a dozen mobile home trailers throughout Clark County' (Raupp, 1987l). Some citizens considered this a threat, especially when the GVC farm manager stated that the company had not given up on a migrant labour camp and the permits would be returned if the camp became a reality (Raupp, 1987m). Opposition continued to mount.

On 2 June, about 200 people filled the courtroom for the meeting to consider GVC's zoning change request. Residents, assisted by local attorneys, successfully argued that water pressure, traffic and humanitarian concerns justified the rejection of the company's plan. They further argued that 'a favorable decision for GVC would set a precedent that would have adverse repercussions locally and throughout the state' (Raupp, 1987n). Despite lengthy explanations by GVC officials, the 6.5-hour meeting resulted in a rejection of the zoning change request (Raupp, 1987n).

Any triumph felt by POP members was short-lived. The company did not surrender plans for a labour camp, and the zoning permits, already awarded, would allow GVC to house Mexican migrant labour in the county. The opposition appealed to the Winchester-Clark County Board of Adjustments and Appeals to have the permits revoked (Raupp, 1987o). At the hearing, the opposition attorney argued that tenant houses, as the trailers were referred to for GVC purposes, traditionally housed only a single tenant farmer or his family and that 'tenant farmers were considered workers who pay rent through a portion of the crop raised' (Raupp, 1987p). GVC's intent varied from the traditionally accepted meanings of both tenant houses and tenant farmers. However, the GVC attorney cited 'court decisions in Michigan and Indiana in which judges ruled that migrants were tenants of the farms on which they worked' (Raupp, 1987p). The hearing resulted in refusal to revoke the permits (Raupp, 1987p).

On 15 June, GVC announced that an agreement was being firmed up with a Lexington motel in adjacent Fayette County to house the first group of migrant workers temporarily. 'That agreement was necessary ... because of delays in securing housing for workers in Clark County' (Raupp, 1987q). Nevertheless, a company official emphasized that this action was temporary and that efforts at securing permanent housing would continue (Raupp, 1987q).

The Mexican Migrant Workers Arrive

The first group of migrants arrived to harvest vegetables on GVC-leased farms in Clark County on 20 June. These 20 migrants were housed in a Lexington motel, and joined approximately 32 local labourers in the harvesting effort. According to GVC representatives, this workforce would be sufficient for a few weeks; however, as the vegetable harvest reached its peak, a second group of migrants, predominantly Mexican or Mexican-American, would be brought in. At this time, housing plans for this second group of migrants were uncertain (Raupp, 1987r).

On 30 July, the GVC farm manager announced that the company had managed to obtain housing for the 200 to 210 migrant workers who had recently arrived in the county; therefore, GVC would not be utilizing the trailer permits at this time. He stated (Raupp, 1987s):

> The majority of GVC's migrants ... are staying at two Lexington motels. ... In addition, about 35 migrants have been living in a Richmond [Madison County] boarding house ... About one third of the workers – approximately 65, ... are being housed in four refurbished tenant houses on the 400 acre, GVC-leased [Becknerville] ... farm.

Although the migrant labour camp did not become a reality, GVC did use migrant labour and a portion of that labour was housed in Clark County.

The Aftermath

There is no doubt that the GVC episode was a bitter one in Clark County history. Some of the emotions remain alive today. Three former POP members refused to be interviewed for this research. Although willing to speak off the record for 30 minutes, one POP member stated that they had received threats at the time of the controversy and was literally afraid of the results of stirring it up again. In a February 1997 interview, Judge Allen stated that there are people who still refuse to speak to him today because of the GVC dispute. Yet, by 8 August of the initial year of the controversy, vocal opposition to GVC had calmed (Marx, 1987). Some citizens considered further efforts futile; others had experienced a gradual transformation of their original fear to an acceptance of GVC and the migrant workers (Guerrant and Guerrant, 1990; B. Blakeman, Kentucky, 1997, personal communication). Whatever the motivation, emotions had quietened, and at its August meeting the Clark County Fiscal Court – for the first time since GVC announced its intent to use migrant workers – did not have a single GVC issue to consider (Marx, 1987).

The dissatisfaction with local officials was demonstrated in the circulation of a petition for a referendum to change Clark County's form of government. The petition called for a vote to decide between the present arrangement, consisting of seven magistrates and a judge executive, and a new commission form, consisting of three commissioners and a judge executive. POP led the movement and quickly gained over twice the number of signatures necessary for the referendum (Anonymous, 1987). Advocates of this change claimed that the present form was not working and that it was 'not responsive to residents ... Because the magistrates are elected from small districts they look out for the interests of their districts rather than for the whole county' (Cohn, 1987). Despite the apparent popularity of the referendum movement in July, it was defeated when it finally came to a vote in the 3 November general election (Daykin, 1987). The measure was defeated in a vote of 2002 for to 2968 against; however, one wonders what the results might have been had the issue come to a vote in May or even June of 1987 (Daykin, 1987).

GVC ceased its Clark County operations after only two seasons, largely as a result of drought conditions in both 1987 and 1988 (Stroud, 1988). On 3 May 1989, the KAMC, facing a debt of more than US$1.2 million and having no resources to cover the deficit, also folded (Stroud, 1989).

Although GVC left Clark County in 1988, Hispanic migrant workers remained. Their initial employment in Kentucky agriculture to harvest vegetables exposed farmers to Hispanic migrant workers as a resource. Subsequently, as Kentucky farms suffered labour shortages, especially in the state's major cash crop – tobacco – they more and more frequently turned to this new labour option. This employment arrangement proved so successful that by 1996 farmers referred to this Hispanic workforce as the 'backbone of the tobacco harvest' (Rios, 1996a).

Even though migrant labour has become highly valued by central Kentucky farmers, Hispanic migrant farm workers are not always held in high esteem by the population as a whole. Some of the challenges faced by these workers, such as difficulty in finding housing, are the direct result of local prejudice and animosity that did not depart with GVC but instead continue to fester today. When asked about his search for an apartment for himself and his family, Juan (Kentucky, 1996, personal communication), a Mexican migrant worker from Puebla, Mexico, explained, 'It is difficult to find an apartment to rent here. People in Winchester [Clark County] do not like Mexicans.' He said that when he had gone to look at apartments, landlords had slammed the door in his face when they discovered that he was Hispanic. Some migrants who have been unable to find adequate housing have been forced to live 'under bridges, in woods, behind buildings and even in dumpsters' (Rios, 1996b). In numerous ways, local prejudice has affected Hispanics living in central Kentucky – both migrant labourers working in the fields and Hispanic immigrants forging a permanent life in the state. And that local prejudice was partially formed or intensified by the community conflict that occurred when GVC introduced Mexican migrant labour into Clark County in 1987.

Conclusion

With the announcement of the intended use of Mexican migrant labour, a predominantly white local population confronted for the first time an influx of Hispanics. Many residents responded to the announcement emotionally and in terms of prejudice and anti-immigrant attitudes. This cannot alone be considered the cause of the controversy, because the immigrants also became a symbol of a bitter struggle against GVC. Many citizens resented the control of local resources, such as water and land rights, by an outside agricultural firm. They were enraged at the potential damage to both the environment and human life that actions such as aerial spraying would produce. There was also an undercurrent of struggle between large and small landowners and perceived attacks on traditional farming practices.

This case study of the GVC introduction of Hispanic migrant labourers into Clark County, Kentucky, demonstrates the multi-faceted nature of community response to a sudden change in farm labour. As Kentucky's Hispanic migrant population continues to grow rapidly, the significance of the case lies in the lessons that can be learned from it. There are lessons for, firstly, white and predominantly white communities in Kentucky and elsewhere that are encountering for the first time an influx of Hispanics and, secondly, those firms contemplating the use of migrant labour in areas where they have not been previously introduced.

For further information on migrant/immigrant experiences in central Kentucky, see Denton (1999). For references on the 'new nativism', see Perea (1997) and Chavez (1998, pp. 192–193). For an excellent history on nativism, see Higham (1988); see also Gutiérrez (1995) and Foley (1997, pp. 40–63). For abuse of the migrant worker by the agricultural system, see Linder (1992). For coverage of migrant workers as competition for domestic labour, see Sosnick (1978, pp. 399–407) and Goldfarb (1981). For community resistance against company-induced change, see Fisher (1993).

References

Anonymous (1987) Clark group seeks referendum to change form of government. *Herald-Leader* (Lexington, Kentucky), 23 July, p. B3.

Ballard, E. (1987a) Fiscal court can't stop migrant use. *The Winchester Sun* 23 April, pp. 1, 14.

Ballard, E. (1987b) POP grills magistrates on aerial spraying. *The Winchester Sun* 14 May, p. 1.

Bureau of the Census (1983) *The County and City Data Book.* US Department of Commerce, Washington, DC.

Chavez, L.R. (1998) *Shadowed Lives: Undocumented Immigrants in American Society,* 2nd edn. Harcourt Brace College Publishers, Orlando, Florida.

Cohn, R. (1987) Clark countians to vote on changing government. *Herald-Leader* (Lexington, Kentucky), 25 October, p. B1.

Daykin, T. (1987) Counties, cities settle issues in referendums. *Herald-Leader* (Lexington, Kentucky), 4 November, p. A8.

Denton, B. (1999) A history of Mexican migrant farmworkers in Central Kentucky, the introduction and early years, 1987–1999. MA thesis, Eastern Kentucky University.

Duke, J. (1987) Clark countians oppose migrant plan. *Herald-Leader* (Lexington, Kentucky), 4 May, pp. B1, B8.

Fisher, S. (1993) *Fighting Back in Appalachia: Traditions of Resistance and Change.* Temple University Press, Philadelphia, Pennsylvania.

Foley, N. (1997) *The White Scourge: Mexicans, Blacks and Poor Whites in Texas Cotton Culture.* University of California Press, Los Angeles, California.

Goldfarb, R.L. (1981) *Migrant Farm Workers: a Caste of Despair.* Iowa State University Press, Ames, Iowa.

Guerrant, W. and Guerrant, L. (1990) Hispanic Ministry Journal. Written record and scrapbook of Hispanic ministry conducted in Winchester, Kentucky. (Journal in author's files with permission of use by the Guerrants.)

Gutiérrez, D.G. (1995) *Walls and Mirrors: Mexican Americans, Mexican Immigrants, and the Politics of Ethnicity.* University of California Press, Los Angeles, California.

Higham, J. (1988) *Strangers in the Land: Patterns of American Nativism, 1860–1925,* 2nd edn. Rutgers University Press, New Brunswick, New Jersey.

Linder, M. (1992) *Migrant Workers and Minimum Wages: Regulating the Exploitation of Agricultural Labor in the United States.* Westview Press, Inc., Boulder, Colorado.

Marx, J. (1987) Controversy over migrants quiets. *Herald-Leader* (Lexington, Kentucky), 8 August, pp. B1, B12.

Nesbitt, R. (1987) Ga. company to raise vegetables in Kentucky. *Herald-Leader* (Lexington, Kentucky), 12 March, pp. A10, A12.

Perea, J. (ed.) (1997) *Immigrants Out! The New Nativism and the Anti-Immigrant Impulse in the United States.* New York University Press, New York.

Raupp, P. (1987a) Board discusses migrant housing request. *The Winchester Sun* 16 April, p. 1.

Raupp, P. (1987b) Firm indicates migrants to be used. *The Winchester Sun* 28 April, pp. 1, 14.

Raupp, P. (1987c) First public hearing on migrants Monday. *The Winchester Sun* 1 May, p. 1.

Raupp, P. (1987d) Officials welcomed news of vegetable processing plant. *The Winchester Sun* 4 May, pp. 1, 16.

Raupp, P. (1987e) Co-op, Georgia vegetable co. depending on each other. *The Winchester Sun* 5 May, pp. 1, 4.

Raupp, P. (1987f) Emotions show during meeting. *The Winchester Sun* 5 May, pp. 1, 3.

Raupp, P. (1987g) Migrant labor used in state previously. *The Winchester Sun* 7 May, pp. 1, 6.

Raupp, P. (1987h) Georgia vegetable has good reputation. *The Winchester Sun* 7 May, pp. 1, 7.

Raupp, P. (1987i) POP airs migrant objections. *The Winchester Sun* 12 May, pp. 1, 16.

Raupp, P. (1987j) POP: Loan should be nullified. *The Winchester Sun* 13 May, p. 1.

Raupp, P. (1987k) City to seek state investigation of KAMC. *The Winchester Sun* 27 May, pp. 1, 20.

Raupp, P. (1987l) GVC gets approval for trailers. *The Winchester Sun* 28 May, pp. 1, 14.

Raupp, P. (1987m) Labour camp application eyed tonight. *The Winchester Sun* 2 June, p. 1.

Raupp, P. (1987n) Planning commission rejects labour camp. *The Winchester Sun* 3 June, pp. 1, 13.

Raupp, P. (1987o) Appeals board to consider revocation of GVC permits. *The Winchester Sun* 15 June, pp. 1, 13.

Raupp, P. (1987p) Board refuses to revoke trailer permits. *The Winchester Sun* 16 June, pp. 1, 16.

Raupp, P. (1987q) Migrants may be housed in Lexington. *The Winchester Sun* 18 June, p. 1.

Raupp, P. (1987r) First group of migrants here in county. *The Winchester Sun* 30 June, p. 1.

Raupp, P. (1987s) Georgia vegetable co. scraps plans for housing migrants in trailers. *The Winchester Sun* 31 July, p. 1.

Raupp, P. (1987t) State probe of KAMC finds no fault. *The Winchester Sun* 18 August, pp. 1, 14.

Rios, B. (1996a) Tobacco growers embrace migrants as vital to the harvest. *Herald-Leader* (Lexington, Kentucky), 20 October, p. A16.

Rios, B. (1996b) Those without a place to stay feel unwelcome. *Herald-Leader* (Lexington, Kentucky), 20 October, pp. A1, A15.

Rosenberg, G. (1992) They are just like family: framing the introduction of Hispanic migrant farmworkers into the Kentucky tobacco harvest. MS thesis, University of Kentucky.

Rosenberg, G. and Coughenour, C.M. (1990) The farm labour situation in Kentucky: the opinions of farmers and community members. *Community Issues* 11, University of Kentucky.

Sosnick, S.H. (1978) *Hired Hands: Seasonal Farm Workers in the United States.* McNally and Loftin, West, Santa Barbara, California.

Stroud, J.S. (1988) Seed remains though firm uprooted. *Herald-Leader* (Lexington, Kentucky), 13 November, pp. B1, B3.

Stroud, J.S. (1989) Winchester agricultural co-op fails to survive. *Herald-Leader* (Lexington, Kentucky), 3 May, pp. B4, B7.

US Department of Agriculture (1994) Hispanic migrant farmworkers in Kentucky. *County Extension Agent Survey,* University of Kentucky.

10 Does Experience as a US Farm Worker Provide Returns in the Mexican Labour Market?

Steven S. Zahniser[1] and Michael J. Greenwood[2]
[1] *US Department of Agriculture, Economic Research Service, Washington, DC, USA;*
[2] *Department of Economics, University of Colorado at Boulder, Boulder, Colorado, USA*

Abstract

This chapter examines the returns in the Mexican labour market to work experience in US agriculture. Using data from the Mexican Migration Project, a standard human capital model is estimated in which the log of monthly income is regressed on a set of explanatory variables, including US migration experience, its square, and controls for business and land ownership. In addition, we measure the impact of an 'occupational match' between the individual's current job in Mexico and most recent job in the USA. Because migration experience is likely to be endogenous, a tobit model is first estimated to generate fitted values for this variable, which are then incorporated within the earnings regression. Also present in the earnings regression are two Mills ratios, which account for the self-selection of persons who were either not employed or in the USA when the data were collected. Separate regressions are estimated for the entire sample, for return migrants, for return migrants who worked only in agriculture while in the USA, and for return migrants with any work experience in US agriculture. For return migrants whose US employment consisted entirely of farm work, the estimated return to the first year of US experience is statistically insignificant and relatively small (3.7%, compared with 7.6% for all individuals). Former US farm workers who are employed in Mexican agriculture enjoy an increase in monthly earnings of some 8–12% by virtue of their occupational match, although this return may originate in US employment outside agriculture.

Introduction

Migration offers tangible economic benefits to its participants, most notably in the form of wages and salaries earned in the destination labour market. Both actual and prospective participants in Mexican migration to the USA are acutely aware of these benefits, as US wages typically exceed the prevailing wage in Mexico by a substantial margin. In June 1999, the manufacturing wage in Mexico averaged US$1.70 per hour, compared with US$13.80 per hour in the USA (INEGI, 1999). Mexicans commonly acknowledge this sharp differ-

 The Dynamics of Hired Farm Labour (eds J.L. Findeis, A.M. Vandeman, J.M. Larson and J.L. Runyan)

ential with the casual observation that 'an hour's work in the USA is worth a day's work in Mexico'.

Migration offers other economic benefits to its participants as well. It was found that US work experience is handsomely rewarded in the Mexican labour market, with an additional year of US experience yielding an 8.9% increase in monthly earnings in Mexico (Zahniser and Greenwood, 1998). Moreover, return migrants who experience an 'occupational match' between their last job in the USA and their current employment in Mexico enjoy an additional 17.7% boost in monthly income.

These results can be attributed to human capital accumulation that occurs during the migrant's US work experience. While working in the USA, migrants presumably learn new skills and may receive additional education and training. For example, an agricultural labourer in Mexico may become an agricultural labourer in the USA, where the individual obtains new and different agricultural skills. In time, this individual may leave agriculture to work in endeavours that have little direct connection to agriculture. In addition, migrants may acquire and improve upon English language skills. If migrants later return to Mexico, they bring with them remunerable knowledge and skills that were acquired in the USA, not in Mexico.

Government policies that channel foreign workers into specific economic sectors are likely to affect the economic rewards that ultimately accrue to these individuals. Policies that enable Mexican migrants to work in occupations with a high level of human capital accumulation would be expected to bolster the returns to migration, if the skills and knowledge obtained in the USA are in demand in Mexico. In contrast, policies that require migrants to work in occupations with a low degree of such accumulation would lessen these returns.

In recent years, US policy makers have contemplated a variety of measures that would allow the entry of additional foreign workers into particular economic sectors. A number of these proposals have targeted the high-tech sector, where human capital accumulation is thought by some to be fast-paced. In October 2000, these efforts culminated with the signing into law of the American Competitiveness in the Twenty-First Century Act, which increased the maximum number of H-1B visas from 115,000 to 195,000 per year for fiscal years 2001, 2002 and 2003 (US Department of Justice, Immigration and Naturalization Service, 2000). These visas allow qualified professionals to enter the USA for up to 6 years, given testimony from their intended employers that their services are needed.

With respect to agriculture, policy makers in the USA have advanced several proposals for a new guestworker programme. For example, Senator Phil Gramm (Republican, Texas) has proposed a programme that would enable undocumented Mexicans already in the USA to obtain permits that would allow them to work legally in the USA. In addition, Senator Bob Graham (Democrat, Florida) has proposed a programme called AgJobs that combines features of a guestworker programme and an immigration amnesty. This proposal would enable undocumented persons who can prove that they worked in US agriculture to receive temporary work permits. If they continued to work in agriculture over a period of several years, they would be entitled to apply for legal immigration status (*Migration Dialogue*, 2001).

In the context of these proposals, the extent to which the returns to US experience in the Mexican labour market vary by the sector of the US economy in which the migrant was employed becomes particularly important. This chapter examines whether these returns may also be obtained through US farm work. Accordingly, this chapter employs roughly the same empirical approach as our previous work on this topic but focuses its attention on the agricultural sector. The chapter is organized as follows: (i) data used in this chapter; (ii) econometric techniques employed; (iii) empirical results; and (iv) conclusions of the analysis.

Data

The data for this chapter are drawn from the Mexican Migration Project, an ongoing study headed by Jorge Durand of the University of Guadalajara and Douglas S. Massey of the University of Pennsylvania. (The data files and their locations are listed in the References section.) This project has its roots in a 1982 study of four communities in the Mexican states of Jalisco and Michoacán (Massey *et al.*, 1987), which have been the origin of much migration to the USA. With the support of grants from the National Institute of Child Health and Human Development and the William and Flora Hewlett Foundation, the project has evolved into an extensive study of the migration experiences of households throughout Mexico. Since 1987, new survey communities have been added to the project's database on an almost annual basis. Although these places are not selected at random, an effort is made to examine communities of different sizes and economic bases, ranging from small rural villages to parts of large metropolitan areas.

Typically, the project interviews 200 household heads, selected at random, in each community. Some household heads from the community were in the USA when interviews were conducted; for this reason, the project uses 'snowball' sampling to locate about 20 such individuals from each Mexican community in the USA and conduct interviews with them as well. Thus, the overall sample contains persons who have never migrated to the USA, migrants who have returned to Mexico, and migrants who had not returned to Mexico by the time that the project visited their respective communities of origin.

Table 10.1. Results of probit models of interview location and employment status.

Variable	Location of interview (Y1)	Employment status (Y2)
Intercept	1.0563*	1.4580***
Age (years)	0.0304	–
Age squared	−0.0002	–
Mexican experience	–	0.0180*
Mexican experience squared	–	−0.0006***
US experience	–	−0.0456**
US experience squared	–	0.0012*
Female	0.6608***	−1.2141***
Education (years)	−0.0377***	0.0144
College educated	0.7898***	0.3254
Married	0.6505***	0.2657
Has relatives in USA	−1.4398***	−0.0573
Legal US resident	−1.7006***	−0.1416
Agricultural land owned (ha)	−0.0016	−0.0009
Number of workers in household	0.0332	0.2245***
Number of non-workers in household	0.0805***	−0.0655***
Spouse works	−0.6410***	−0.2494*
Business owner	0.3285***	0.0589
n	5273	5273
Interviewed in USA (Y1 = 0)	389	389
Interviewed in Mexico (Y1 = 1)	4884	4884
Not employed (Y2 = 0)	523	523
Employed (Y2 = 1)	4750	4750
Log likelihood	−727.47	−934.50

Results for dummy variables denoting communities of origin are not reported. Results of chi-squared test: ***passes at 99% level; **passes at 95% level; *passes at 90% level.

For the purposes of the analysis, observations on household heads over 65 years of age are excluded. The resulting data set contains 5273 observations, 4884 drawn from 43 communities in Mexico and 389 drawn from the USA. Of the 4884 household heads interviewed in Mexico, 4367 were employed at the time of the interview. These 4367 persons are the focus of the research, but the information provided by the other respondents is also used in the analysis (see Zahniser and Greenwood, 1998, for additional information on the sample and its communities).

Econometric Approach

The econometric model is a standard human-capital function in which the log of monthly earnings in Mexico is regressed on a vector of explanatory variables. These variables include Mexican work experience and its square, US experience and its square, and years of schooling. Mexican experience is defined as the individual's age (in years) minus the following terms: 6 years (to account for early childhood), years of schooling, and US experience (when relevant). Experience in the USA is the number of years spent in the USA, which is calculated by dividing the total months of US experience by 12. Unfortunately, the survey instrument does not indicate the number of months that the individual actually worked in the USA: it is assumed that all months spent in the USA entailed working. In addition, the measure of US experience does not distinguish among different occupations in the USA.

The contours of the data set and the nature of the economic question at hand necessitate the incorporation of several special features into the econometric approach. Firstly, Mexican income is not observed for household heads who were in the USA at the time of the interview or for persons who were in Mexico at the time of the interview but were not working. To correct for possible selection bias, two probit models are estimated, one for the probability of being interviewed in Mexico and one for the probability of working (Table 10.1). Inverse Mills ratios are derived from these probit models and then included in the earnings models.

Secondly, the amount of US experience may influence the expected wage differential between Mexico and the USA and thus affect the migrant's decision about whether to return to Mexico. Consequently, the expected Mexican wage may influence how long the migrant works in the USA. For this reason, US experience and the log of monthly earnings are treated as endogenous variables. Specifically, a tobit model is used to estimate fitted values for US experience (Table 10.2) and then substitute these values in the earnings regression. A series of Hausman-Wu tests supports this decision.

Thirdly, the population of the survey communities varies greatly, from less than 1000 to nearly 2.9 million (based on 1990 census data). Observations from larger communities may have greater variance and consequently cause heteroskedastic errors. A chi-squared test [$\chi^2(42) = 859.12$, $P < 0.005$, in the most basic earnings model] easily rejects the null hypothesis of a homoskedastic error structure. Thus, to correct for heteroskedasticity, the earnings models are estimated using maximum likelihood rather than ordinary least squares.

In total, four specifications of the earnings model are estimated. Model 1 utilizes all 4367 observations in the data set. Model 2 includes only return migrants (1599 observations); it is not possible to estimate the model for owners of farmland only, as the number of observations in that subgroup is too small for the model to converge. Model 3 focuses on return migrants whose US work experience included farm work (920 observations), a subset that includes persons whose US employment consisted entirely of farm work as well as individuals who performed both agricultural and non-agricultural work in the USA. Finally, Model 4 examines those return migrants who worked only in agriculture while in the USA (583 observations). Tables 10.3 and 10.4 contain descriptive statistics for the observations corresponding to these models.

Table 10.2. Results from first-stage tobit model used to generate fitted values for US migration experience.

Vector/variable	Parameter estimate
Intercept	−2.1430
Individual characteristics	
Mexican experience (years)	−0.1633***
Mexican experience squared	−0.0008
Female	−6.0484***
Education (years)	−0.4951***
College educated	0.5413
Married	1.5676**
Migration networks	
US migration experience of father precedes that of individual	1.8814***
US migration experience of sibling precedes that of individual	−1.0431***
Household characteristics	
Agricultural land owned (ha)	0.0047
Number of workers in household	0.9829***
Number of non-workers in household	−0.0494
Spouse works	−3.0323***
Business ownership	
Street vendor of food	2.1945***
Street vendor of goods	1.5294*
Grocery store	3.8777***
Cattle trader	0.8394
Agricultural goods trader	−0.6807
Manufactured goods trader	2.6642***
Tortilla mill	0.5505
Butcher shop	2.5389*
Restaurant	0.5692
Workshop	2.1226***
Repair shop or garage	−2.6355
Small assembly shop	4.1532**
Other type of business	2.1819
Occupation	
Irregularly employed	4.7324***
Technician or non-manual labourer	−3.4277**
Office worker	−0.7933
Salesperson	0.6303
Industrial owner or supervisor	1.1224
Skilled manual worker	0.2427
Unskilled manual worker	0.6816
Agricultural worker	1.7832***
Service worker	0.4894
Mills ratios	
Interview location	10.6165***
Employment status	10.3724***
Scale factor	6.0631
n	4367
Has US migration experience	1600
Lacks US migration experience	2767
Log likelihood	−6218.76

Results for dummy variables denoting community of origin are not reported. Results of chi-squared test: ***passes at 99%; **passes at 95%; *passes at 90% level.

Table 10.3. Means and standard deviations for observations used in earnings regressions (Models 1 and 2).

	All individuals (Model 1) (n=4367)		Return migrants (Model 2) (n=1599)	
Vector/variable	Mean	Std dev.	Mean	Std dev.
Monthly income (in 1994 new pesos)	1384.90	2672.23	1653.42	4132.67
Individual characteristics				
Age (years)	41.5661	11.6666	42.2233	11.5822
Mexican experience (years)	28.2442	14.1155	27.0982	14.0378
US experience (actual, in years)	1.5119	4.0673	4.1269	5.8648
US experience (fitted, in years)	0.8179	2.2426	1.9725	3.2879
Number of US trips	1.5748	3.8533	4.2470	5.3863
Subsequent internal migration	0.1120	0.3154	0.2989	0.4579
Recent migrant to USA (< 3 years past)	0.1566	0.3635	0.4278	0.4949
Female	0.0753	0.2640	0.0313	0.1741
Education (years)	5.8099	4.7279	4.9981	4.0949
College educated	0.1049	0.3064	0.0607	0.2388
Married	0.9020	0.2974	0.9450	0.2281
Legal US resident	0.0985	0.2980	0.2620	0.4399
Migration networks				
Has relatives in USA	0.6648	0.4721	0.8361	0.3703
US migration experience of father precedes that of individual	0.1800	0.3842	0.2827	0.4504
US migration experience of sibling precedes that of individual	0.3151	0.4646	0.3327	0.4713
Household characteristics				
Agricultural land owned (ha)	1.6375	18.1035	1.9443	16.2162
Number of workers in household	1.8425	1.2406	1.8774	1.2607
Number of non-workers in household	3.6011	2.1080	3.7448	2.1102
Spouse works	0.1523	0.3593	0.1488	0.3560
Business ownership				
Street vendor of food	0.0341	0.1816	0.0381	0.1916
Street vendor of goods	0.0206	0.1421	0.0156	0.1241
Grocery store	0.0414	0.1993	0.0532	0.2244
Cattle trader	0.0055	0.0739	0.0063	0.0789
Agricultural goods trader	0.0071	0.0840	0.0063	0.0789
Manufactured goods trader	0.0137	0.1164	0.0138	0.1165
Tortilla mill	0.0055	0.0739	0.0038	0.0612
Butcher shop	0.0053	0.0724	0.0075	0.0863
Restaurant	0.0080	0.0892	0.0069	0.0827
Workshop	0.0371	0.1890	0.0363	0.1870
Repair shop or garage	0.0085	0.0917	0.0038	0.0612
Small assembly shop	0.0041	0.0641	0.0050	0.0706
Other type of business	0.0664	0.2490	0.0607	0.2388
Occupational match	0.1337	0.3404	0.3596	0.4800
Agricultural match	0.0815	0.2737	0.2176	0.4128
Occupation				
Irregularly employed	0.0433	0.2035	0.0325	0.1774
Professional or administrative	0.0939	0.2917	0.0544	0.2269
Technician or non-manual labourer	0.0121	0.1095	0.0038	0.0612
Office worker	0.0302	0.1712	0.0188	0.1357
Salesperson	0.1216	0.3269	0.1182	0.3229
Industrial owner or supervisor	0.0048	0.0692	0.0031	0.0558
Skilled manual worker	0.2114	0.4083	0.1995	0.3997

Table 10.3. (*Continued.*)

	All individuals (Model 1)		Return migrants (Model 2)	
Vector/variable	Mean	Std dev.	Mean	Std dev.
Unskilled manual worker	0.1161	0.3204	0.1232	0.3288
Agricultural worker	0.2386	0.4263	0.3377	0.4731
Service worker	0.1280	0.3341	0.1088	0.3115
Mills ratios				
Interview location	0.0870	0.1825	0.1493	0.2605
Employment status	0.1122	0.2329	0.1090	0.2250

Empirical Results

As previous research indicated (Zahniser and Greenwood, 1998), US work experience has a significant payoff in the Mexican labour market (Tables 10.5 and 10.6). In the basic model (Model 1), an additional year of US experience yields a 7.6% increase in monthly earnings. Estimated at the mean income of those who never migrated, this amounts to 93 new pesos per month at 1994 prices, or about US$28. The figure in dollars is calculated using the period-average exchange rate for 1994 found in International Monetary Fund (1999). This return is greater than the return to an additional year of Mexican experience (1.1%) and the return to an additional year of schooling (3.7%).

For return migrants who performed any farm work in the USA, the return to US experience is substantially higher (10.7%, Model 3). However, this increase may stem from work experience outside agriculture. In the regression for return migrants whose US employment consisted entirely of farm work (Model 4), the coefficient for US experience is positive but insignificant, while the coefficient for the squared term is negative and significant. Moreover, the coefficient for US experience (3.9%) is less than the corresponding coefficients in the other models. These results suggest that the returns to US experience are substantially smaller (and may even be negative) for return migrants whose US employment was limited to agriculture, especially for those individuals with extensive experience as a US farm worker.

Two occupational-match variables are used to identify return migrants whose occupation in Mexico is the same as their last job in the USA. One variable indicates the presence of such a match in any occupation, while a second identifies persons with an occupational match in agriculture. These individuals enjoy the distinct advantage of accumulating skills and knowledge in the same occupation over an extended period of time, uninterrupted by their spells as migrants.

The individual with an occupational match enjoys a 26.5% increase in monthly earnings, using the estimates of Model 1. However, the additional attribute of an agricultural job match reduces monthly earnings by 15.2%. Although the net effect of an agricultural job match is still positive (11.2%), it is substantially less than that resulting from a non-agricultural job match. In addition, agricultural workers in Mexico constitute the lowest paid occupational group in Model 1. According to the coefficient for this group, the monthly income of agricultural workers in Mexico is 52.6% less than the income of professionals and administrators, a finding that is strongly significant.

For return migrants whose US employment consisted entirely of farm work, the parameter estimates for the occupational match variables are statistically insignificant (Model 4). Thus, for these individuals, the earnings model generates no evidence of a positive return to US experience, through either the fitted value for US experience (and its square) or the occupational match variables.

Table 10.4. Means and standard deviations for observations in earnings regressions (Models 3 and 4).

	US experience includes farm work (Model 3) (n=920)		Only US experience is farm work (Model 4) (n=583)	
Vector/variable	Mean	Std dev.	Mean	Std dev.
Monthly income (in 1994 new pesos)	1716.25	5130.11	1400.57	2491.05
Individual characteristics				
Age (years)	44.5913	11.9848	44.8799	11.9937
Mexican experience (years)	29.9360	14.5143	31.1908	14.5661
US experience (actual, in years)	4.6749	6.3889	3.8709	5.9124
US experience (fitted, in years)	2.3855	3.7040	2.1118	3.7091
Number of US trips	5.4793	6.2623	4.6518	5.9384
Subsequent internal migration	0.3261	0.4690	0.3019	0.4595
Recent migrant to USA (< 3 years past)	0.4261	0.4948	0.3585	0.4800
Female	0.0109	0.1037	0.0103	0.1010
Education (years)	3.9804	3.5226	3.8182	3.6224
College educated	0.0293	0.1689	0.0309	0.1731
Married	0.9652	0.1833	0.9674	0.1777
Legal US resident	0.2707	0.4445	0.2144	0.4108
Migration networks				
Has relatives in USA	0.8163	0.3874	0.7770	0.4166
US migration experience of father precedes that of individual	0.2946	0.4561	0.2607	0.4394
US migration experience of sibling precedes that of individual	0.3033	0.4599	0.2590	0.4385
Household characteristics				
Agricultural land owned (ha)	2.1739	11.6502	1.9005	10.1511
Number of workers in household	1.9989	1.3581	2.1218	1.4480
Number of non-workers in household	3.7543	2.1496	3.8268	2.2212
Spouse works	0.1413	0.3485	0.1355	0.3426
Business ownership				
Street vendor of food	0.0337	0.1805	0.0292	0.1684
Street vendor of goods	0.0120	0.1087	0.0137	0.1164
Grocery store	0.0587	0.2352	0.0583	0.2345
Cattle trader	0.0087	0.0929	0.0069	0.0826
Agricultural goods trader	0.0054	0.0736	0.0034	0.0585
Manufactured goods trader	0.0130	0.1135	0.0120	0.1090
Tortilla mill	0.0033	0.0570	0.0034	0.0585
Butcher shop	0.0054	0.0736	0.0034	0.0585
Restaurant	0.0033	0.0570	0.0034	0.0585
Workshop	0.0337	0.1805	0.0343	0.1822
Repair shop or garage	0.0022	0.0466	0.0034	0.0585
Small assembly shop	0.0054	0.0736	0.0051	0.0716
Other type of business	0.0413	0.1991	0.0429	0.2028
Occupational match	0.4446	0.4972	0.5026	0.5004
Agricultural match	0.3750	0.4844	0.4940	0.5004
Occupation				
Irregularly employed	0.0380	0.1914	0.0240	0.1532
Professional or administrative	0.0283	0.1658	0.0274	0.1635
Technician or non-manual labourer	0.0033	0.0570	0.0034	0.0585
Office worker	0.0130	0.1135	0.0172	0.1300
Salesperson	0.1033	0.3045	0.0961	0.2949
Industrial owner or supervisor	0.0022	0.0466	0.0017	0.0414

Table 10.4. (*Continued.*)

Vector/variable	US experience includes farm work (Model 3)		Only US experience is farm work (Model 4)	
	Mean	Std dev.	Mean	Std dev.
Skilled manual worker	0.1761	0.3811	0.1595	0.3665
Unskilled manual worker	0.1076	0.3101	0.0806	0.2725
Agricultural worker	0.4500	0.4978	0.5129	0.5003
Service worker	0.0783	0.2687	0.0772	0.2671
Mills ratios				
Interview location	0.1479	0.2585	0.1262	0.2259
Employment status	0.1084	0.2055	0.1146	0.2125

Conclusion

The rewards to US experience in the Mexican labour market that are observed for all return migrants do not appear to extend to persons who worked entirely in agriculture while in the USA. These persons enjoy neither a direct return to their US experience nor an additional benefit through an agricultural job match. In contrast, return migrants whose US work included both agricultural and non-agricultural employment obtain a positive return to their US experience in the Mexican labour market, but these returns may very well originate in work performed outside agriculture.

These results suggest that any guest-worker programme that would channel additional Mexican workers into US agriculture at the exclusion of other sectors of the US economy would limit the long-term benefits of migration obtained through human capital accumulation during the migrant work experience. Obviously, the intent of agricultural guestworker proposals is to legally secure the services of farm workers from outside the USA, not to foster human capital accumulation by migrants. Still, the question of whether human capital accumulation accompanies migrant employment is important for the USA, especially since some migrants remain in this country for an extended period of time.

The return migrants who would seem to be in the best position to apply skills and knowledge acquired during US agricultural employment are members of Mexican farm households. The results show that persons with an agricultural job match enjoy an increase in monthly earnings in the neighbourhood of 8–12%. This result disappears when return migrants whose US employment consisted entirely of farm work are considered in isolation, so again it is not clear whether this return originates in farm work, non-agricultural employment, or some combination of the two. Moreover, agricultural workers are among the lowest paid occupational group in Mexico, according to our data on monthly incomes. However, US earnings may enable some Mexican farm households to remain in agriculture. If so, then the impact of US experience for them may lie not in increased earnings, but in continued participation in Mexican agriculture.

Table 10.5. Parameter estimates from earnings regressions: Models 1 and 2.

Vector/variable	All individuals (Model 1)	Return migrants (Model 2)[a]
Intercept	7.0147***	7.3442***
Individual characteristics		
Mexican experience (years)	0.0106***	0.0024
Mexican experience squared	−0.0001***	−0.0001
US experience (fitted, in years)	0.0762***	0.0884***
US experience squared	0.0005	−0.0027
Number of US trips	0.0060	0.0100**
Subsequent internal migration	−0.0144	0.0773*
Recent migrant to USA (< 3 years past)	0.0485	0.0839**
Female	−0.1923***	−0.6052***
Education (years)	0.0370***	0.0163*
College educated	0.1633***	0.5238***
Married	0.1321**	0.0901
Household characteristics		
Agricultural land owned (ha)	0.0017***	0.0029**
Number of workers in household	−0.0005	−0.0025
Number of non-workers in household	0.0077	0.0040
Spouse works	0.0310	−0.0458
Business ownership		
Street vendor of food	−0.0846	−0.0780
Street vendor of goods	0.0381	−0.1049
Grocery store	−0.0655	0.1804**
Cattle trader	0.6042***	−0.0124
Agricultural goods trader	0.1859*	0.1231
Manufactured goods trader	0.1925**	0.2745*
Tortilla mill	0.3018**	0.5889*
Butcher shop	0.1505	−0.0061
Restaurant	0.1961	0.1290
Workshop	0.0610	0.2347**
Repair shop or garage	0.4117***	0.6073**
Small assembly shop	−0.3978***	−0.3261*
Other type of business	0.1276***	0.1014
Occupational match	0.2648***	0.2731***
Agricultural match	−0.1519**	−0.1557*
Occupation		
Irregularly employed	−0.4532***	−0.4010**
Technician or non-manual labourer	−0.2000*	−0.4684
Office worker	−0.2927***	−0.4433***
Salesperson	−0.2769***	−0.2726**
Industrial owner or supervisor	0.5150***	−0.0902
Skilled manual worker	−0.2859***	−0.3411***
Unskilled manual worker	−0.3262***	−0.1760
Agricultural worker	−0.5256***	−0.5073***
Service worker	−0.2977***	−0.2920**
Mills ratios		
Interview location	−0.1451	−0.0908
Employment status	−0.4218***	0.1122
n	4367	1599
Log likelihood	−4644.36	−1858.71

[a]To ensure convergence, one observation was dropped from Model 2. Results for dummy variables denoting community of origin are not reported.

***indicates $|t| >= 1.96$; **indicates $1.67 <= |t| < 1.96$; *indicates $1.29 <= |t| < 1.67$.

Table 10.6. Parameter estimates from earnings regressions: Models 3 and 4.

Vector/variable	US experience includes farm work (Model 3)	Only US experience is farm work (Model 4)
Intercept	6.9168***	6.7614***
Individual characteristics		
Mexican experience (years)	0.0167**	0.0099
Mexican experience squared	−0.0002*	−0.0001
US experience (fitted, in years)	0.1071***	0.0387
US experience squared	−0.0033*	−0.0077***
Number of US trips	0.0064	0.0028
Subsequent internal migration	0.1135**	0.1145**
Recent migrant to USA (< 3 years past)	0.1558***	−0.0212
Female	−0.6342**	−0.4740
Education (years)	0.0385***	0.0493***
College educated	0.4170**	−0.0654
Married	0.1860	0.1274
Household characteristics		
Agricultural land owned (ha)	0.0071**	−0.0005
Number of workers in household	−0.0112	−0.0132
Number of non-workers in household	−0.0003	0.0029
Spouse works	0.0596	−0.1065
Business ownership		
Street vendor of food	−0.1084	−0.0489
Street vendor of goods	−0.2638	0.0380
Grocery store	0.0259	0.4497***
Cattle trader	−0.5670	−0.6069
Agricultural goods trader	−0.4258*	−0.3390
Manufactured goods trader	0.4452**	0.4221**
Tortilla mill	0.0527	0.5002*
Butcher shop	−0.5830	0.0668
Restaurant	0.4012	0.0825
Workshop	0.3834***	0.1757
Repair shop or garage	1.1485**	1.0803**
Small assembly shop	−0.3270	−0.9392***
Other type of business	−0.0750	−0.1156
Occupational match	0.4051***	0.1097
Agricultural match	−0.3273**	−0.4025
Occupation		
Irregularly employed	−0.4836**	−0.4601*
Technician or non-manual labourer	−0.3126	−0.0361
Office worker	−0.4861**	−0.3872
Salesperson	−0.1419	−0.1484
Industrial owner or supervisor	1.9829***	1.0052*
Skilled manual worker	−0.1476	−0.1041
Unskilled manual worker	−0.1771	−0.2121
Agricultural worker	−0.3624*	−0.1765
Service worker	−0.1973	−0.2743
Mills ratios		
Interview location	−0.4053**	0.1858
Employment status	−0.0135	−0.1947
n	920	583
Log likelihood	−975.09	−503.33

Results for dummy variables denoting community of origin are not reported.
***indicates $|t| >= 1.96$; **indicates $1.67 <= |t| < 1.96$; *indicates $1.29 <= |t| < 1.67$.

References and Further Reading

INEGI (1999) Indicadores de Competitividad, Salarios en la Industria Manufacturera en Varios Países. Instituto Nacional de Estadística, Geografía, e Informática, Banco de Información Económica. http://dgcnesyp.inegi. gob.mx/cgi-win/bdi.exe (accessed 18 October).

International Monetary Fund (1999) *International Financial Statistics Yearbook 1999*. IMF, Washington, DC.

Massey, D.S., Alarcón, R., Durand, J. and González, H. (1987) *Return to Aztlán: the Social Process of International Migration from Western Mexico*. University of California Press, Berkeley, California.

Mexican Migration Project (1997) HOUSFILE (machine-readable downloaded file). Population Studies Center, University of Pennsylvania, Philadelphia, Pennsylvania, March.

Mexican Migration Project (1997) LIFEFILE (machine-readable downloaded file). Population Studies Center, University of Pennsylvania, Philadelphia, Pennsylvania, March.

Mexican Migration Project (1997) MIGFILE (machine-readable downloaded file). Population Studies Center, University of Pennsylvania, Philadelphia, Pennsylvania, March.

Mexican Migration Project (1997) PERSFILE (machine-readable downloaded file). Population Studies Center, University of Pennsylvania, Philadelphia, Pennsylvania, March.

Mexican Migration Project (1997) Table 1: Sample information. Population Studies Center, University of Pennsylvania, Philadelphia. http:// lexis.pop.upenn.edu / mexmig / sampletable. html

Migration Dialogue (2001) University of California at Davis (2001) Congress: guest workers. *Migration News* 8(5). http://migration.ucdavis. edu / mn / archive_mn / may_2001-04mn.html (May).

US Department of Justice, Immigration and Naturalization Service (2000) Questions and answers: changes to the H-1B program. http:// www.ins.usdoj.gov / graphics / publicaffairs/ questsans/h1bchang.htm (21 November).

Zahniser, S.S. and Greenwood, M.J. (1998) Transferability of skills and the reward to US migration experience in the Mexican labour market. In: Mexican Ministry of Foreign Affairs and US Commission on Immigration Reform, *Migration Between Mexico and the United States, Binational Study, Volume 3, Research Reports and Background Materials*. Morgan Printing, Austin, Texas, pp. 1133–1152.

11 Income Distribution and Farm Labour Markets

Robert D. Emerson[1] and Fritz Roka[2]

[1]*Food and Resource Economics Department, University of Florida, Gainesville, Florida, USA;* [2]*Southwest Florida Research and Education Center, University of Florida, Immokalee, Florida, USA*

Abstract

This chapter examines the extent to which increased income inequality in the USA may have altered the supply of labour to agriculture. An aggregate model of the hired farm labour market is estimated, incorporating minimum wage effects in addition to the changes in income distribution. The results suggest that the widening income distribution has not significantly affected the supply of labour to agriculture. The level of non-farm wages is found to have a strong negative effect on the supply of farm labour. Farm labour supply is highly elastic with respect to the farm wage rate, and farm labour demand is highly inelastic.

Introduction

The US labour force has experienced a dramatic increase in the dispersion of earnings since the early 1980s. While highly skilled labour has experienced increased compensation, unskilled labour has maintained a nearly constant rate of compensation in real terms. It is well established that there is significant interaction between farm and non-farm labour markets with mobility between the sectors strongly influenced by relative wage rates in the two sectors. A question addressed in this chapter is the extent to which the increased dispersion of compensation in the economy alters the interactions between farm and non-farm labour markets.

Recent wage and employment outcomes on farm and non-farm labour markets are examined and contrasted with earlier time periods. Various aggregate models of the farm labour market are updated with more recent data, and modified for the current policy structure and question of interest in this chapter. Particular attention is given to appropriate inclusion of the minimum wage within the model. Equations explaining wage and employment levels are estimated both as reduced forms and as structural supply and demand equations. The results show that equilibrium employment and wages continue to be strongly influenced by the non-farm wage rate. In addition, supply response is highly elastic, both with respect to own wage and (negatively) with respect to the non-farm

(eds J.L. Findeis, A.M. Vandeman, J.M. Larson and J.L. Runyan)

wage. Although the demand equation estimates are not very robust, one interpretation is that demand response is highly inelastic. It is concluded that the increased wage dispersion has not had a significant impact on the farm labour market.

Farm and Non-Farm Labour Markets

Labour market interaction

The literature on farm labour markets over the past half-century has emphasized the importance of the linkage between farm and non-farm labour markets in understanding the operation of farm labour markets. Johnson (1951) was an early contributor to the literature, emphasizing the role of labour markets in equilibrating farm with non-farm incomes. With relatively low returns to resources in agriculture at the time, much of the attention was directed to farm households rather than hired farm labour. The questions at the time were why relatively higher non-farm incomes were not drawing sufficient labour from agriculture to equalize returns, and what factors were preventing the labour mobility. The subsequent massive labour mobility through the 1960s was in two forms: permanent movement to the non-farm sector, and off-farm work by members of the farm family. Following the exodus of labour from agriculture, incomes of farm households eventually reached the level of non-farm household incomes in the 1980s and 1990s.

Early econometric models of the farm labour market treated the non-farm wage rate as a measure of the alternative wage for farm labour. Schuh (1962) focused on the hired farm labour market, estimating a supply and demand model for the market. Non-farm income, adjusted for the unemployment rate, was a key variable representing alternative opportunities for farm labour in the non-farm economy. The effect of non-farm income on farm labour supply was statistically significant and negative, having an elasticity larger in absolute value than the own supply elasticity (Schuh, 1962). Subsequent papers by Tyrchniewicz and Schuh (1966, 1969) extended the model to regions of the USA, and then to a joint model for farm operator, family labour and hired farm labour. Non-farm income opportunities remained a strong shifter, both statistically and numerically, in each of the labour supply equations.

A more recent aggregate supply and demand model is by Duffield and Coltrane (1992). Their hired farm labour model, patterned after the Schuh model, was designed to examine potential shortages of farm labour in the USA due to the 1986 Immigration Reform and Control Act (IRCA). The supply side of the model again incorporated non-farm earnings as the alternative wage rate, finding that it continued to be a strong negative shifter of the supply of labour to agriculture.

Barkley (1990) examined the labour mobility out of US agriculture in response to non-farm labour returns relative to returns in agriculture. His results again provided strong support for the response of farm operators to non-farm earnings.

Gardner (1972, 1981) took a different approach to modelling the labour market in examining the effect of minimum wages on the equilibrium quantity of labour and the equilibrium wage. Rather than attempting to estimate the structural supply and demand equations for agricultural labour, he specified the reduced form equations. Increases in the non-farm wage rate were found to increase the equilibrium real wage in agriculture, and to reduce the equilibrium quantity of hired farm labour.

Richards and Patterson (1998) take a novel approach to examining the farm and non-farm labour market interaction with a particular emphasis on the question of a shortage of farm labour. They attempt to estimate the real option value of remaining employed in non-farm employment rather than switching to agriculture when the relative wages alone suggest that switching would be optimal. (The option value is a measure of the value of waiting for additional information before switching employment sectors rather than switching immediately as the wage ratio moves in favour of one sector rather than the other.) Their empirical analysis was based on regional wages for agriculture and selected non-agricultural industries in Washington State. They concluded that there was a positive real option value to non-farm employment, and that its existence implied that a shortage of agricultural labour was more likely than a surplus for their data set, although Department of Labor local area statistics suggested a surplus of labour. In the context of this chapter, the implication of their results is that mobility between the sectoral labour markets is somewhat more retarded than much of the prior existing work would suggest.

Another set of literature examining the responsiveness of farm labour to non-farm opportunities has utilized cross-sectional data on individual workers. Sumner (1982), using a sample of Illinois farmers, found a positive response of labour to off-farm work with higher non-farm wages. Tokle and Huffman (1991), utilizing Current Population Survey (CPS) data, obtained similar findings. Perloff (1991) examined the probability of farm work by non-farm workers using CPS cross-sectional data, and found an increased probability of farm work the higher was the farm wage rate relative to the potential non-farm wage rate. Gisser and Dávila (1998), using Census Public Use Microdata Sample (PUMS) data, presented evidence that once adjustments for differences in worker endowments and price levels are made, the wage differential between farm and non-farm work largely disappears, particularly for younger workers, who are the dominant group in agriculture. Their result suggested equilibrium between the farm and non-farm labour markets.

Income inequality

Considerable attention has been given to the widening distribution of income for the US economy in the 1980s and 1990s. Nevertheless, Goldin and Margo (1992) presented evidence that the recent extent of labour market inequality was merely a return to the level of inequality in the early 1940s, and that the interim reduction in inequality was the result of a confluence of temporary short-term phenomena. There is a strong consensus in the literature that the present widening of the income distribution has resulted from an increasing premium for skilled over unskilled labour in the labour market (Berman *et al.*, 1994; Welch, 1999). Welch's calculations with the CPS data for white men employed full-time suggested an increase in the variance of the log of weekly wages by a factor of 2 from 1970 to 1993. Whether this was due to increasing technology or to a further opening of the economy and increased international trade remains an open question.

Farm Labour Market Characteristics

The number of hired farm workers has declined dramatically in the USA. Figure 11.1 illustrates this, using annual averages of hired farm workers from 1940–1998 as reported in the *Farm Labor* series published by the US Department of Agriculture. The decline is from just over 2.5 million in 1940 to just under 1 million throughout the 1990s. The reasons for the decline are familiar: increased mechanization and technology, and the shifting of some activities previously done on the farm to non-farm firms.

Figure 11.2 illustrates the hired farm wage rate in constant 1982–1984 dollars. There was a substantial increase from around US$3.00 per hour to US$4.74 per hour from 1948 to 1978. Although there is a major gap in the data series through most of the 1980s, the limited data available show no significant departure from the 1980–1989 annual average wage rates and the farm wage has yet to return to the 1978 level in real terms. The 1998 average real wage was US$4.58 per hour.

The minimum wage was extended to agriculture in 1967, although with limited coverage and initially with lower rates than for the covered non-farm sector. In real terms, the minimum wage reached a peak at the same time as the average farm wage rate peaked in 1978. The 1998 minimum in real terms is nearly the same as the initial minimum wage introduced in 1967. It is noteworthy that the largest part of the rise in the average farm

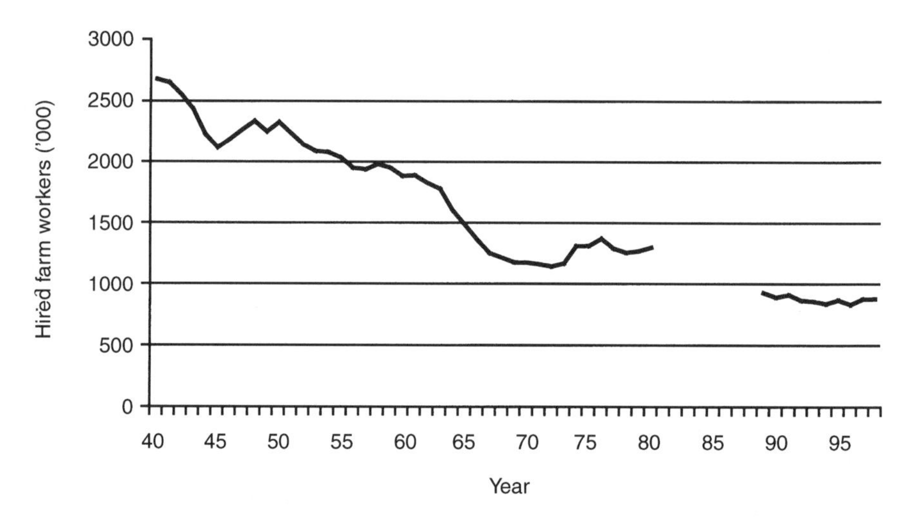

Fig. 11.1. US hired farm workers. Data used for the number of farm workers are workers directly hired by farms. Agricultural service workers were only separately tabulated starting in July 1978. In addition, the survey methodology for agricultural service workers changed in 1982. While it would be desirable to include the number of agricultural service workers, the least bad choice appears to be to exclude agricultural service workers due to major changes in agricultural service worker data series over time. (Source: *Farm Labor*, US Department of Agriculture.)

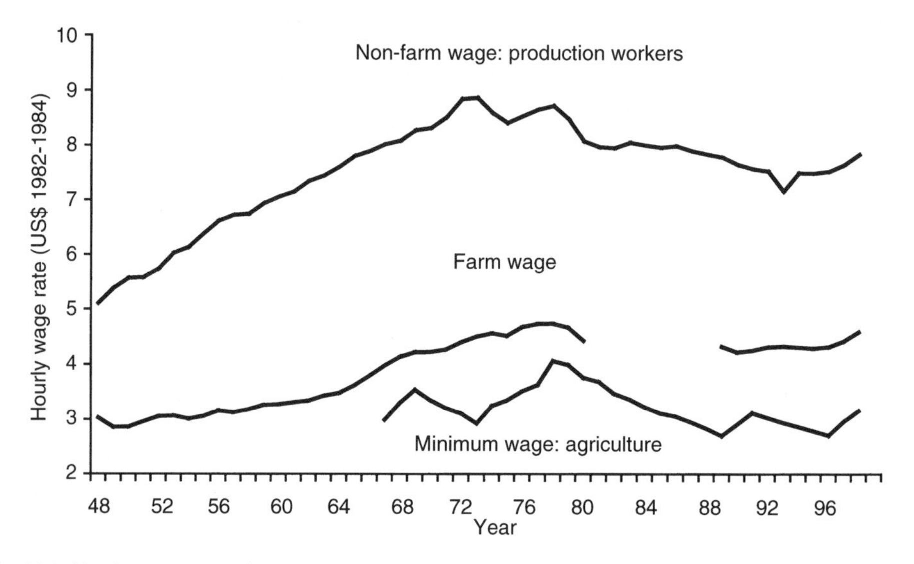

Fig. 11.2. Hourly wage rates (see text for the definition of production workers). (Source: *Farm Labor*, National Agricultural Statistics Service, US Department of Agriculture; Bureau of Labor Statistics, US Department of Labor.)

wage rate was coincident with the initial years of the minimum wage for agriculture.

Figure 11.2 also illustrates the pattern of non-farm wages for production workers in the US economy. The non-farm wage rate used is the average hourly earnings for all private production and non-supervisory workers. This includes production workers in mining and manufacturing (those up through working supervisors, excluding executive, managerial and clerical workers), construction workers in the construction industry (those up through working supervisors) and non-supervisory workers in the remaining private sector industries (US Bureau of Labor Statistics, 1997, pp. 16–17).

While the overall pattern of wage rates in real terms is similar for non-farm production workers and hired farm workers (Fig. 11.2), the levels are dramatically different. The ratio of farm to non-farm wage rates was 0.59 in 1948, the highest of any year throughout the 1948–1998 period. There was a gradual decline in the ratio to 0.45 in the early 1960s, and then a continuing increase to a level of 0.58 in 1998, virtually where it started 50 years earlier. Contrasting the patterns of relative wage rates with the reduction in the hired farm workforce, most of the decline in the hired farm workforce shown in Fig. 11.2 took place as the farm to non-farm wage ratio was declining throughout the 1950s and early 1960s.

One indicator for the economy-wide distribution of income is illustrated in Fig. 11.3. The graph illustrates the ratio of the average household income for the top quintile of the US population to the average household income of the bottom quintile of the population. The data are readily available for 1967–1998 and show the lowest ratio to be just under ten in 1974, then rising rather steadily to just under 14 in 1998.

As noted above, there is a strong presumption that the widening of the income distribution is a result of the increased demand for skilled relative to unskilled labour. The median education for farm workers in the USA is reported as the eighth grade level (US Department of Labor, 1993), clearly an educational level limiting participation in most skilled occupations in the US economy. The question of interest is to what extent the increased demand for highly skilled labour has resulted in an increased availability of low-skill workers to industries such as agriculture that employ high concentrations of low-skill workers.

Empirical Model

Schuh's (1962) original framework for the analysis of aggregate supply and demand for farm labour is a convenient starting place. The demand for labour (Q) is assumed to be a function of the real wage rate for farm work (W), prices paid by producers for purchased inputs (other than labour) (PP) and prices received for farm products (PR); Gardner (1981) adds the quantity of land (L) (the data source and variable definitions are in Table 11.1). The

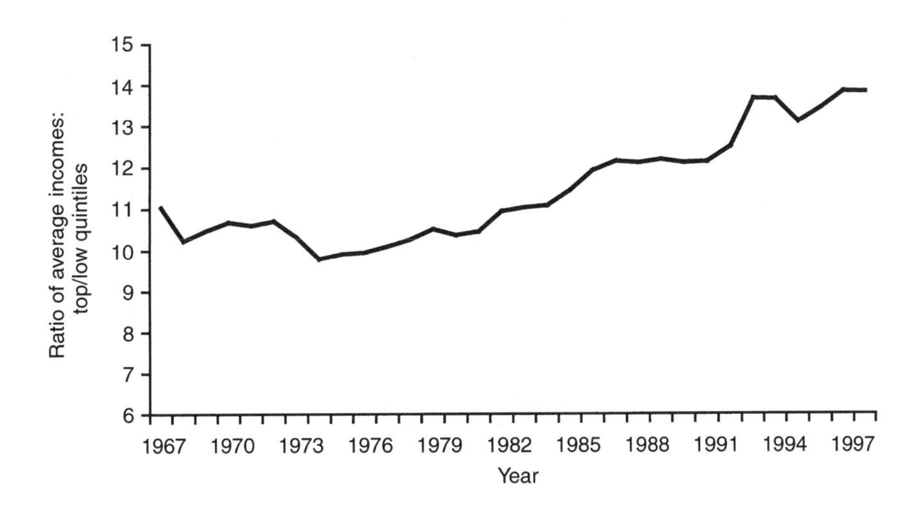

Fig. 11.3. US household incomes. (Source: US Census Bureau.)

Table 11.1. Variable definitions and data sources.

Variable	Definition	Source
Workers	Annual average number of farm workers hired directly by farms	US Department of Agriculture, *Farm Labor*
Wage	Average wage of all hired farm workers, deflated by consumer price index (CPI)	US Department of Agriculture, *Farm Labor*
NFW	Hourly earnings of all private production or non-supervisory workers, deflated by CPI, times one minus the unemployment rate	US Bureau of Labor Statistics
PP	Prices paid index for purchased inputs except labour	US Department of Agriculture, *Agricultural Statistics*
PR	Price received for farm products by farmers (index)	US Department of Agriculture, *Agricultural Statistics*
MinWg	Legal minimum wage for agriculture	US Department of Labor (2001)
L	Input index for farm real estate	US Department of Agriculture, *Agricultural Statistics*
LF	US civilian labour force 25 years and older with 8 or fewer years of education	US Bureau of Labor Statistics
DIST	Average household income of the top US quintile divided by average household income of the bottom US quintile	US Bureau of the Census
CPI	Consumer price index	US Bureau of Labor Statistics

supply of labour (Q) is typically specified to be a function of the real wage rate for farm work (W), an alternative non-farm real wage rate (NFW) and a measure of the magnitude of the labour force (LF). The quantity of labour (Q) and the real farm wage (W) are assumed to be simultaneously determined for the aggregate model. The remaining variables are argued to be exogenous in the context of the farm labour market. Purchased inputs are sufficiently non-specific to the industry that variations in agriculture can have little effect on their prices. Similarly, changes in the agricultural sector are unlikely to influence the non-farm alternative wage rate (NFW) or the labour force (LF). Gardner (1972) argued that prices received may be simultaneously determined with the outcome of the farm labour market; in addition, product prices are not known for most products until the end of the production period. Lagging prices received (PR) by one period handles both problems. Land is also sufficiently inflexible that there is unlikely to be any influence by the labour market on the contemporaneous quantity of land (L).

It should be noted that, due to the seasonal nature of much farm work, the earnings discrepancy between farm and non-farm workers is considerably larger than the wage discrepancy. However, since earnings are the product of hours and the wage rate, a 10% increase in the wage rate (holding hours constant) results in a 10% increase in earnings. If, for example, non-farm workers have 2000 hours of work, and farm workers have 1000 hours of work, the percentage change in earnings remains the same as the percentage change in the respective wage rate.

The papers by Schuh (1962) and by Tyrchniewicz and Schuh (1966, 1969) were based on data from a time period prior to the institution of a minimum wage in agriculture. Farms became subject to the minimum wage in 1967. Initially, the rate was lower than for workers in other industries, but was gradually increased to become the same as for non-farm workers in 1978. There are special exemptions for agricultural employers, excluding from coverage those employers with fewer than 500 man-days of labour in any quarter of the previous year. Although many employers are not covered under the minimum wage, most workers are employed by large employers who are covered by the minimum wage. Agricultural employers are also exempt from paying higher overtime wages, and there are other minor differences for agriculture (Runyan, 2000a). Using CPS data, Runyan (2000b, p. 32) estimated that 25% of all hired farm workers earned less than the federal minimum wage in 1998. Data on crop workers for the period 1996–1998 indicate that 12% of the workers earned less than the minimum wage (US Department of Labor, 2000, p. 33).

Gardner (1981) argues that the minimum wage significantly complicates the specification of the structural supply and demand equations for labour. The complication has two components. The first is that skilled workers are largely unaffected since their equilibrium wage exceeds the minimum wage prior to its implementation. The second concerns the incomplete coverage of unskilled workers in the sector. To the extent that some unskilled workers are covered by the minimum wage, there will be wage and employment effects. He argued that the remaining unskilled, uncovered workers will only be affected by the minimum wage to the extent that the demand for unskilled workers in the uncovered market shifts. Gardner (1981) further argues that this combination of effects in the three labour market components (skilled, uncovered unskilled and covered unskilled) results in an inward distortion of the aggregate demand curve, depending on the relative importance of the adjustment in the unskilled component of the labour market. As a result, the observed aggregate wage rate and employment will be on the distorted demand curve, and the supply function displaced by the extent of the employment effect in the unskilled market. With these problems, Gardner chose to estimate reduced form equations for the labour market to evaluate the effects on equilibrium employment and wage rates.

A major change in the farm labour market since the earlier papers estimating aggregate labour supply and demand models is the increased role of workers from Mexico in the labour market. For example, among crop workers surveyed in the National Agricultural Workers Survey (NAWS) for 1997–1998, 77% were Mexican born (US Department of Labor, 2000). Moreover, 42% of the workers considered Mexico to be their home base. This characteristic suggests that variations in Mexican economic conditions may have a strong influence over the supply of farm workers in the USA. Initial efforts to estimate the reduced form and structural models included Mexican economic variables. The variables had no statistically significant effects as long as the US non-farm wage rate was included. An interpretation is that although Mexican economic variables may indeed be influential, their effect is transmitted through the US non-farm wage rate. Since non-farm industries also employ large numbers of workers from Mexico, it appears that the non-farm wage rate absorbs the variations in the Mexican economy, and that they do not have an independent effect on the farm labour market beyond the variations in the US non-farm wage rate.

Reduced form estimates

Although the focus of this chapter is not on minimum wages, the effects that minimum wage policies have on the observed employment and wage rate must be accounted for in the empirical specification. A variant of Gardner's (1972) reduced form specification will serve as a starting point for the empirical analysis. The two variables to be explained are the annual average number of workers employed on farms (Q) and the real farm wage (W). The exogenous variables in terms of the model are the non-farm wage rate (NFW), prices paid for non-labour purchased production inputs (PP), lagged index of prices received for farm goods (PR), and land (L), using the US Department of Agriculture input index for farm real estate. The nominal farm wage, the non-farm wage, prices paid and prices received are all deflated by the consumer price index (CPI). The non-farm wage rate is, in addition, multiplied by 1 minus the percentage of the labour force unemployed, to better represent expected earnings from non-farm work.

Gardner's (1981) data set ended with 1978 – before it was apparent that there were significant changes in both the farm labour market and the overall US labour market (Figs 11.1–11.3). Starting in the early to mid-

1970s, real wages stopped rising, the decrease in the number of farm workers slowed, and the distribution of income in the USA started to widen. To examine whether or not there is a significant difference in the worker and wage relationships before and after 1973, standard tests for structural change are done for the separate versus the combined time periods. Equality of the parameters in the two time periods for both the worker and wage equations is strongly rejected. The test statistic for the worker equation is 11.23, with a critical value of $F(6,30)_{0.01} = 3.47$; the test statistic for the wage equation is 4.70, and the $F(6,29)_{0.01} = 3.50$. The conclusion in each case is that there are significant differences in the parameters in the two time periods.

Lines 1 and 2 of Table 11.2 give the worker equation estimates for the separate time periods; line 3 gives the estimates for the combined time periods. The contrast of most interest between the two time periods is the minimum wage effect. In the first period, including the introduction of the minimum wage and the early years of the minimum wage, there was a very strong negative effect on employment in agriculture. The second time period, starting with 1973, shows a non-significant effect on employment. The pooled equation (3) in Table 11.2 completely masks this difference. The non-farm wage rate appears to have a negative effect on agricultural employment, although it is not very strong in the latter time period.

Lines 5 and 6 of Table 11.2 give the wage equation estimates for the separate time periods; line 7 gives the estimates for the combined sample. The non-farm wage has a strong positive effect on the real farm wage for both time periods. The minimum wage also has a positive effect for both time periods and is highly significant in each case. The contrasts between the two time periods are not as dramatic in the case of the wage rate as they are for employment.

Lines 4 and 8 of Table 11.2 augment the two reduced form equations with a labour force variable (LF) and an income distribution variable (DIST) to introduce changes in the distribution of income in the overall economy to the farm labour market model. The LF variable is the civilian labour force aged 25 years and older with an education of 8 or fewer years – the portion of the labour force from which much of the farm workforce would be drawn. (Although the CPS may underrepresent unconventional households characteristic of undocumented workers, it remains the standard source for labour force measurement.) The most commonly used labour force measure used is persons aged 16 years and over, employed or looking for work. However, the education level is readily available only for those aged 25 years and over. Given the emphasis in this chapter on low-wage workers, the latter was used to reflect the dwindling portion of the labour force with eight or fewer years of education.) Between 1967 and 1998, this group has declined from 31% to 7% of the labour force aged 25 years and older (US Bureau of Labor Statistics, 1999). The absolute number has declined as well, from 32.6 million to 12.8 million persons. These are dramatic reductions, reflecting the reduced supply of unskilled labour as an increasing portion of the population has remained in school beyond the eighth grade.

The distribution variable (DIST) is the ratio of the average household income of the top quintile to the average household income of the bottom quintile, as illustrated in Fig. 11.3. There is some consensus that the widening income distribution is due in part to an increased premium for skill (Berman *et al.*, 1994; Welch, 1999). Suppose there is a skilled labour market and an unskilled labour market: agriculture is expected to employ primarily unskilled workers. An outward shift in the demand for skilled labour would generate the increased skill premium with a movement along the upward sloping supply schedule for skilled labour. Given two types of labour, the increased equilibrium quantity of skilled labour due to the shift in the demand schedule results in a leftward shift in the supply of unskilled labour, implying a higher equilibrium unskilled wage rate with a lower quantity of unskilled labour. Although the available income distribution variable includes unearned as well as earned income, there is little reason to believe that the changes would be dramatically different

Table 11.2. Reduced form estimates[a].

Equation	NFW	PP	PR_{t-1}	MinWg	L	LF	DIST	Constant	P	*n*
(1) Workers	−419.9***	6.134	−5.659***	−45.58***	2.177	–	–	4583	0.984	26
1947–1972	(48.53)	(5.585)	(1.607)	(14.63)	(8.485)			(1203)		
(2) Workers	−54.20	13.28***	1.505	33.65	−6.796	–	–	755.5	0.986	16
1973–1996	(42.24)	(3.92)	(0.986)	(37.67)	(5.674)			(490.3)		
(3) Workers	−131.4***	12.76***	1.033	−106.4***	21.46***	–	–	−1011	0.975	42
1947–1996	(23.19)	(3.16)	(1.264)	(14.19)	(5.25)			(496.4)		
(4) Workers	−61.89	7.896***	1.242	49.27	−1.191	0.011**	−12.69	615.9	0.983	22
1967–1996	(38.20)	(2.479)	(1.193)	(31.95)	(4.653)	(0.004)	(21.06)	(625.5)		
(5) Wage	0.431***	0.0045	0.0057***	0.126***	−0.021*	–	–	1.596	0.985	25
1948–1972	(0.059)	(0.0068)	(0.0020)	(0.017)	(0.011)			(1.551)		
(6) Wage	0.214***	0.0090	−0.0018	0.227***	−0.028**	–	–	4.399	0.917	16
1973–1996	(0.087)	(0.0081)	(0.0020)	(0.078)	(0.012)			(1.010)		
(7) Wage	0.211***	0.013***	−0.0016	0.191***	−0.043***	–	–	5.694	0.984	41
1948–1996	(0.024)	(0.003)	(0.0013)	(0.015)	(0.005)			(0.520)		
(8) Wage	0.366***	0.0038	0.0019	0.194***	−0.014*	−0.00003***	−0.011	2.851	0.948	22
1967–1996	(0.070)	(0.0045)	(0.0022)	(0.058)	(0.008)	(0.000008)	(0.038)	(1.139)		

[a]The numbers in parentheses are estimated standard errors.
Estimates are statistically different than zero at the 10%, 5%, or 1% significance levels as indicated by *, **, or ***, respectively.

than for a distribution variable based solely on earned income.

The results for the employment equation comparing lines 2 and 4 in Table 11.2 are not dramatically different. While the labour force variable is positive and strongly significant, the income distribution variable appears to have no effect on farm employment. Virtually the same statement may be made for the wage equation contrast between lines 6 and 8 in Table 11.2. Reductions in the labour force have a strong positive effect on the real wage in agriculture, and again the income distribution variable appears to have no discernible effect on the real wage. The remaining parameters are largely consistent between the two specifications.

Structural equations

A concern noted above with estimating the structural equations in the presence of a minimum wage is that the programme may distort the demand equation when there is only partial coverage of the labour market. One way to incorporate this distortion directly into the demand equation is to include the product of the wage rate and the minimum wage in addition to the minimum wage by itself. This allows a change in the minimum wage rate to alter the slope of the aggregate demand curve as well as the intercept. Gardner's (1981) framework also suggests that the supply curve revealed by the data is an inward displacement of the true supply curve as a result of the minimum wage reducing unskilled employment. This is readily handled with the inclusion of the minimum wage in the supply equation.

The estimates in Table 11.3 are based on such a specification. The underlying model is the basic specification introduced by Schuh (1962), adjusted for the minimum wage considerations and incorporating the income distribution variable in the supply

Table 11.3. Structural labour supply and demand equations[a].

Variable	Demand[b]	Supply[c]
Farm wage	31.29 (518.8)	886.3*** (263.4)
Non-farm wage	–	−340.2*** (106.5)
Prices paid	4.438 (3.084)	–
Prices received$_{t-1}$	3.460*** (0.989)	–
Minimum wage	154.5 (727.5)	−46.32 (58.58)
Farm wage * minimum wage	−22.27 (160.9)	–
Land	6.728* (4.086)	–
Labour force	–	0.0440*** (0.0124)
Distribution	–	−23.16 (34.67)
Constant	−677.7 (2331)	−766.2 (1067)
$\partial Q/\partial W$[d]	−40.64 (97.28)	–
Wage elasticity[d]	−0.161 (0.385)	3.504*** (1.042)
$\partial Q/\partial MinWg$[d]	56.91 (42.21)	–
Minimum wage elasticity[d]	0.166 (0.123)	–
Non-farm wage elasticity[d]	–	−2.352*** (0.736)
n	22	22

[a]The data used are for 1967–1980 and 1989–1996. Numbers in parentheses are estimated standard errors. Estimates are statistically different than zero at the 10%, 5%, or 1% significance levels as indicated by *, **, or ***, respectively.
[b]The instruments used for the demand equation are all exogenous variables in the system, the squares of all exogenous variables, and the first four principal components of the pair-wise products of all exogenous variables.
[c]The instruments used for the supply equation are all exogenous variables in the system.
[d]Evaluated at the means of the data.

equation. Since the wage and quantity of labour are assumed to be jointly determined, the equations are estimated with instrumental variable techniques. The demand equation is nonlinear in the wage rate; an extended set of instruments is necessary to obtain consistency of the parameter estimates (Kelejian, 1971). Kelejian suggests using all instruments plus higher order terms in the instruments. With limited observations, incorporating all second order terms was not feasible. As an alternative, all exogenous variables, their squares, and the first four principal components of the cross-products of the exogenous variables were included as instruments. The first four principal components represented 98% of the variance of the 21 cross-products.

Estimation of the parameters of the demand equation is not very robust. The prices received index has a strong positive effect on the quantity of labour demanded, and the land quantity effect is positive and marginally significant. The effect of the wage rate ($\partial Q/\partial W$), although carrying a negative sign, is not significantly different than zero. The implied wage elasticity of −0.161 is likewise not significantly different than zero, suggesting at best a highly inelastic demand schedule. The minimum wage effect ($\partial Q/\partial$Min wage) has a sign opposite of that expected but the estimate is not significantly different than zero.

The supply equation performs somewhat better. The farm wage, non-farm wage and the labour force variables are all statistically significant. The farm wage is highly significant, implying an elasticity of supply of 3.5 at the sample means. The non-farm wage rate has the expected negative sign and is also highly significant. The minimum wage variable, although insignificant, does have the correct sign. The reduced form equations for the same time period (lines 4 and 8 of Table 11.2) indicate a positive wage effect but no employment effect. The reduced form finding is consistent with a negative shift in the supply schedule and an inelastic demand schedule, suggesting some degree of correspondence between the (unrestricted) reduced form estimates and the structural equation estimates.

The income distribution variable again has no statistically significant effect, suggesting that it does not shift the supply schedule for farm labour. This result is consistent with an increased demand for highly skilled labour in the economy, earning a commensurately higher return on their skills, while relatively unskilled labour continues to earn a constant return on their labour. The relatively flat non-farm wage for production workers since the late 1960s is also consistent with a constant return for relatively unskilled labour. Alternatively, a falling non-farm wage rate for production workers would suggest an excess supply of unskilled labour. One interpretation of these results is that with a constant return to unskilled labour in the economy, the non-farm wage rate incorporates all available information about the relative availability of labour to the agricultural sector. In this case, the fact that the income distribution has been widening has no direct bearing on the farm labour market other than through the non-farm wage rate; the widening income distribution is merely a phenomenon of increased returns to highly skilled labour in the overall economy.

Summary and Conclusion

This chapter extends the existing empirical analysis of the interaction of farm and non-farm labour markets with a simple supply and demand model at the aggregate US level. Equilibrium levels of farm wages and employment are found to be responsive to changes in the non-farm wage rate. Results for the effect of variations in the minimum wage are mixed for a more recent time period than examined by Gardner (1981). While the minimum wage had a strong negative effect on employment and a positive effect on wages up through 1972, as Gardner found, analysis of the 1973–1996 period does not reveal any systematic employment effect, though there is a positive wage effect.

The supply response to own wages is found to be quite elastic, suggesting relative ease of adjustment in the quantity supplied to shocks in labour demand. The quantity supplied is also highly responsive to changes in the non-farm wage rate; a 1% increase in the non-farm wage rate would result in a 2.4% reduction in the quantity of labour supplied, *ceteris paribus*. Both the own and non-farm wage responses are considerably larger than reported in the early work by Schuh (1962), as well as the more recent estimates by Duffield and Coltrane (1992). Agricultural employers are concerned about competition for workers with other industries, and the estimated response of labour to the non-farm wage is consistent with this concern. With higher non-farm wage rates, the supply curve shifts to the left, resulting in higher farm wage rates. Referring to Fig. 11.2, real non-farm wage rates appear to have increased slightly in the past 2 years, a change that would imply pressure to increase farm wage rates in real terms. Note, however, that the non-farm wage rates remain about US$1 in 1982–1984 dollars (US$1.63 in 1998 dollars) below non-farm wage rates in the early 1970s.

Predictions about the future supply of farm labour require assumptions about the future pattern of non-farm wage rates. Assuming that the returns to unskilled labour remain relatively constant, there is little reason to believe that non-farm wage rates for production workers would rise significantly. In this case, there would be little change expected in the availability of labour for the farm labour market. Significant reductions in the flow of labour into the country or in the enforcement of restrictions on alien employment could be expected to raise the non-farm wage rate and would, in turn, impact the farm labour market with higher wage rates.

The increased dispersion between skilled and unskilled labour since the early 1970s has apparently had little or no effect on the farm labour market. The primary reason is that the return to unskilled labour has remained relatively constant while the return to highly skilled labour has increased, generating the increased wage dispersion. As long as the return to unskilled labour remains near its present level, there is little reason to suggest an impact on the farm labour market.

The estimates of the demand equation are not particularly robust. To the extent that they do have any meaningful interpretation, it is that the labour demand is highly inelastic. As a result, the wage effects of any shocks to supply or demand will simply be the movement along the supply curve to equilibrate the fixed demand. Given the highly elastic supply response, however, the wage effects will be rather muted in comparison with the employment changes.

Continuation of the *Farm Labor* series from the US Department of Agriculture is essential for monitoring and analysis of the farm labour market. Unfortunate lapses in the series, as occurred during the 1980s, leave major gaps in our knowledge and ability to determine potential structural changes in the labour market. Other types of information that would be worth considering are employment and wage information for production and non-production worker categories in agriculture. Presently there are field and livestock worker categories but, even for these, the numbers in each of the categories do not appear to be published. As agriculture continues to adapt to new technological developments, there is likely to be a movement toward more highly skilled labour in the hired farm workforce. The present extent of this is unknown and our knowledge deficit is likely to increase in its continued absence.

References

Barkley, A.P. (1990) The determinants of the migration of labor out of agriculture in the United States, 1940–85. *American Journal of Agricultural Economics* 72, 567–573.

Berman, E., Bound, J. and Griliches, Z. (1994) Changes in the demand for skilled labor within US manufacturing: evidence from the annual survey of manufactures. *Quarterly Journal of Economics* 109, 367–397.

Duffield, J.A. and Coltrane, R. (1992) Testing for disequilibrium in the hired farm labor market. *American Journal of Agricultural Economics* 74, 412–420.

Gardner, B. (1972) Minimum wages and the farm labor market. *American Journal of Agricultural Economics* 54, 473–476.

Gardner, B. (1981) What have minimum wages done in agriculture? In: Rottenberg, S. (ed.) *The Economics of Legal Minimum Wages.* Amer-

ican Enterprise Institute, Washington, DC, pp. 210–232.
Gisser, M. and Dávila, A. (1998) Do farm workers earn less? An analysis of the farm labor problem. *American Journal of Agricultural Economics* 80, 669–682.
Goldin, C. and Margo, R.S. (1992) The great compression: the wage structure in the United States at mid-century. *Quarterly Journal of Economics* 107, 1–34.
Johnson, D.G. (1951) Functioning of the labor market. *Journal of Farm Economics* 33, 75–87.
Kelejian, H. (1971) Two-stage least squares and econometric systems linear in parameters but nonlinear in the endogenous variables. *Journal of the American Statistical Association* 66, 373–374.
Perloff, J.M. (1991) The impact of wage differentials on choosing to work in agriculture. *American Journal of Agricultural Economics* 73, 671–680.
Richards, T.J. and Patterson, P.M. (1998) Hysteresis and the shortage of agricultural labor. *American Journal of Agricultural Economics* 80, 683–695.
Runyan, J.L. (2000a) *Profile of Hired Farmworkers, 1998 Annual Averages.* Agricultural Economic Report No. 790. Economic Research Service, US Department of Agriculture, Washington, DC.
Runyan, J.L. (2000b) *Summary of Federal Laws and Regulations Affecting Agricultural Employers, 2000.* Agricultural Handbook No. 719. Economic Research Service, US Department of Agriculture, Washington, DC.
Schuh, G.E. (1962) An econometric investigation of the market for hired labor in agriculture. *Journal of Farm Economics* 44, 307–321.
Sumner, D. (1982) The off-farm labor supply of farmers. *American Journal of Agricultural Economics* 64, 499–509.
Tokle, J.G. and Huffman, W.E. (1991) Local economic conditions and wage labor decisions of farm and rural nonfarm couples. *American Journal of Agricultural Economics* 73, 652–670.
Tyrchniewicz, E.W. and Schuh, G.E. (1966) Regional supply of hired labor to agriculture. *Journal of Farm Economics* 48, 537–556.
Tyrchniewicz, E.W. and Schuh, G.E. (1969) Econometric analysis of the agricultural labor market. *American Journal of Agricultural Economics* 51, 770–787.
US Bureau of Labor Statistics (1997) *BLS Handbook of Methods.* US Department of Labor. US Government Printing Office, Washington, DC.
US Bureau of Labor Statistics (1999) US Department of Labor. http://stats.bls.gov
US Bureau of the Census (1996) A brief look at postwar US income inequality. *Current Population Reports P60-191.* Washington, DC. Updated with http://www.census.gov/hhes/income/income98/in98dis.html
US Department of Agriculture (various years) *Agricultural Statistics.* National Agricultural Statistics Service. US Government Printing Office, Washington, DC.
US Department of Agriculture (various years) *Farm Labor.* National Agricultural Statistics Service, Washington, DC.
US Department of Labor (1993) *US Farmworkers in the Post-IRCA Period.* Research Report No. 4. Office of the Assistant Secretary, Office of Program Economics, Washington, DC.
US Department of Labor (2000) *Findings from the National Agricultural Workers Survey (NAWS) 1997–1998.* Research Report No. 8. Office of the Assistant Secretary, Office of Program Economics, Washington, DC.
US Department of Labor (2001) *History of Changes to the Minimum Wage Law.* Employment Standards Administration, Washington, DC. http://www.dol.gov/dol/esa/public/minwage/chart.htm
Welch, F. (1999) In defense of inequality. *American Economic Review: Papers and Proceedings* 89, 1–17.

12 Rural Deprivation and Farm Worker Deprivation: Who's at the 'Sharp End' of Rural Inequalities in Australia?

Jim McAllister
School of Psychology and Sociology, Central Queensland University, Rockhampton, Queensland, Australia

Abstract

While it is acknowledged that there is devastating poverty within certain sectors of Australia's urban areas, low incomes are concentrated in rural areas, where rural sociology has been concerned mostly with (owner-operator) farmer lifestyles and livelihoods. It cannot be denied that commodity prices and seasonal conditions frequently create 'bad times' for farm owners or commodity 'growers'. Nevertheless, this chapter disaggregates the 'rural' epithet – used for all that goes on outside the cities – and separates farm workforces from other categories of rural and urban workers, bringing the farm wage and salary earner (employee) category to the fore. Overall, this chapter addresses two major aspects of agricultural labour: (i) how the employees' conditions differ from those of their farmer employers and of farm owner-operators who live and work in the same production environment but appear to employ only immediate family members; and (ii) the conditions of life and work of agricultural employees. Data of two types are presented, and the analysis is in three sections. Census data from 1996 (most recent) confirm farm workers as belonging to three class categories (employers, self-employed and employees). Farm workers in aggregate are first compared with other producers, then compared among classes of farm producers. This cross-sectional analysis is enhanced with data from a small national attitude survey, and then two case studies are used to elaborate the limited census information.

Conditions of work and living are shown to be poorer for farm employees than for other farm work categories (owner-operator and employer farmers), and their appreciation of their opportunities for economic and social improvement reflects limited capacity for change within their work milieu. Particularly of concern here is farm, local community and legislative response to providing a labour force for harvesting sugarcane and mandarin oranges (tangerines).

Changing Policy for a Changing Agriculture

Legislation, and the policy it embodies, sometimes follows public thinking and sometimes sets the pace for it. Since the early 1980s, Australian agriculture has witnessed a variety of legislation specifically directed to altering various agricultural industries. At the same time, agriculture has had to adjust to

(eds J.L. Findeis, A.M. Vandeman, J.M. Larson and J.L. Runyan)

the consequences of a variety of other policies – especially economic ones – not intended to directly impact farm decision making or the environment in which farm decisions are taken, but which have consequences too wide-ranging for farming people to avoid.

In the early 1980s, during a period of economic stagnation in the world economy, markets for Australian agricultural products expanded rapidly and then, following the stock market crash of 1988, shrank almost as rapidly. The chief effect of this economic fluctuation has been the disruption of the expansion of the Asian economies, Australia's primary hope for increasing its food export markets. Farm decisions made during this 'climate of optimism' in the early 1980s have been regretted more recently as markets contracted; farm populations, as well as non-farm rural and urban people, have had to live with the consequences of the national and international economic policy decisions of the 1980s. The consequences have not been experienced uniformly and different segments of the farming sector have differentially borne the brunt of the full range of production and policy decisions.

Paid Agricultural Workers in Australia

In Australia, the discipline of rural sociology has been principally concerned with three issues: (i) the political economy of family farming (Lawrence and Gray, 1999); (ii) land and water degradation and/or sustainability (Lockie and Vanclay, 1997); and (iii) country life, with towns as farm service centres and seats of local power (Gray *et al.*, 1993; Higgins, 1997). Recently, women have also begun to find their place in this story (James, 1989; Alston 1992), and increasing prominence has also been given to concern with agricultural commodities as human food (Parsons, 1996; Lockie and Collie, 1999) and with corporate agriculture (Nankervil, 1980; Hungerford, 1996). Nevertheless, it would appear that an additional issue – who are the hired agricultural workforce and what do they do? – still needs to be considered within the realm of rural sociology. Until very recently, family farmers (who are presumed to include husband–wife and parent–offspring, as well as sole-proprietor, business arrangements) have represented by far the most numerous category of producers in Australian agriculture. Additionally, whenever agricultural labour is discussed, there is a tendency to conflate farm-family work and farmwork-for-payment (without ownership), and view it from the farmer's side.

One stimulus to reconsider this issue is that a small number of larger-than-family farms control a very large proportion of the land area of Australia. One should bear in mind the complication of 'carrying capacity', which becomes more important in less developed and low rainfall areas. That said, the farms in this grouping are important land controllers. Indeed, some of the wealthiest people in Australia control a great proportion of the land, deriving very large incomes and high social prestige from involvement in sheep and cattle grazing. For example, Jones (1998) reported that the top 70 or so 'Barons of the Bush' qualified as 'barons' because they ran 'more than 15,000 head of cattle' on their properties. The circumstances of these 'barons' contrast sharply with those of the folkloric 'rural battlers' and serve to illustrate the quantity of hard, physical work and sheer numbers needed to make the productivity and wealth of these corporate 'barons' a reality.

A specific example of a corporate producer is AMP's Stanbroke Pastoral Company, headquartered in the State of Queensland. In 1998 it was Australia's largest beef producer, with 535,000 head (Jones, 1998), and also 'the world's single largest landholder with 13.5 million hectares' (in 31 properties) of which 3.9 million ha are located in Queensland. 'The nation's largest private cattle producer, the McDonald family, ... [is] the second largest private landholder in Australia [and] now runs an estimated 116,000 head of cattle on 3.37 million hectares' (Jones, 1998, p. 23). One telling example is of the New Zealand owned dairy company in Victoria that milks 2000 head.

The pressing question raised by the existence of corporate agriculture is: where does the workforce come from to support this form of agriculture? Obviously, farm employees are a significant component of these types of operations. Some (smaller) farms still maintain one or two workers, nevertheless, and the consequences for them of structural change are also a research concern. Overall, this chapter is particularly concerned with two major aspects of agricultural labour: (i) the conditions of life and work of agricultural employees on both large-scale farms and in small operations; and (ii) how the employees' conditions differ from those of their farmer employers, and from those of the farm owner-operators who live and work in the same production environment but appear to employ only immediate family members.

Types of paid farm workers

Finemore and McAllister (1999) have proposed that there are three types of paid farm workers: on-farm employees ('farm hands'), 'industrial agriculture' employees (including intensive-livestock workers) and 'harvest labour'.

The small amount of literature on paid employment of all types that does exist suggests that farm hands are few in number and decreasing in importance (Dempsey, 1990; Karmel *et al.*, 1991). Nevertheless, because of their increasing marginality to agricultural industries (Nalson, 1977; Lawrence, 1987), their (probable) older age and likely retirement into rural towns (Hudson, 1984; Dempsey, 1990) and their individual, intimate contact with their employer (Newby, 1977; Barlett, 1986), their case seems likely to present interesting and policy-relevant aspects of industrial relations, and insights into the labour process in farming (Thomas, 1981; Willis, 1998).

In the USA, Heffernan (1972) and Constance (1988), for example, proposed that the egg and broiler poultry industries and, possibly, cattle feedlotting and pig production are producing an expanding population of paid workers (factory farm employees), whose work situation might be expected to be increasingly like that of other factory employees, and whose workplace is often just one more 'factory' – vertically integrated within a food production and marketing company (Nankervil, 1980; Lawrence, 1987). In Australia, intensive-stock production appears to have developed in two directions. Vanclay and Lawrence (1995) described an industrial labour force in beef feedlotting but, equally, Rickson and Burch (1996) and Lyons (1996) recognized, for poultry, the 'commodity contracting' highlighted by Heffernan (1972) and Constance (1988).

Gill (1981) contrasted two categories of wool employees, showing the difference in lifestyle, work habits and deference of sedentary farm hands as compared with those of the itinerant shearing labour force: the latter, organized industrially and consequently working strict hours in spartan, but controlled, conditions defined by a relatively favourable industrial agreement; and the former poorly remunerated, isolated and subject in their private lives and working conditions to the 'suasion' of their employers. Ruben (1992) investigated itinerant fruit pickers in rural Victoria, finding that – unorganized, by contrast with the shearers – they are open to much more exploitation in their varying workplaces where:

> Picking is undisputedly an unattractive job ... Work is hard and low paid. There are some other problems that contribute to the unappealing character of the job, namely, inadequate accommodation, lack of efficient transport, poor working conditions and deficient work organization.

Analysis of the farm workforce from the 1996 Australian Census indicates that paid workers (wage, piecework and contract) constitute about 60% of the farm workforce in Australia. But the statement by Garnaut and Lim-Applegate (1998, p. 3) that 'family labour ... total[s] 85% of all farm labour' suggests that 60% of farm labour does only 15% of the farm work, and this implies a high proportion of part-time work. It might even be assumed that a considerable proportion are cobbling together a living from a number of jobs, one of which is farm work. Duff (1999) calls this 'portfolio work' and refers to many family farmers obtaining a living in this way. There possibly is a trend toward farmers and their families contributing to this workforce as well.

Finding Social Categories in the Census Data

The designation of the population in the Australian National Census is crude, but it does allow adults to be 'classified' (after the manner of Eric Olin Wright) in terms of the work they do and the control they have over resources, their declared incomes, and their personal demographic, family and dwelling characteristics. Census data permit comparison of various segments of the farm labour force (employers, self-employed and employees) with each other, and with comparable classes producing in non-primary industrial activities. In general, the data show that the designated classes represent income hierarchies, farm and non-farm; but while owner-operatorship is strong among Australian farmers, this category hardly exists any longer among non-farming industries.

Changes in the proportions of these classes and in their characteristics have occurred as major aspects of government policy have acted to deregulate both economy and labour market. While trends in numbers of workers are of demographic interest, what is much more interesting sociologically is what changes in work roles there have been over time, and what workplace relationships are established within various agricultural structures. Duff (1999) captures this meaning well:

> Although the most common economic indicators are essential to an analysis of the social effects of economic restructuring, there is a continuing need to supplement them with social research which reveals the way in which restructuring affects people's lives in different parts of rural and regional Australia.

The census categories investigated here are sociologically poor, but the analysis is deepened by incorporating other research – a subsample from national data on attitudes, and field investigation of sugar and citrus growing – to elaborate perception and realization of class differences and of lifestyles.

This chapter argues that, while there is poverty in Australia within urban areas – some of it devastatingly oppressive and attacking the most vulnerable members of our society – nevertheless, low incomes are concentrated in rural areas, where rural sociology has tended to show most interest in (owner-operator) farmer lifestyles and livelihoods. While it cannot be denied that commodity prices and seasonal conditions frequently create 'bad times' for farm owners and commodity growers, this chapter aims to show that a class disaggregation of the rural population brings the farm employee category to the fore. Their objective conditions of living are shown to be poorer than those of the other farm classes, and their appreciation of their opportunities for economic and social improvement reflects limited capacity for change within their work milieu. Farm employees are at the 'sharp end' of rural poverty in Australia and its consequences for their lives show up not only in low incomes, but also in the conditions of their living and their expectations from life.

Data Analysis

Data of two types are presented, and analysis is in three sections: (i) the most recent census data from 1996 are used to compare members of the farm workforce with all other workers, and then to investigate the varied conditions of the farm classes; (ii) a small subsample of farm people is used to deepen this analysis to include group attitudes; and (iii) fieldwork case study data are used to elaborate the census information. Particularly of concern in the final section is farm, local community and legislative response to providing a labour force for harvesting sugarcane and mandarin oranges.

Census data for Australia

The census enumeration for Australia in regard to a number of population characteristics – economic, demographic, familial and level of living – is shown in a set of tables that allows comparison of types of workers. The right side of each table shows a comparison between farm and non-farm categories of worker in regard to each characteristic of the population shown in preliminary analysis to be appropriate. On the left side of the tables, three farm classes are compared. The data are generally presented in very few categories; this is an outcome of some prior work associated with comparing these data with data from previous censuses. The 1986 Census was particularly poorly differentiated and has restricted some of the subsequent categorizations. All values of chi-square (χ^2) are significant at the 1% level.

The proportions of the various categories of individual workers who earn no (or a negative) income, from Aus$1 to Aus$599 per week, and Aus$600 or more are shown in Table 12.1. Compared with non-agricultural workers, agricultural workers cluster predominantly in the two categories below Aus$600 per week. Only half the proportion of agricultural, as compared with non-agricultural, workers earn Aus$600 or more (17.2% compared with 35.8%).

Comparing within the farming classes, a low proportion of employees declare negative or no income, this being a result of the taxation system only really available to property owners who are permitted to lose money by

Table 12.1. Incomes and hours of work for surveyed individuals[a]. (Source: Australian Bureau of Statistics, 1966 Census (data not randomized).)

	Agriculture and service to agriculture					
	Employer	Self-employed[b]	Employee	Total	All non-agricultural	Overall total
Weekly incomes of surveyed individuals (in Aus$)[c]						
Negative or none	2,288	8,106	4,462	14,856	37,174	52,030
	(9.0)	(10.0)	(2.7)	(5.4)	(0.5)	(0.7)
$1–599	15,670	60,457	135,979	212,106	4,472,854	4,684,960
	(61.8)	(74.4)	(81.1)	(77.3)	(63.7)	(64.2)
$600 or more	7,378	12,688	27,240	47,306	2,511,506	2,558,812
	(29.1)	(15.6)	(16.2)	(17.2)	(35.8)	(35.1)
Total	25,336	81,251	167,681	274,268	7,021,534	7,295,802
Incomes of households in which surveyed individuals reside (in Aus$)[d]						
Negative or none	993	3,565	1,745	6,303	8,424	14,727
	(4.6)	(4.9)	(1.2)	(2.6)	(0.1)	(0.2)
$1–599	5,848	30,942	53,253	90,043	1,095,329	1,185,372
	(27.0)	(42.9)	(36.5)	(37.5)	(17.5)	(18.3)
$600 or more	14,834	37,607	91,008	143,449	5,145,104	5,288,553
	(68.4)	(52.1)	(62.3)	(59.8)	(82.3)	(81.5)
Total	21,675	72,114	146,006	239,795	6,248,857	6,488,652
Hours worked per week by surveyed individuals[e]						
1–34 hours	3,177	12,472	36,651	52,300	1,940,489	1,992,789
	(12.6)	(15.6)	(22.2)	(19.3)	(28.4)	(28.0)
35 hours or more	22,031	67,723	128,414	218,168	4,898,297	5,116,465
	(87.4)	(84.4)	(77.8)	(80.7)	(71.6)	(72.0)
Total	25,208	80,195	165,065	270,468	6,838,786	7,109,254

[a]Actual numbers and column percentages (percentages shown in parentheses).
[b]Includes only those self-employed farmers (owner-operators) with no employees.
[c]Farm/non-farm comparison: $\chi^2 = 118,822.6$; df = 2; significant at 0.01 level. Within farm-categories comparison: $\chi^2 = 9556.5$; df = 4; significant at 0.01 level.
[d]Farm/non-farm comparison: $\chi^2 = 128,308.6$; df = 2; significant at 0.01 level. Within farm-categories comparison: $\chi^2 = 2731.1$; df = 4; significant at 0.01 level.
[e]Farm/non-farm comparison: $\chi^2 = 10,534.7$; df = 1; significant at 0.01 level. Within farm-categories comparison: $\chi^2 = 736.5$; df = 4; significant at the 0.01 level.

depreciating their assets while living fairly comfortably; they can also 'average' their income over a number of years for taxation purposes. By contrast, employees who have no income and few assets to trade on are poor if they say so. Even so, farm employees fall predominantly in the Aus$1–599 range; four out of every five earn weekly incomes in this range. At the same time, a high proportion of employer farmers earn incomes in excess of Aus$600 per week, as might be expected. As such, Aus$600 is not a high income: in June 1996, the average weekly earnings of employees in Australia was Aus$564 (Australian Bureau of Statistics, 1997), while for households the median income was Aus$637 (Australian Bureau of Statistics, 1996). Saunders (1996) proposed that, at the same time, the poverty line for a single person paying rent was Aus$215.60; for a couple it was Aus$288.40; for a couple with two children, Aus$404.90; and for a single parent with two children, Aus$335.00.

Some workers claim to have no (or negative) earnings (Table 12.1). Although individual employees without income might be able to pool their poverty with their families (and/or co-residents), in the absence of more detailed data it seems necessary to assume that the 1.2% of farm employees with no or negative income drew on savings or 'the kindness of strangers' to make ends meet that year. By contrast, the self-employed worker and employer farmers may have been able to 'average' their income in subsequent better seasons to recover from apparent present poverty.

Table 12.1 also shows the accumulation of income within households, using the same categories as shown for individual earnings. Compared with non-agricultural workers, agricultural workers again cluster predominantly in the two categories below Aus$600 per week whereas more than four-fifths of non-agricultural households earn Aus$600 or more. Comparing within farming classes, again a low proportion of employees declare negative or no income, but this time self-employed farmers predominate in the low-income category. However, a high proportion of employee households also accumulate incomes in excess of Aus$600 per week. Here there seems to be the combined effect of more than one earner contributing to raise the proportion of relatively high incomes among employees, whereas (again) the relatively high proportions of employers and self-employed workers with no or negative income can be taken to be an artifact of the taxation system. It must be said, however, that among farm workers, farm employees are less likely to have high incomes than their employers, but that self-employed or 'own account' farmers are less likely still to declare high incomes. Later in this chapter, there will be further discussion about ways in which farm employees in Australia piece together an income from a variety of activities.

Finally, Table 12.1 shows the percentages of the various categories of individual workers who work only a short working week, compared with those who work full-time (35 hours or more). Compared with non-agricultural workers, higher proportions of agricultural workers work the longer week. Comparing within the farming classes, a high proportion of employees work a week of less than 35 hours – suggesting, as surmised earlier, that many of them are part-time workers.

Table 12.2 shows the age at which the worker who filled out the census form left school. Compared with non-agricultural workers, workers in agriculture have a tendency to leave school early. Comparing within the farming classes, self-employed farmers in Australia have been the ones who predominantly leave school at an early age, as is well known from the farming literature.

Table 12.2 also shows the percentage of men and women in the agricultural and non-agricultural workforces. Compared with non-agricultural workers, agricultural workers are much more predominantly men. Farm employees tend to be the most 'masculine' workforce, whereas farm employers are either taking advantage of tax concessions available to men and women in partnership or else women farmers may be more likely to be employers, hiring the labour to do the traditionally 'male' jobs on the farm. In addition, Table 12.2 shows the proportions of the various categories of workers who have households of various sizes. Compared with

Table 12.2. Age left school, gender, and number of individuals in dwelling of surveyed individuals[a]. (Source: Australian Bureau of Statistics, 1996 Census.)

	Agriculture and service to agriculture				All	Overall
	Employer	Self-employed[b]	Employee	Total	non-agricultural	total
Age at which surveyed individual left school[c]						
Up to 16 years	16,562	57,595	112,427	186,584	3,450,471	3,637,055
	(64.2)	(70.2)	(67.8)	(68.2)	(50.6)	(51.2)
17 years and older	9,225	24,426	53,441	87,092	3,373,945	3,461,037
	(35.8)	(29.8)	(32.2)	(31.8)	(49.4)	(48.8)
Total	25,787	82,021	165,868	273,676	6,824,416	7,098,082
Gender of surveyed individual[d]						
Men	16,969	56,326	123,932	197,227	3,968,237	4,165,464
	(64.5)	(67.3)	(71.7)	(69.7)	(55.6)	(56.1)
Women	9,340	27,410	48,830	85,580	3,173,793	3,259,373
	(35.5)	(32.7)	(28.3)	(30.3)	(44.4)	(43.9)
Total	26,309	83,736	172,762	282,807	7,142,030	7,424,837
Number of people in dwelling of surveyed individuals[e]						
1	1,258	5,282	14,468	21,008	548,128	569,136
	(0.6)	(7.2)	(9.6)	(8.6)	(8.7)	(8.7)
2	7,594	29,485	45,254	82,333	1,793,958	1,876,291
	(34.2)	(40.0)	(30.4)	(33.5)	(28.4)	(28.6)
3	4,365	14,331	30,369	49,065	1,331,462	1,380,527
	(19.6)	(19.4)	(20.3)	(20.0)	(21.1)	(21.0)
4	4,192	12,724	30,399	47,315	1,543,123	1,590,438
	(18.8)	(17.2)	(20.4)	(19.3)	(24.4)	(24.2)
5	3,040	7,836	18,196	29,072	751,301	780,373
	(13.7)	(10.6)	(12.2)	(11.9)	(11.9)	(11.9)
6+	1,792	4,145	10,555	16,492	350,007	366,499
	(8.1)	(5.6)	(7.1)	(6.7)	(5.5)	(5.6)
Total	22,241	73,803	149,241	245,285	6,317,979	6,563,264

[a]Actual numbers and column percentage (percentages shown in parentheses).
[b]Includes only those self-employed farmers (owner-operators) with no employees.
[c]Farm/non-farm comparison: $\chi^2 = 355.3$; df = 1; significant at 0.01 level. Within farm-categories comparison: $\chi^2 = 32{,}692.6$; df = 2; significant at 0.01 level.
[d]Farm/non-farm comparison: $\chi^2 = 22{,}201.9$; df = 1; significant at 0.01 level. Within farm-categories comparison: $\chi^2 = 242.6$; df = 2; significant at 0.01 level.
[e]Farm/non-farm comparison: $\chi^2 = 5502.9$; df = 5; significant at 0.01 level. Within farm-categories comparison: ($\chi^2 = 1470.0$; df = 10; significant at 0.01 level.

non-agricultural workers, agricultural workers are slightly overrepresented within two-person households, and again slightly more in very large households. Comparing within the farming classes, very few employer farmers live alone, whereas one in ten farm employees does. Forty per cent of owner-operator (self-employed) farmers live in two-person households, slightly more than either of the other classes.

The percentages of the various categories of individual workers who fully own or are purchasing their own homes, compared with those who are renting or have other similar housing arrangements, are shown in Table 12.3. Compared with non-agricultural workers, workers in agriculture have a slightly higher propensity either to be involved in or to aim to achieve private property in housing. A very low percentage of farm employees is in the owner/purchaser category (two-thirds compared with nine-tenths for the other two classes) and a high proportion (one in three) of the former pay rent.

The size of house lived in is also indicated in Table 12.3. Agricultural workers are slightly overrepresented in very small living quarters compared with non-agricultural workers, and they also seem to have more than their share of large houses (with four bedrooms, or more), but by contrast they take fewer than their 'fair share' of one- and two-

Table 12.3. Dwelling characteristics and household motor vehicle availability among surveyed individuals[a]. (Source: Australian Bureau of Statistics, 1996 Census.)

	Agriculture and service to agriculture					
	Employer	Self-employed[b]	Employee	Total	All non-agricultural	Overall total
Tenure type of dwelling[c]						
Fully owned or being purchased	15,589 (90.4)	51,328 (90.4)	89,177 (68.4)	156,094 (76.4)	3,062,211 (71.4)	3,218,305 (71.6)
Rented or other	1,647 (9.6)	5,454 (9.6)	41,201 (31.6)	48,302 (23.6)	1,228,063 (28.6)	1,276,365 (28.4)
Total	17,236	56,782	130,378	204,396	4,290,274	4,494,670
Number of bedrooms in dwelling[d]						
None[e]	46 (0.3)	258 (0.5)	932 (0.7)	1,236 (0.6)	19,250 (0.4)	20,486 (0.5)
1	186 (1.1)	856 (1.5)	3,965 (3.0)	5,007 (2.4)	140,240 (3.3)	145,247 (3.2)
2	1,210 (7.0)	5,308 (9.3)	17,050 (13.0)	23,568 (11.5)	780,318 (18.1)	803,886 (17.8)
3	7,559 (43.7)	28,652 (50.3)	66,843 (51.0)	103,054 (50.2)	2,169,137 (50.4)	2,272,191 (50.3)
4+	8,308 (48.0)	21,873 (38.4)	42,245 (32.2)	72,426 (35.3)	1,198,733 (27.8)	1,271,159 (28.2)
Total	17,309	56,947	131,035	205,291	4,307,678	4,512,969
Number of motor vehicles per household[f]						
No motor vehicle	93 (0.5)	287 (0.5)	3,551 (2.7)	3,931 (1.9)	198,491 (4.7)	202,422 (4.6)
1	2,010 (11.7)	8,423 (14.9)	31,896 (24.1)	42,329 (20.9)	1,474,860 (34.8)	1,517,189 (34.2)
2 or 3 vehicles	10,959 (64.1)	38,724 (68.7)	77,953 (60.4)	127,636 (63.1)	2,370,136 (56.0)	2,497,772 (56.2)
4+	4,048 (23.7)	8,978 (15.9)	15,624 (12.1)	28,650 (14.1)	192,569 (4.5)	221,219 (5.0)
Total	17,110	56,412	129,024	202,546	4,236,056	4,438,602

[a]Actual number and column percentages (percentages shown in parentheses).
[b]Includes only those self-employed farmers (owner-operators) with no employees.
[c]Farm/non-farm comparison: $\chi^2 = 2391.9$; df = 1; significant at 0.01 level. Within farm-categories comparison: $\chi^2 = 6190.1$; df = 2; significant at 0.01 level.
[d]Farm/non-farm comparison: $\chi^2 = 9215.9$; df = 4; significant at 0.01 level. Within farm-categories comparison: $\chi^2 = 615.0$; df = 8; significant at 0.01 level.
[e]Including bedsitters.
[f]Farm/non-farm comparison: $\chi^2 = 51,599.8$; df = 3; significant at 0.01 level. Within farm-categories comparison: $\chi^2 = 1965.9$; df = 6; significant at 0.01 level.

bedroom houses. Comparing within farming classes, employees are the category who make the most use of 'bedsitters, etc.', and they predominate in their use of small houses as well. Again, employer farmers seem to have the highest standard of living and their use of large houses is almost twice as frequent as for non-farm workers.

Finally, Table 12.3 shows the proportions of households that have a number of motor vehicles available to them (probably, but not clearly, as farm establishments, where appropriate). Relative to non-agricultural workers, agricultural workers are less likely to have no or one vehicle, whereas higher proportions have more than one vehicle per household. A fair proportion of households in agriculture has upward of four vehicles. Almost one in four farm employees has one vehicle available within their household, as compared with 11.7 and 14.9% of the other two categories in agriculture. Owner-operators have two or three vehicles available to more than two-thirds of them, and again, the employer farmers fare especially well, with almost one in four having four or more vehicles at the disposal of their household – more than four times the proportion that prevails within the Australian population overall.

On the basis of these data, it seems fair to make several generalizations. Farm people as a group are disadvantaged compared with the non-farm population, but not all farm workers endure the low standards. In Australia many farm employers do well all the time, and a lot do well most of the time. A proportion of the farm owner-operators are 'poorer off' than the national non-farm standard. Nevertheless, if there is a group that systematically bears the brunt of rural deprivation, it is the farm employees. Future analysis of farm deprivation clearly needs to disaggregate farm worker categories.

In addition, this analysis of objective class differences suggests the need for a consideration of subjective concerns about the consequences of these differences. The census data are unsuitable for investigating interpersonal interactions and relationships, but this chapter now argues that further study of paid agricultural employees from the point of view of their place in the social structure has potential for illuminating an underresearched category of Australian workers, and for elucidating aspects of workplace interaction within a rapidly changing agriculture. To do that, it is necessary to step away from the demographic level of explanation, and relate attitudes and commitments to features of respondents' workplace situation in an effort to deepen our understanding.

Attitudes and commitments

Investigation is now made of data collected in 1984–1986 for 'a sample selected in two components: an urban sample of localities of 10,000 or more persons as of the 1981 census; and a complementary rural sample using the electoral roles as the sample frame' (Kelley *et al.*, 1987). The urban sample was randomly selected. Within each dwelling, one person aged 18 years or older was interviewed (in English) to represent the household. The rural sample covered the rest of the Australian population, by sampling electoral divisions in clusters in proportion to the size of their rural population.

The surveys were carried out to examine public attitudes and beliefs about current social, political and economic events, about life and the way respondents experience it, and about sources of personal and social support and satisfaction (Kelley *et al.*, 1987). Information covered attitudes toward how much money is spent on foreign aid, defence, unemployment, medical and social services, among other topics; other issues such as taxation, inflation, crime and punishment, business affairs, and so on; confidence in institutions such as banks, police and government; and perceived economic and social priorities for Australia over the next 10 years. Other questions related to the respondents' personal feelings about life; health, the need for medical services, and availability of trustworthy friends; religious beliefs and priorities; and political and economic data. Background variables included family history, education, birthplace and other similar variables (Kelley *et al.*, 1987). Only a selection of these responses is used in the analysis.

Again, a person's work status is characterized as employer, self-employed and employee, and by industry. Having established distinct and objectively different categories, the analysis now investigates their subjective consciousness of differences among themselves. A total of 140 farm people from among 6203 respondents provided data. The exposition has two parts: (i) their political and philosophical affiliations; and (ii) their personal satisfactions.

Some questions probe respondents' beliefs with such questions as, 'Do you favour government involvement in [redistributive activity]?' An agreeable response indicates a social concern for less privileged members of Australian society. Direction and intensity of beliefs about these issues were assessed with a fixed-format response framework ranging from 'strongly agree' to 'strongly disagree'. A representative selection of responses (Table 12.4) depicts higher percentages of employees than self-employed or employer farmers in favour of redistributive social programmes, keeping unemployment under control rather than reducing inflation (if a choice between these two must be made), protecting local industry, and maintaining (or even increasing) government spending (Table 12.5). When this ideological commitment is pushed a little further, again higher proportions of paid agricultural employees wish the government to own electricity supply utilities, the steel industry and local public transport; higher proportions of paid agricultural employees also wish the government to control prices charged by these providers, while lower proportions of paid agricultural employees wish for regulation, or for privatization. No choice of government ownership was offered for banks or the automobile industry, but, among the choices that were offered, here too the pattern is the same.

In contrast, Table 12.6 reveals little difference between the classes in terms of those proportions who are satisfied with life or feel happy, though lower proportions of paid agricultural employees classify themselves as competitive or ambitious. Considering satisfaction with income, the most significant figure is the 47% of self-employed farmers who are unsatisfied. Also, a slightly lower proportion of paid agricultural employees reports having good health and a slightly higher percentage describes their health as poor.

Table 12.4. Percentage in favour of government involvement in social policy, compared by farmer class. (Source: National Social Science Survey, 1987.)

	Employer	Self-employed, no hired workers	Employees
Wealth redistribution	24	20	40
Undecided	12	20	30
Tax the rich more heavily			
Strongly agree	13	12	27
Agree	18	21	30
Decrease government spending	74	73	57
Protect industry			
Agree	57	64	74
Disagree	40	22	8
Inflation control *or*	70	48	43
Unemployment reduction	14	19	40
Favour increased government spending on:			
Health	24	33	40
Education	16	16	23
Old people	35	47	68
Unemployed	22	16	17

Table 12.5. Percentage in favour of government ownership, compared by farmer class. (Source: National Social Science Survey, 1987.)

	Employer	Self-employed, no hired workers	Employees
Electricity supply			
Own it	27	26	38
Control prices	47	35	50
Regulate	13	13	6
Promote private ownership	13	26	6
Steel industry			
Own it	7	0	25
Control prices	14	13	31
Regulate	14	26	6
Promote private ownership	66	61	38
Local public transport			
Own it	13	4	25
Control prices	30	30	44
Regulate	17	26	12
Promote private ownership	40	39	19
Automobile industry			
Own it	–	–	–
Control prices	27	13	47
Regulate	13	22	6
Promote private ownership	60	65	47
Banks			
Own it	–	–	–
Control prices	33	35	56
Regulate	20	22	0
Promote private ownership	14	26	6

Certain consistencies in responses are evident. These results provide ample justification for believing that the three classes lead lives that are materially different and are motivated by opinions that derive from fundamentally different social philosophies. Employees live in poorer physical and financial circumstances, and their responses to their own lives and the possibilities for change tend to derive from what are regarded in Australia as radical opinions, and as 'more progressive' policy prescriptions for social ills. On the face of it, this throws the Deferential Worker thesis (Newby, 1977) into some disarray.

This class distinctiveness is usually lost when resorting to the use of a catch-all expression such as 'country people', and a strong case can be made for the need to stretch beyond the demographics of farm worker categories to an empirical investigation of paid agricultural employees, in an attempt to position them socially, both within the rural context (where their special concerns tend to become lost in the equation of 'rural dweller' with farmer and family of a farmer) and within the wider society of wage earners, especially those also involved in transporting the products of agriculture and transforming them into food.

The principal issues at stake here are two. This study stresses the reality of paid agricultural employees as a category of agricultural workers who respond to life and work in unique ways that differentiate them from the owners and operators of farms, and whose distinctiveness is usually lost in expressions like 'country people'. No attempt is made or implied to deny the validity of considering social relationships among agricultural producers and their families, but it seems too simple, on the face of it, to believe that a category defined simply as 'farm persons' should be taken for granted, or, for

Table 12.6. Personal satisfactions, compared by farmer class (percentage). (Source: National Social Science Survey, 1987.)

	Employer	Self-employed, no hired workers	Employees
'I'm satisfied with life'			
Yes	48	43	48
No	43	48	38
'I feel happy'			
Very	17	16	20
Considerably	75	75	77
Somewhat	8	9	3
'I'm competitive'			
Yes	53	36	33
No	31	34	30
'I'm ambitious'			
Yes	61	41	32
No	18	27	37
'I'm satisfied with my income'			
Yes	63	37	57
No	18	47	29
'My health is...'			
Good	69	71	60
Fair	7	3	10
Poor	24	26	30

example, that it should be anticipated that 'agrarianism' (Craig and Phillips, 1983) would be its ideology. The responses to surveys of farm employees are just as relevant to a clear description of Australian society, and their concern for matters of agricultural policy just as significant as those of the more frequently canvassed farm owners. Their apparently increasing numbers make their case even stronger.

There is a matter of industrial policy at issue, as well. As Winter (1984) pointed out, a study of the rural-dwelling population can also be turned to 'an understanding of social stratification and property relations in society at large'. Admittedly, as Havens and Newby (1986, p. 293) remarked, 'In reality, the state can follow only a limited number of effective policies without causing general socio-economic disruption,' but sometimes governments need approval for the policies they enact, and some groups (in this case farm employees just as much as their employers) need to be offered the chance to identify themselves as a constituency. The results of this analysis suggest that a class categorization, as specified by Wright (1978), generates a model of relationships, social structural categories and policy preference and outcomes with some empirical relevance, well deserving further study.

Evidence from case studies: sugarcane and citrus fruit in Queensland

Notwithstanding the argument about 'class' above, it cannot be assumed that all segments of farm employees are alike, nor that policy made 'in their interest' would affect everyone in the same way. Field study provides further depth into understanding who works as paid employees on Australia's farms. Different employee categories can be differentiated, for example, with the sugarcane and fruit industries, as described below.

Canecutters

In a recent study of the sugar industry around the cities of Bundaberg and Mackay on Australia's northeast coast, Finemore and McAllister (1999) quoted changes within the harvest labour force for that industry:

> [In the hand-harvesting era], canecutters generally worked under a piecework system, i.e.: they contracted to cut a set number of ton[ne]s in a season and were paid at a fixed rate for each ton[ne] cut. In the far northern region ..., a gang of up to eight men would sign a contract with one or more farmers in the one area to cut an agreed tonnage. The earnings would be shared equally amongst members (Balanzategui, 1990, p. 78).

Recently, agricultural contracting has become an alternative route for capital's subsumption of agricultural surplus: cane harvest contracting has come to function like the more traditional sheep-shearing contracting. As the Mackay producers characterize them, contractors are seldom merely 'bright young entrepreneurs'; they are 'farmers' sons, farmer/contractors, [and/or] farmer-run' (Mackay focus group member). In short, contractors may be farmers (with varying sized enterprises) supplementing or diversifying their income through contracts with neighbours; 'second sons' (*sic*) of farmers, financed by a lien on the parental property; or independent business people seeing the possibility of a livelihood; and so on. Their harvest labour employees are recruited either locally or from among itinerant labourers. Sugar has traditionally also provided employment for full-time farm hands, who tend to reside locally and make their lives in the country towns, which seems to be becoming more the case for harvest workers as well. Within the citrus fruit industry in a nearby irrigated area, it was noticed that the harvest labour force consisted of both local-living and itinerant pickers (McAllister, 1998).

Local citrus pickers

Within the towns where we interviewed people and photographed (McAllister and Griffin, in preparation), a certain proportion of the population worked in agricultural pursuits. For most of the year, their work is farm labour, which is particularly poorly remunerated at less than Aus$10 an hour. When the fruit-picking season comes, many of them see the possibility of 'tipping over' into the piece-work of fruit picking and making much more money than they normally would in their 'hourly rate' jobs. There is a stabilizing influence associated with becoming fruit pickers in the season: a person might be prepared to accept Aus$9 per hour on a daily basis for (say) 7 months of the year if they can become a piece-work fruit picker and earn considerably more for the other 5 months. Thus, over the year, income averages to a tolerable life-support. From the point of view of the employer, this procedure has all the advantages for the home farm associated with people who work in the district and are controlled by the local norms of what communities do particularly well: social control, provision of housing and social networks, and so on.

Workers from around the local town also become fruit pickers in season. They are tradespeople and seasonal builders, 'backyard' mechanics, house painters and 'timber workers', they look after livestock, and do various other jobs locally. When the fruit season comes, they drop those jobs and go off to make what they persistently called 'good money' in fruit picking, without having to leave home; they travel back and forth to work, while using all the amenities of their own homes. They are single males and females, and couples. A few young men even said that they had done this kind of work in their weekends when they were schoolchildren, so that they could make pocket money and begin their working lives within their own district.

We found that in an effort to ensure a stable supply of farm workers who live in town and to make their farms attractive to people as a place to work full-time, some growers were attempting to produce crops for as much of the year as possible. Although the growers were not offering full-time work as such, they were offering year-round work, although it was always 'for hire on a daily basis', except in the fruit-picking season when the hire was by 'contract' by piece-work.

About a third of the 30 or so interviewees mentioned that they had been on the itinerancy circuit when they made the calculation that there was enough work in Gayndah to allow them to remain in the region, working year-round and becoming fruit pickers in that part of the year when harvesting was available to them. On the other hand, the manager of the large farm at Emerald pointed out that he is attempting to install an 11-month harvesting regime on his property – not just of citrus, but also of grapes and other crops – so that he will be able to attract a full-time 'dependable' labour force to live in the town. This dependability seems to be closely associated with small community living.

ITINERANT FRUIT PICKERS. Fruit picking is a hard physical job but it does not require much education nor, on the face of it, a specialist training. Thus, people can come into this occupation at any age and with any amount of other skill. Itinerant pickers fall predominantly into the following three categories:

- People now working full-time in fruit picking because they have left some other activity, for all sorts of reasons: they got tired of their prior work, they had lost their job, or they had resigned. Without any pejorative being intended, these people might be called 'drop-outs' from their previous work and 'drop-ins' to fruit picking.
- A 'contingent' of people who appear to have 'fallen' into this line of work. They have no particular ambition in regard to their fruit picking – it is just 'what they do', and they make a living from it. Their income equals their expenditure. At the end of a season they have no savings and no recourse but to start the circuit over again.
- The deliberate full-time fruit pickers. This is the work that they have decided on, and they did so because they had heard there was good money in it. They set about putting together a 'nest egg', working extremely hard in conditions uncongenial to human existence, living rough and travelling precariously, but, while they work, they save as much as they possibly can towards the later fulfilment of some particular dream. So, in a sense, for them the work they are currently doing is a 'postponement of life'. This is their work; their life is something else. It is part of their 'dream'. People talked about buying their own business, buying a block of farm land, or retiring on their savings.

There are two additional categories of itinerant workers:

- Immigrants to Australia, with poor English skills. In general, the labour market is particularly restricted for them, but they are able to find in fruit picking a style of work that does not require English capability (this needs to be qualified by a realization that workplace health and safety incidents are increased by poor communication skills on the job). Provided they can move around the country, they can find remunerative work. There is for them the danger that their limited facility with English will be exploited by their employers and others.
- Those full-time workers whose ambition was to put aside enough money to buy themselves a small block of ground but were unsuccessful. A number of pickers said they came from places where they have land of their own, or have contracts to buy. These 'Blockies' (as they are known locally) settled on their 'block' from trade jobs, and on pensions and so on, at various times within the last 15 years, thinking (they recalled) that they would be able to find or create work locally (maybe working part-time at their trade) and develop their farmlet so that they might run a 'pleasure horse' or livestock farm of various sorts, but where they would be independent of the need to go off to work. The unavailability of work in rural subdivisions (especially in drought times) made it necessary for them to go further afield to find a means of sustenance, particularly if they had (young) families. Fruit picking has represented one possibility of leaving home for a short time (a season in the Gayndah

area might be 5 months), accumulating their earnings while working away and then supporting their family on that money either for the rest of the year or until it ran out. Their experience suggests that those of the full-time pickers who planned to set themselves up on a small retirement block have this kind of 'rude awakening' still in store.

BACKPACKERS. Young hikers tend to be also called 'blow-ins' by the more established pickers in the industry. Their intention is generally to make just a little money and then go, either to continue with their tourist activities or on to some other fruit-picking area. Demographically, they tend to be young and the 'backpack' describes their mode of travel. They are usually either overseas visitors to Australia or young Australians, out for the big adventure. Among the established pickers, the backpackers were not well-liked, as a general rule, nor were they favourites with farmers. A number of farmers pointed out that they had had raids on their properties by inspectors from the Department of Immigration and Ethnic Affairs (DIEA), who saw it as their responsibility to prevent undocumented backpackers and certain other people from working. That is, DIEA employees made farm visits to check people's documentation to ensure either that they were not wanted by the law or that they were permitted to work in Australia.

Backpackers have a reputation for being prepared to accept a lower piece-rate than the majority of fruit pickers want. Needing to make only a certain amount of money, they are prepared to accept a low return for their labour.

RETIREE PICKERS. Retiree pickers fall into two categories. Some are victims of the recent recession – people who have lost their jobs at a relatively mature age, perhaps from about 40 onwards. Many seem to have been fairly heavily committed financially, to the point where they just could not afford to live on social security, and would be required to sell their assets. Working as couples in fruit picking, they might be able to save the price of a caravan and move to where their work is, or otherwise to camp in 'caravan parks' while on the fruit-picking circuit.

The other category of retirees have sold their existing businesses. They bought a Land Rover-type vehicle and a caravan, invested the 'nest egg' that they had from selling their business, and living frugally, went off to tour the country. They discovered that travelling is not inexpensive, but fortunately also discovered that fruit picking was available and decided that an appropriate way to see the country, for a little while at least, would be to 'work the fruit circuit'.

These two categories have the similarities that they are well into middle age; perhaps the former are from a lower socio-economic grouping. Since they are not attempting to accumulate large amounts of money, they can probably work at a pace that suits people of their more mature years and 'cobble together' a reasonable style of living. Being prepared to work at a sedate pace makes them popular with the farmers. The backpackers (in particular) and other itinerants who want to work as hard as they can for many hours, and to take away from the job as much money as can be made, are disinclined to be especially responsible for the crop they pick. By contrast, the older and more financially secure pickers are glad to work more slowly and can take more detailed instructions about the quality of crop the farmer wants. They point out that when they came to the job at the beginning, they were not necessarily particularly adept at it: they were not especially fit, they were unable to work the kind of pace and hours that younger people were, but nevertheless they found that they were able to become quite fit and numbers of them spoke about enjoying the work very much.

Discussion and Conclusion

It is clear that legislation, and the policy it embodies, may both follow and set public thinking. Since the early 1980s, Australian legislators have passed a variety of laws and changes specifically targeting various agricultural industries. At the same time, the agricultural sector has had to adjust to the consequences of a variety of other policies that have unavoidable consequences for farming people.

Overall, this chapter has addressed two major aspects of agricultural labour: (i) the conditions of life and work of agricultural employees on both large-scale farms and in small operations; and (ii) how the employees' conditions differ from those of their farmer employers, and from those of the farm owner-operators who live and work in the same production environment but appear to employ only immediate family members. The chapter has argued that, while it is acknowledged that there is devastating poverty within certain sectors of Australia's urban areas, nevertheless low incomes are concentrated in rural areas, where rural sociology has been concerned mostly with (owner-operator) farmer lifestyles and livelihoods. It cannot be denied that commodity prices and seasonal conditions frequently create bad times for farm owners/commodity growers but this chapter has attempted to show that a class disaggregation of the rural population brings needed attention to the farm wage and salary earner category. Their objective living conditions are shown to be poorer than those of the other farm work categories (owner-operator and employer farmers), and their appreciation of their opportunities for economic and social improvement reflects limited capacity for change within their work environment.

Nevertheless, field study has shown that the lives they create can be rich and varied and that, given the necessity to work, this work is something of a choice and many employees are content to make a living at their work. It is also observed that a variety of subcultures of work (and attendant lifestyles) exist between industries and often within farm commodity frameworks. The panorama of work situations and workforce responses has been sketched only in outline but the argument serves to focus on an under-theorized category of Australian workers, and to develop further the picture of agricultural and rural life emerging within Australian rural sociology at the present time (see, for example, *Rural Sociology*, 1999, 64(2)). Clearly, further research is warranted for a cross-section of farm products and work situations, and more up-to-date attitude data would show how farm employees have responded to more recent changes to social policy and recruitment initiatives within the commodity systems in which they work.

Acknowledgements

This chapter makes use of data supplied by the Australian Bureau of Statistics, and these are acknowledged within it. Data used in part of this analysis are from the National Social Science Survey (Kelley *et al.*, 1987), supported by grants from the Research School of Social Sciences at The Australian National University and the Australian Research Grant Scheme, and distributed by the Social Science Data Archives at the Australian National University. Field research was performed by Michael Finemore and by the author. Many thanks to colleagues from the School of Psychology and Sociology at Central Queensland University for assistance with funding to attend the conference, *The Dynamics of Hired Farm Labour: Constraints and Community Responses*.

References

Alston, M. (1992) *Rural Women*. Centre for Rural Welfare Research, Charles Sturt University, Wagga Wagga, New South Wales.

Australian Bureau of Statistics (1996) *Census of Population and Housing: Selected Family and Labour Force Characteristics*. ABS Bulletin No. 2017.0. Australian Government Publishing Service, Canberra.

Australian Bureau of Statistics (1997) *Year Book Australia, No.* 79 (ABS Catalogue No. 1301.0). Australian Government Publishing Service, Canberra.

Balanzategui, B.V. (1990) *Gentlemen of the Flashing Blade*. Department of History and Politics, James Cook University, Townsville, Queensland.

Barlett, P. (1986) Profile of full-time farm workers in a Georgia county. *Rural Sociology* 51, 78–96.

Constance, D. (1988) Obstacles or detours: the broiler industry revisited. Paper presented at the 1998 Rural Sociological Society Conference, Athens, Georgia.

Craig, R.A. and Phillips, K.J. (1983) Agrarian ideology in Australia and the United States. *Rural Sociology* 48, 409–420.

Dempsey, K. (1990) *Smalltown: a Study of Social Inequality, Cohesion and Belonging*. Oxford University Press, Melbourne.

Duff, J. (1999) Unemployment in rural Australia: the uneven social consequences of economic restructuring. Paper presented at the Australian Sociological Association Conference, Melbourne, Australia.

Finemore, M. and McAllister, J. (1999) Hiring labour for sugar harvesting: farmers, farm workers and sub-contractors. In: Burch, D., Goss, J. and Lawrence, G. (eds) *Restructuring Global and Regional Agricultures: Transformations in Australasian Agri-food Economies and Spaces*. Ashgate, Aldershot, UK, pp. 237–252.

Garnaut, J. and Lim-Applegate, H. (1998) *People in Farming*. Australian Bureau of Agriculture and Resource Economics, Canberra.

Gill, H. (1981) Land, labour or capital: industrial relations in the Australasian primary sector. *Journal of Industrial Relations* 23, 139–162.

Gray, I., Lawrence, G. and Dunn, A. (1993) *Coping with Change: Australian Farmers in the 1990s*. Centre for Rural Social Research, Charles Sturt University, Wagga Wagga, New South Wales.

Havens, A.E. and Newby, H. (1986) Agriculture and the state: an analytical approach. In: Havens, A.E., Hooks, G., Mooney, P.H. and Pfeffer M.J. (eds) *Studies in the Transformation of US Agriculture*. Westview Press, Boulder, Colorado, pp. 287–304.

Heffernan, W.D. (1972) Sociological dimensions of agricultural structures in the United States. *Sociologia Ruralis* 12, 481–499.

Higgins, V. (1997) Breaking down the divisions? A case study of political power and changing rural local government representation. *Rural Society* 7, 51–58.

Hudson, P. (1984) Provision of services to the aged in small rural communities. *Urban Policy and Research* 2, 40–41.

Hungerford, L. (1996) Australian sugar and the global economy. In: Burch, D., Rickson, R. and Lawrence, G. (eds) *Globalization and Agri-Food Restructuring: Perspectives from the Australasia Region*. Avebury Press, London, pp. 127–138.

James, K. (1989) *Women in Rural Australia*. University of Queensland Press, St Lucia.

Jones, J. (1998) Corporate barons in major rural property reshuffle. *Australian Farm Journal* 8, 20–44.

Karmel, T., Andrews, L., Ryan, C., Unglers, P., Dearson, L., Whittingham, B., Pawsey, A. and O'Reilly, B. (1991) *Australia's Workforce in the Year 2001*. The Economic and Policy Analysis Division of DEET, Australian Government Publishing Service, Canberra.

Kelley, J., Cushing, R.G., Headey, B. and associates (1987) *National Social Science Survey; SSDA 423*. Social Science Data Archives, Australian National University, Canberra.

Lawrence, G. (1987) *Capitalism and the Countryside: the Rural Crisis in Australia*. Pluto Press, Sydney.

Lawrence, G. and Gray, I. (1999) The restructuring of Australia's farm population. *Agri-food VII*, The Australia and New Zealand Agrifood Association 1999 Conference. University of Sydney, New South Wales.

Lockie, S. and Collie, L. (1999) 'Feed the man meat': gendered food and the conceptualisation of consumption in agri-food research. In: Burch, D., Goss, J. and Lawrence, G. (eds) *Restructuring Global and Regional Agricultures: Transformations in Australasian Agri-food Economies and Spaces*. Ashgate, Aldershot, UK, pp. 255–274.

Lockie, S. and Vanclay, F. (1997) *Critical Landcare*. Centre for Rural Social Research, Charles Sturt University, Wagga Wagga, New South Wales.

Lyons, K. (1996) Agro-industrialisation and rural restructuring: a case study of the Australian poultry industry. In: Lawrence, G., Lyons, K. and Momtaz, S. (eds) *Social Change in Australian Agriculture*. Rural Social and Economic Research Centre, Central Queensland University, Rockhampton, Queensland, pp. 167–177.

McAllister, J. (1998) The human face of harvest work: citrus picking in central Queensland. Paper presented at the Australian Sociological Association Conference, Queensland University of Technology, Gardens Point, Brisbane.

Nalson, J. (1977) Rural Australia. In: Davies, A.F., Encel, S. and Berry, M.J. (eds) *Australian Society: a Sociological Introduction*. 3rd edn. Longmans Cheshire, Melbourne, pp. 304–333.

Nankervil, P. (1980) Australian agribusiness: structure, ownership and control. In: Crouch, G., Wheelwright, T. and Wiltshire, T. (eds) *Australia and World Capitalism*. Penguin Books, Ringwood, Victoria, pp. 160–168.

Newby, H. (1977) *The Deferential Worker*. Allen Lane, Harmondsworth, UK.

Parsons, H. (1996) Supermarkets and the supply of fresh fruit and vegetables in Australia: implications for wholesale markets. In: Burch, D., Rickson, R. and Lawrence, G. (eds) *Globalization and Agri-Food Restructuring: Perspectives from the Australasia Region.* Avebury Press, London, pp. 251–270.

Rickson, R. and Burch, D. (1996) Contract farming, organizational agriculture and its effects upon farmers and the environment. In: Burch, D., Rickson, R. and Lawrence, G. (eds) *Globalization and Agri-Food Restructuring: Perspectives from the Australasia Region.* Avebury Press, London, pp. 173–202.

Ruben, A.M. (1992) A study of seasonal rural workers in an advanced capitalistic society: the fruit pickers in Victoria, Australia. PhD Dissertation, La Trobe University, Bundoora, Victoria.

Saunders, P. (1996) Poverty and deprivation in Australia. *Year Book Australia, No. 78* (ABS Cat. No. 1301.0). Australian Bureau of Statistics, Australian Government Publishing Service, Canberra, pp. 226–240.

Thomas, R. (1981) The social organization of industrial agriculture. *The Insurgent Sociologist* 10, 20–26.

Vanclay, F. and Lawrence, G. (1995) *The Environmental Imperative: Socioeconomic Concerns for Australian Agriculture.* Central Queensland University Press, Rockhampton, Queensland.

Willis, E. (1998) *Technology and the Labor Process: Australasian Case Studies*. Allen and Unwin, Sydney.

Winter, M. (1984) Agrarian class structure and family farming. In: Bradley, T. and Lowe, P. (eds) *Locality and Rurality.* Geo Books, Norwich, UK, pp. 115–128.

Wright, E. (1978) *Class, Crisis and the State*. New Left Books, London.

13 The Role of the State in Manitoba Farm Labour Force Formation

Avis Mysyk
Department of Anthropology, University of Manitoba, Winnipeg, Manitoba, Canada

Abstract
This chapter examines the Canadian state's role in the development of Manitoba agriculture and specifically in the formation of ethnically based class fractions of the Manitoba farm labour force through a combination of agricultural, Indian and immigration policy. It argues that the state acts primarily in the interests of monopoly capital (agribusiness) and only secondarily in the interests of competitive capital (farmers) and the working class (farm labourers).

Introduction

> The maintenance of a surplus of workers at the bottom [of the employment ladder] is extremely costly to the nation in unemployment benefits, welfare assistance, loss in tax collections and in weakening the productive capacity of the country, to say nothing of the social evils created by idleness.
> (Manitoba Department of Agriculture and Immigration, 1959, p. 88.)

According to O'Connor (1973), the state must fulfil two basic and sometimes contradictory functions: accumulation and legitimization. Panitch (1980) would add a third: coercion. The task is far more difficult than it first appears. To fulfil the functions of accumulation and legitimization, the state must manage relations not only within and between various classes ('monopoly capital', 'competitive capital', the working class) and between various classes and the state, but also between various levels (federal, provincial, municipal) of the state itself. O'Connor (1973) defines monopoly capital as a high ratio of capital to labour and high productivity. In agriculture, suppliers of farm inputs (machinery, fertilizer, herbicides, seeds), wholesalers, retailers and food processors would be considered 'monopoly capital'. One might also include those groups who lobby for the interests of monopoly capital. O'Connor (1973, p. 13) also defines 'competitive capital' as a low ratio of capital to labour and low productivity. In agriculture, farmers would be considered competitive capital.

Although the state claims to act in the interests of society as a whole, it primarily acts in the interests of monopoly capital (O'Connor, 1973). To a lesser degree, it also acts in the interests of competitive capital

(eds J.L. Findeis, A.M. Vandeman, J.M. Larson and J.L. Runyan)

(in this case farmers), by creating and maintaining a reserve army of labour despite its expressed concern for the surplus of workers at the bottom of the employment ladder. The purpose of this chapter is to examine the state's role in the development of Manitoba agriculture and specifically in the formation of ethnically based class fractions (Native, Japanese, Mexican Mennonite, Mexican, Laotian) of the farm labour force through a complex combination of agricultural, Indian and immigration policies. 'Native' is a general term for those aboriginal groups – Indian, Métis and Inuit – that are officially recognized by the Canadian government. The historical misnomer, 'Indian', is slowly being replaced by the term 'First Nations'.

Canadian agricultural, Indian and immigration policies have not always coincided in the past, nor do they today. None the less, some generalizations can be made.

A Brief Overview of Manitoba Agriculture

Over time, Canadian agriculture has become regionally specialized. Atlantic Canada is dominated by horticulture, especially potatoes, fruit and vegetables; Quebec and Ontario by dairy and pork production and horticulture; the Prairies by grains (particularly wheat) and cattle. Their historical importance was captured succinctly by Morton (1985, p. 28): 'The successor to prairie grass and buffalo was wheat and beef.' British Columbia is the most diversified.

Compared with the Prairie provinces of Saskatchewan and Alberta, Manitoba has also become fairly diversified (Fig. 13.1). Wheat and cattle are still the most important agricultural commodities but canola and pork production are increasing. The dairy, poultry and egg industries are important in the provincial market, while the sugarbeet industry

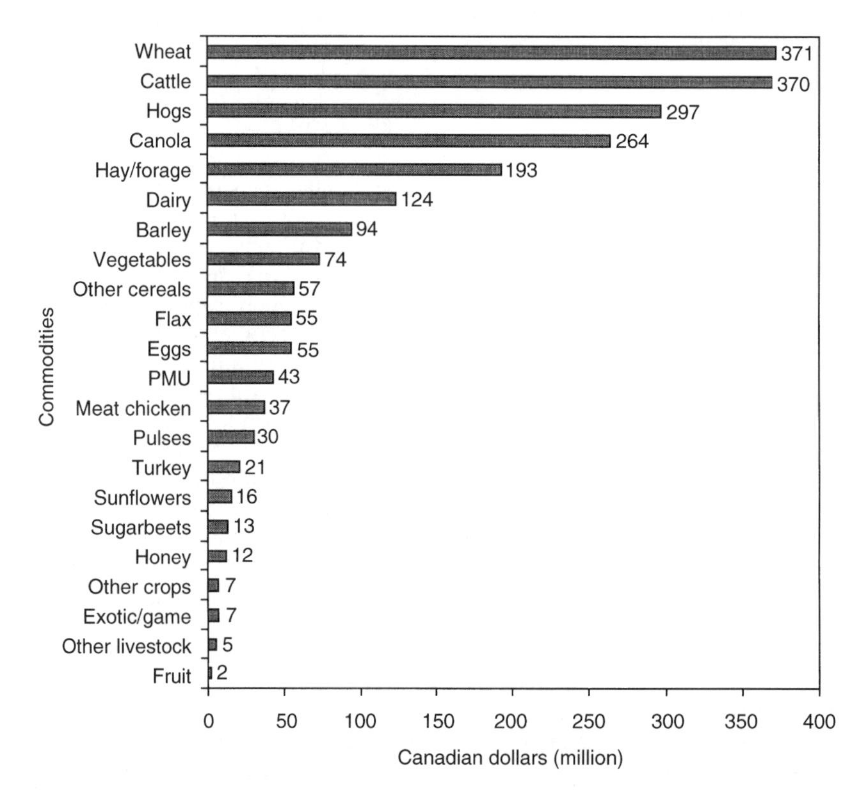

Fig. 13.1. Value of farm production in Manitoba, Canada, 1993. (Source: Agriculture and Agri-Food Canada and Manitoba Agriculture, 1994.)

and horticulture (primarily potatoes) enjoy wider, even international markets. Vegetable production, which is concentrated in south central Manitoba, is a comparatively small but robust industry (Table 13.1) comprising ten major growers, six of whom hire Mexicans. The industry's success has always depended upon access to 'elusive' markets and a plentiful supply of low-wage labour (Table 13.2). The map in Fig. 13.2 illustrates the regions listed in Table 13.2.

The Early Farm Labour Force in Manitoba

From 1867 until 1929, the federal government's primary goals were: (i) to settle the West; and (ii) to find an agricultural staple that could support financial and industrial interests in eastern Canada. The first goal was achieved through an immigration policy of 'white only or white if possible' (Hawkins, 1977, p. 78) which, incredible as it may seem, remained in place until 1962. Indian policy was closely tied to this goal as well. Initial attempts to isolate Indian peoples on reserves and/or assimilate them were institutionalized in the Indian Act of 1876. Its policy of 'coercive tutelage', which made Indian peoples wards of the state, lasted until 1969 (Frideres, 1999, p. 118). The second goal was achieved by focusing on the production of wheat, which seemed the best-suited of all crops to Prairie soils. Once a staple crop had been found, the federal government imposed tariffs on all agricultural supplies imported into western Canada from the USA in order to protect monopoly interests in eastern Canada. This imbalance of power between monopoly and competitive capital, which continues today, is known as the cost-price squeeze.

The actual origins of Manitoba agriculture are humble. Not until 1827 did it become firmly established in the Red River Colony where, until the 1930s, grains, root crops and summer crops were grown for local needs on river lots. To a greater or lesser extent, provincially sponsored societies and institutes such as the Brandon Experimental Farm (1888), the Manitoba Agricultural College (1905) and the Morden Research Station (1915) worked with farmers to increase agricultural productivity but left marketing to them.

During this time, one source of labour upon which farmers relied was that of the Ojibwa from Sandy Bay, Long Plain, Indian

Table 13.1. Value (Can$ '000)[a] of selected horticultural crops in Manitoba, 1936–1991. (Source: Manitoba Department of Agriculture, 1982, 1989, 1990, 1991.)

		Commercial vegetables		
Year	Potatoes	Fresh vegetables	Processing vegetables	Non-commercial vegetables
1936	13,670	na	na	na
1941	19,415	na	na	na
1946	15,021	na	na	na
1951	16,279	11,029	1,226	6,863
1956	12,209	14,444	4,444	7,778
1961	8,965	14,428	6,219	8,955
1966	17,658	5,255	1,590	8,993
1971	21,407	8,133	885	7,375
1976	26,504	5,739	752	5,639
1981	44,672	8,419	na	na
1986	33,577	10,150	na	na
1991	na	na	na	na

[a]In constant 1986 dollars.
na = not available.

Table 13.2. Hired agricultural labour by census agricultural region in Manitoba. (Source: Statistics Canada, 1982, 1987, 1992, 1997.)

	Year-round		Seasonal	
Census regions[a]	Number of farms reporting	Total weeks worked/workers per farm[b]	Number of farms reporting	Total weeks worked
Province				
1981	2,347	133,373/1.1	7,611	114,354
1986	2,617	167,867/1.2	10,798	174,839
1991	3,487	208,745/1.2	8,813	145,326
1996	3,992	290,052/1.4	7,430	146,805
Region 7				
1981	302	18,963/1.2	927	18,393
1986	306	24,062/1.5	1,313	33,171
1991	413	35,843/1.7	1,098	26,193
1996	528	45,880/1.7	915	25,774
Region 8				
1981	426	25,942/1.2	1,174	20,990
1986	479	34,379/1.4	1,819	31,575
1991	623	36,506/1.1	1,494	27,362
1996	740	53,570/1.4	1,214	24,150
Region 9				
1981	340	26,689/1.5	854	13,504
1986	418	32,310/1.5	1,198	19,058
1991	538	39,374/1.4	948	17,399
1996	554	52,641/1.8	870	20,697
Region 11				
1981	194	11,764/1.2	441	6,914
1986	200	13,293/1.3	629	9,602
1991	208	13,693/1.3	464	6,217
1996	230	19,509/1.6	384	8,013

[a]For a map showing these regions, see Fig. 13.2.
[b]Statistics Canada does not distinguish between the numbers of year-round and seasonal workers. The number of year-round workers per farm can be calculated in the following manner: (total number of weeks ÷ 52 weeks) ÷ number of farms = number of workers per farm. The number of seasonal workers cannot be calculated, since the number of weeks worked in a year is unknown.

Gardens, Swan Lake and Roseau River, and of the Dakota from the Dakota Tipi and Long Plain Sioux (Fig. 13.3). In the late 19th century, the federal government attempted to 'civilize' these primarily hunting peoples by settling them on reserves and transforming them into farmers. Aware that a scarcity of game would eventually force them to pursue other means of subsistence, some bands enthusiastically took up farming – as documented in the archival Canada Sessional Papers (CSP).

The Indian Gardens and Swan Lake bands initially showed great interest in growing cereal and root crops on their fairly fertile lands. The Long Plain band also tried to cultivate its lands even though the soil was very sandy. The Roseau River band tried to farm but the Dominion Government soon expropriated its best lands for the benefit of white farmers. Almost from the start, the band collected seneca root and worked in the harvest off-reserve, both of which proved more lucrative than farming (CSP, 1888, No. 15, p. 48; CSP, 1889, No. 16, p. 44). Two bands that did not farm were the Dakota Tipi and Sandy Bay. The Dakota Tipi, refugees of the Minne-

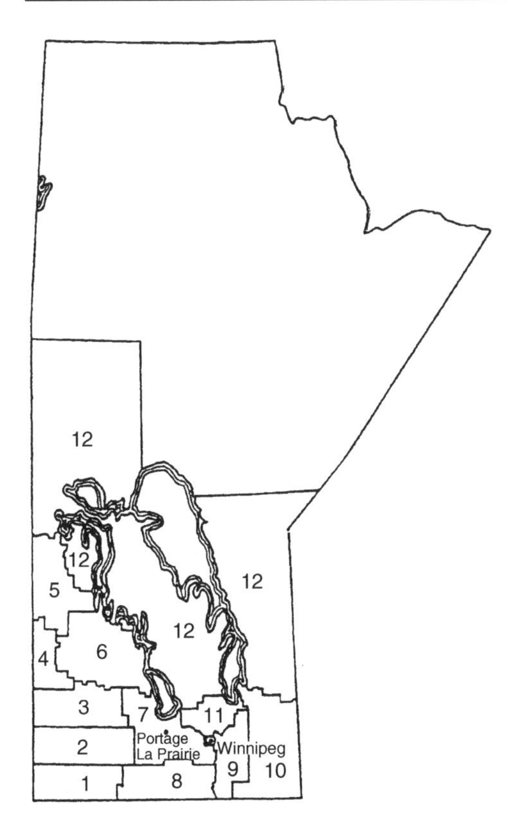

Fig. 13.2. Census agricultural regions in Manitoba, Canada, 1991. (Source: Manitoba Department of Agriculture, 1992.)

sota Uprising (1863–1864), were never given a reserve. Instead, they supported themselves as casual workers in Portage la Prairie or as farmhands in its vicinity, where their labour was much in demand and well remunerated for the time (Howard, 1984; Elias, 1988). The lands of the Sandy Bay band, located on the southwest shore of Lake Manitoba, proved unsuitable for agriculture. Even so, the band supported itself so successfully by fishing, trapping, digging seneca root, raising cattle and working on grain fields in southern Manitoba that, in 1900, Indian Agent Swinford (CSP, 1901, No. 27, p. 89) was able to report: 'They are always well dressed and fat, which is the best proof that their resources and occupations are manifold and profitable.'

Despite the willingness of some bands to farm, the federal government was often slow to provide assistance beyond what was stipulated in Treaties 1 and 2, i.e. one plough and one harrow per family and one ox per band (National Archives of Canada, RG 10 Black, v. 3721, f. 23715). What supplies the bands did receive were often of inferior quality (CSP, 1879, No. 7, p. 55). In spite of these problems, Indian agriculture advanced to such a degree throughout the 1880s that white farmers began to complain about unfair competition. In response, the federal government began to enforce a policy of 'peasant farming' of 4 acres (1.6 ha) and one cow per Indian family, thereby restricting their land base, their access to labour-saving technology and, ultimately, their ability to compete with white farmers (Carter, 1990).

The peasant farming policy had the desired effect. By 1900, agriculture no longer formed the basis of the reserve economy. Extensive tracts of reserve lands in and around Portage la Prairie lay idle for lack of agricultural machinery (CSP, 1901, No. 27, p. 85), and the Dominion Government had sold more than half of the fertile lands of the Swan Lake and Roseau River bands to white farmers (CSP, 1909, No. 27, pp. 108–109). Indians, including those from Sandy Bay and Dakota Tipi, were proclaimed to be a 'general nuisance' who would fare better if they were relocated to areas more suitable to their 'natural mode of life' rather than interfere with the settlement of western Canada (CSP, 1907–1908, No. 27, p. 102; CSP, 1912, No. 27, p. 106). The government had created what would become a permanent reserve army of labour for Manitoba farmers.

The Wartime Farm Labour Force

The Depression and war years (1930–1945) necessitated emergency measures on the part of the federal government, but only when the market for wheat collapsed in the 1930s did it realize that farmers were not operating on a level playing-field with monopoly capital and make an effort to assist them through subsidy payments and price supports. During the Depression, there was no change in Indian policy because it was assumed that Indian

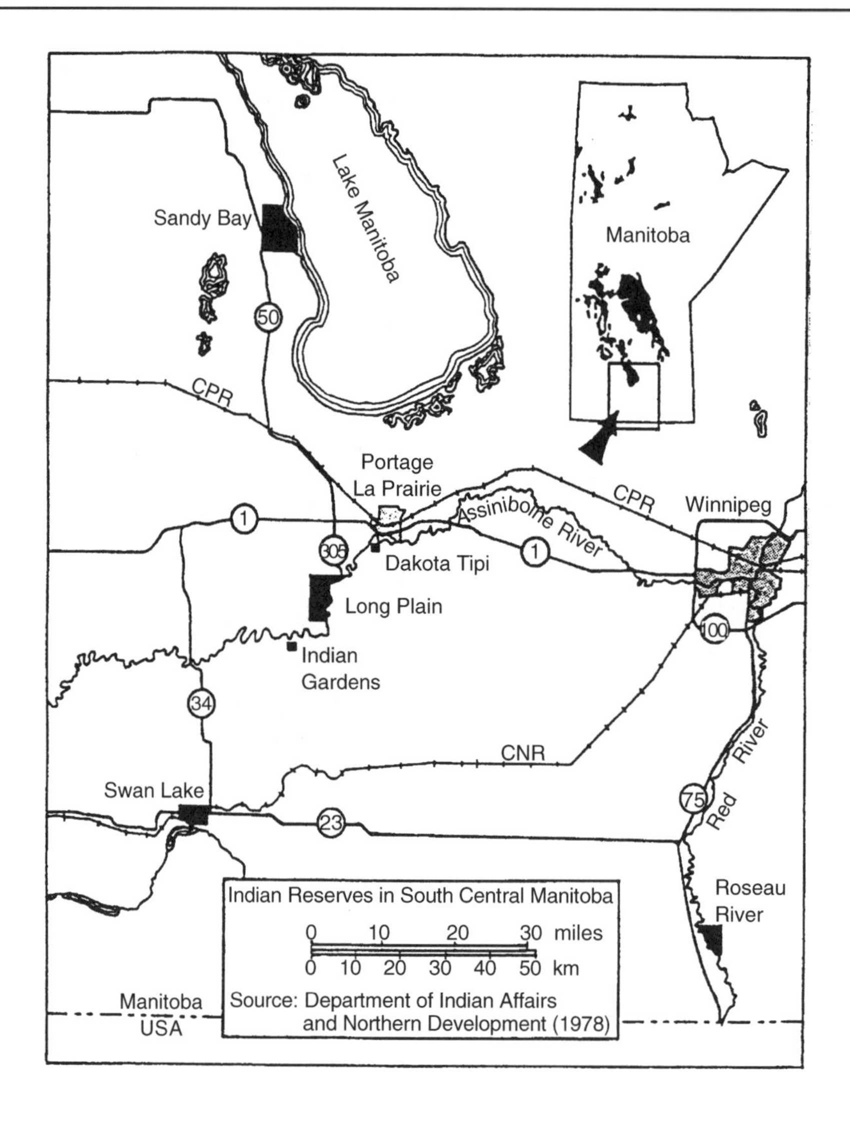

Fig. 13.3. Indian reserves in south central Manitoba, Canada.

peoples would be able to eke out a living on their reserves. From 1939 to 1945, however, both agriculture and labour, including Indian labour, were mobilized for the war effort.

During this time, Canadian agriculture was faced with its most serious labour shortages to date as able-bodied men and women flocked into the war industries and the military. In Manitoba, farmers were granted exemption from military service and normal labour market relations were suspended, so that high-school students and city-dwellers could be mobilized to help in the grain and sugarbeet harvests. Indian peoples from both southern and northern reserves were also expected to assist. Those from the south were experienced in farm work but, to the surprise of farmers and government alike, the inexperienced northerners 'adapted quickly' to harvest work and were offered Can$4.00 a day for stooking (stacking sheaves of grain) and Can$4.50 a day for threshing, as opposed to

Can$3.00 and Can$3.50, offered to the average worker, whether Native or white (*Winnipeg Tribune*, 1943).

The most controversial wartime labour force consisted of Japanese workers. Their immigration had been curtailed since 1930 (Weinfeld and Wilkinson, 1999) but existing immigrants were evacuated from British Columbia by the federal government and forced to work on sugarbeet farms in the Prairie provinces and Ontario until the end of the war. On behalf of the Manitoba Sugarbeet Growers Association and the Manitoba Sugar Company, the British Columbia Security Commission (BCSC) sought to place more than 1000 Japanese on sugarbeet farms in Manitoba (La Violette, 1948; Adachi, 1976). The Manitoba government agreed to the placements only under the condition that the federal government assume all supervisory and financial responsibilities for them. The evacuees hoped that farm work would allow them to retain some semblance of family units rather than to be dispersed across Canada and quickly regrouped to meet the requirement that 'families include at least 80% workers and number approximately six in order to fit the available housing' (Roy *et al.*, 1990, p. 142).

The experience was overwhelmingly disappointing to the Japanese, as described in the Provincial Archives of Manitoba Oral History Collection, *The Japanese in Manitoba (JM)*. Upon their arrival in Winnipeg, they felt that farmers chose them 'like slaves in a slave market' (*JM*, Tapes C842, Archival C860). Once in rural Manitoba, some evacuees discovered that, despite recommendations by the BCSC that farmers should provide them with access to potable water and shelter suitable for year-round occupancy, their sources of drinking water were miles away and their accommodations were small, cold and filthy. Some families were housed in insect-infested barns and old granaries, the floors of which were covered with mouldy grain (*JM*, Tapes C846, C864). One family piled first dirt and then snow around the bottom of their house to keep the floor warm; another heaped horse manure around the house, 'right up to the windowsills', to insulate it from the cold (*JM*, Tapes C842, C852).

To improve their living quarters, the BCSC provided them with building materials but, according to one evacuee, 'some farmers were not above holding up the supplies for their own use' (Takata, 1983, p. 138). Working conditions were as abysmal as living conditions. During the harvest, men worked from dawn to dusk but were paid only half of what they earned if they were paid at all (*JM*, Tapes C846, C864), an obviously illegal practice.

Unable to support themselves over the winter and unhappy with their housing, the Japanese petitioned the BCSC to allow them the freedom to move to other farms. When growers tried to block the petition, the Japanese threatened not to honour their contracts for the following year. The growers relented (Takata, 1983). This collective action was initiated by the Manitoba Japanese Joint Council (MJJC), which formed clandestinely in 1942 but, because the BCSC forbade the Japanese to congregate, was not recognized until 1943, when its presence was finally welcomed by the Federal Department of Labour's Japanese Division (formerly the BCSC) as a mediator between the government, sugarbeet interests and the Japanese. In 1943 alone, the MJJC successfully resolved no fewer than five work stoppages on the part of the Japanese who demanded that growers adhere to the terms of the contracts (Dion, 1991).

By the autumn of 1943, it had become obvious that the Japanese would not be able to survive another winter on their seasonal wages and would either have to seek off-farm employment or else receive relief. Fearing that the Japanese would not return to the farms once they left, sugarbeet interests tried in vain to argue that the BCSC originally had 'frozen' workers to the industry. The Japanese Division, however, was unwilling to shoulder the expense of relief. In an attempt to satisfy all parties, it limited its job-search programme for the Japanese to seasonal employment in order to force them back to work on sugarbeet farms in the spring.

Despite monetary incentives by the government to offset a sugar scarcity in 1945, farmers were reluctant to increase sugarbeet production as the availability of alternative employment for the Japanese increased. As a result, the MJJC was able to negotiate better contracts for those Japanese who remained on the farms until the end of the war (Dion, 1991). Farmers willingly complied.

The Post-War Farm Labour Force

It became clear that agriculture had to be protected, at least minimally, from the vagaries of the free market (Drummond *et al.*, 1966). Provincial governments, in particular, actively encouraged the creation of grower-controlled marketing boards in an attempt to increase farm-gate prices. In response to the post-war boom, Canada's immigration policy shifted to one of 'tap on, tap off' in 1962 (Parai, 1975). On paper, this meant that immigration would be increased in times of labour shortages and decreased in times of labour surpluses. In reality, it meant opening Canada's borders to a low-wage and ethnically diverse reserve army of labour on behalf of competitive capital which either could not or would not pay higher wages to Canadian farm workers. The new immigration policy directly contradicted Canada's Indian policy of tutelage which, at least in theory, was responsible for Native peoples' economic well-being.

The immediate post-war years marked the beginning of another cost-price squeeze in Canadian agriculture. The appearance of larger but fewer farms, increasing expenses and, for many, decreasing income were correlated with the division of farmers into small, medium and large farmers who both owned and operated (i.e. worked) their farms, and even the formation of a group of small capitalist farmers who owned and managed, but did not work their farms. Manitoba farmers in general and market gardeners (i.e. small, medium and large farmers) in particular were not immune to these trends.

From 1945 to 1960, commercial market gardening expanded out of the Winnipeg area and into rural Manitoba. The move was encouraged by the Manitoba Department of Agriculture in its attempt to promote diversification in the province's wheat economy. The initial results were so successful that provincial agricultural experts were certain that the region had the potential to supply at least ten processors with fresh produce (*Winnipeg Tribune,* 1958). But initial success did not come without financial risk. Once growers had taken the risk, Manitoba Department of Agriculture officials had to admit that the province's total acreage of vegetables had stalled (*Manitoba Co-operator* [hereafter *MC*], 1960). Potato production had stalled as well. By 1968, the province was producing only a fraction of what it was capable of and Manitoba Department of Agriculture officials further agreed that '[increased] production will take place only as additional market outlets for this production, be it fresh or processed, are developed' (Stone, 1968, p. 47).

In 1982, medium producers, represented by the Vegetable Growers Association of Manitoba (VGAM), finally won the right to sell their produce through the grower-controlled Manitoba Vegetable Producers Marketing Board. The right was not won easily. They were faced with strong opposition from powerful buyers who formed the Manitoba Fruit and Vegetable Wholesalers Association, to keep downward pressure on farm-gate prices, and from the Manitoba Vegetable Processors Association, which argued that marketing boards interfered with its right to buy from the cheapest source possible.

Once growers had some protection in marketing, they turned their attention to the farm labour force. Sugarbeet and potato producers in southwestern Manitoba had long had their own source of labour in those referred to as Mexican Mennonites. The original members of this Anabaptist sect from Russia had settled in Manitoba in the late 19th century, seeking group settlement, freedom of language and religion and exemption from military service, in an attempt to keep its ideals of strict conformity to sacred and secular matters intact. Within 10 years of the sect's arrival, the Manitoba Government began to back out of its agreement with the Mennonites. This, together with land shortages in the province, persuaded many to emigrate to Mexico. Not only did their emigration cause serious rifts within the Mennonite community itself (Redekop, 1969), but their experience in Mexico did not solve their problem of land shortages: instead, it further exacerbated differences between the landed *wirte* and the landless *anwohner* (Sawatzky, 1971). Those *anwohner* who could afford to would return illegally to Manitoba on a seasonal basis to work the land of their more wealthy brethren.

When the federal government relaxed its immigration policy after the Second World War and allowed Mexican Mennonites born to Canadian parents in or after 1947 to apply for Canadian citizenship, even more made the seasonal trip to Manitoba. So desperate were they to escape the poverty in which they found themselves in Mexico that in 1966 it came to the attention of the federal government that some were being transported 2000 miles non-stop to work in sugarbeets and potatoes at minimum wage or less in southwestern Manitoba. In one case, 29 people were crowded into a camper unit designed for four or five on top of a half-ton truck. Several were reported to have been suffering from dysentery (*Winnipeg Tribune,* 1966). Although they are accused by their Manitoba brethren of taking advantage of opportunities they once had rejected (Sawatzky, 1971), Mexican Mennonites have since become a low-wage floating-surplus population for farmers in southwestern Manitoba. Traditionally against unions, Mexican Mennonites tolerate low wages and unsafe working conditions in their repeated attempts to earn enough to buy a small parcel of land in Mexico.

Because both sugarbeet and potato production have become capital intensive, a Mexican Mennonite labour force is usually sufficient. Fruit and vegetable production, however, remains highly labour intensive. In 1966, the powerful Ontario farm lobby convinced Canada Manpower to allow it to import workers under the Canada–Commonwealth Caribbean Seasonal Agricultural Workers Program (SAWP). Because of high unemployment rates in Canada at the time, the federal government justified the programme to the public as one of 'temporary development aid' to the Caribbean (Bogacz and Forsyth, 1990). But growers resented the fact that, under the programme, they were required to pay the Caribbean workers' return airfare; they preferred, instead, to pay the less expensive brokerage fees to recruiters of Mexican farm labourers who followed the harvest north in the USA. However, Canada Manpower officials warned growers that only after they had exhausted all Canadian sources would Mexican workers be brought in (*Winnipeg Tribune,* 1973).

The federal government agreed, under pressure from Ontario growers, to extend the Foreign SAWP to include Mexico as well. When Manitoba commercial market gardeners heard of these arrangements, they, too, wanted to participate. In an effort to standardize the working and living conditions of Canadian farm labourers, Canada Manpower established 30 Agricultural Employment Services (AES) offices across Canada and introduced employer–employee agreements which stipulated both the terms of employment and the type and cost to the employee of accommodation.

Although the details of the Canada–Mexico SAWP are almost identical to the former US–Mexico Bracero Program, its success to date has depended upon continual coordinated efforts between the Canadian and Mexican governments. Potential workers must meet strict requirements before they are allowed to enter Canada for no longer than 8 months a year to work only in agriculture. The employer must be able to guarantee at least six 40-hour weeks of work for each employee and to provide acceptable living accommodation at no cost to the worker. Certain costs such as health insurance premiums may be deducted from each worker's pay. The programme is valid for a 3-year term but is reviewed annually and may be terminated at any time. After 3 years, it may be renegotiated and renewed.

Despite the occasional difference of opinion between federal and provincial levels of government as to whether or not Mexican labourers are actually needed in Manitoba, the arrangement is very cost-effective for Canada as a whole. The state and the employer pay only for the cost of maintaining the worker during his stay in Canada while the cost of supporting the worker's family is borne either by Mexico or by the family itself

(Burawoy, 1976). The arrangement also acts as a safety-valve for the Mexican government, which is unable to provide employment for the whole of its own country's population. The participants in the Canada–Mexico SAWP are either unemployed or underemployed in Mexico, and some derive from states such as Puebla and Tlaxcala, whose peasant economies are now on the decline.

In 1975, 30 Mexican farm workers arrived in Manitoba and growers were eager to share their opinions with the public. 'The Indians and Mexicans work together with no bad blood,' said one grower (Hunter, 1975, p. 5). The Mexican presence might have been forgotten by the public had not one unfortunate incident made the headlines that summer. It was discovered that living accommodation for Native workers on certain sugarbeet farms in southern Manitoba were so inadequate that the Children's Aid Society removed five children from the shacks and old milk trucks that served as their homes. One journalist commented: 'It is ironic that the Mexican migrants, who have a history of being exploited by California land owners, live in comfort while the Indians, from reserves in rural Manitoba, exist in filth and squalor' (Hunter, 1975, p. 1).

In 1976, some 450 Indian and Métis farm workers, assisted by a labour consultant to the provincial New Democratic Party (NDP), which was in power at the time, announced that they were organizing themselves into the Manitoba Farm Workers Association (MFWA). Highest on their list of demands (none of which were included under Manitoba's Employment Standards Act) were recognition of the MFWA as the workers' bargaining agent, mandatory job classifications and recommended wage rates. Other demands included: a grievance procedure; sick leave, higher safety standards and workers' compensation; first-aid kits, drinking water, and toilets in the fields; and rest periods and lunch facilities. Housing and transportation were also pressing issues, since a lack of these amenities put MFWA members at a distinct disadvantage compared with the Mexican workers. Some Native workers had to leave their communities at 5.00 a.m. to arrive at work by 8.00 a.m.

Despite assurances from Manitoba's Agriculture Minister that the MFWA was not a union but a lobbying group and thus did not have the right to strike, many growers were furious. Some refused to negotiate or to sign a written agreement; others refused to hire Native workers any more (*Winnipeg Tribune,* 1976a). Their opinion of the MFWA's demands was summed up by one grower: '[Once] you have a set of rules ... the flexibility is lost, it dehumanizes' (McCook, 1977). The NDP urged farmers to be more progressive in their thinking; the MFWA accused the growers of racism. Talks between the MFWA and growers and mediated by the Agriculture Minister soon broke off, with the MFWA threatening not to assist in that year's harvest if it had no written agreement. Several days later, 20 members of the MFWA picketed the main Canada Manpower office in Winnipeg to protest against the importation of Mexican workers (*Manitoba Co-operator,* 1976).

As tension escalated between growers on the one hand and the provincial government and the MFWA on the other, the NDP commissioned a study into the farm labour situation in Manitoba. The study (Marcoux, 1976, unpublished paper) concluded that, since the available number of Indian and Métis farm workers (435) far exceeded growers' requirements (104 full-time and 126 part-time workers), the importation of Mexican workers should be phased out by 1980. The study also recommended a Local Employment Assistance Program (LEAP) by which Native workers, with government assistance, would be mobilized for the growers' benefit. The strategy, which included alcohol education, a nutrition programme, a basic life skills programme, and a youth programme to foster pride in Native identity, also included worker retraining and upgrading and ongoing government support of the MFWA. Two growers signed as sponsors.

By June 1976, the 24 Mexican labourers who had been expected that year were working quietly alongside their Indian and Métis counterparts. Disputes had been minimal and there had been no work stoppages. Nonetheless, the Mexican Consul General at the time commented, 'I warned them [the

Mexican labourers] when they arrived, the first fight, the first riot, I'd send them back to Mexico' (*Winnipeg Tribune*, 1976b, p. 15). One Mexican worker admitted that he detected 'some bitterness' on the part of local labourers that summer but another said they just ignored it: 'We say we don't understand Indian' (*Winnipeg Tribune*, 1976b, p. 15). Undercurrents of hostility against Native workers could be detected in the comment of one grower, however, who said, 'Mexicans may be more productive in comparison with local labour because their need is greater ... Their Canadian colleagues ... are accustomed to more of a hand-to-mouth existence' (*Winnipeg Tribune*, 1976b, p. 15). Another commented that Mexicans demonstrated to others that 'you have to *work* in this country to make a living' (McCook, 1977, p. 5, emphasis added).

As a result of Marcoux's (1976, unpublished paper) study, it was discovered that 312 of the 450 MFWA members went without work in 1975. The discovery surprised no one. A spokesman for the growers had already informed the media that he knew there were enough locals for his own labour needs but that he simply preferred to hire Mexicans (Rosner, 1976). The NDP decided to terminate the Canada–Mexico SAWP for 1977.

Growers were upset. One claimed that, due to labour shortages, his vegetables were rotting in the fields. Another claimed that the ban on Mexican workers was crippling the industry. All blamed 'bureaucrats' in Ottawa for taking decisions with no first-hand knowledge of the situation. Mexicans created more jobs than they took away, growers argued, especially since they were 'more adapted to menial labour' (Wall, 1977, p. 4). Canada Manpower officials were unmoved by growers' complaints. 'I find it difficult to understand why farmers can't find 24 suitable labourers out of [a] pool of 300,' said one (*Winnipeg Tribune*, 1977, p. 4).

Although the NDP lost the 1977 provincial election to the Progressive Conservative (PC) Party, the proposed LEAP to train local labour and to educate growers in employer–employee relations went ahead. The results did not impress growers. The programme was a waste of money, they complained, because locals with any ambition would seek better jobs elsewhere. Those who did stay were incapable of retaining the information given them. One PC member of the Legislative Assembly went so far as to blame the MFWA's employment problems not on its members' lack of training or education but on their lack of initiative (*Winnipeg Free Press*, 1978). By early 1978, the MFWA conceded that, if it could be proven that employment opportunities for Canadians would increase as growers claimed they would, some Mexican workers should be brought in.

Eighteen Mexicans were brought in that year, a number so small, scoffed one grower, 'as to be ridiculous in terms of the national employment situation' (Francis, 1978, p. 53). The obvious question then might be: why bother at all? Instead, the grower concluded that 100 Mexicans should be imported the following year. Another complained that, since the formation of the VGAM, the number of growers had declined due to the combination of low crop prices and a shortage of experienced labour (Francis, 1978). He did not mention the warning that small producers had issued in the late 1960s that they would be forced out of the industry because they would not be able to compete with medium growers. Nor did he mention the preference of Canadian retailers and processors for cheaper American and Mexican produce. According to growers, the problem had to be the labour force.

The Farm Labour Force in the 1990s

After the global recession of the 1970s, the federal government's interests turned to the liberalization of agricultural trade, first through the Canada–US Free Trade Agreement (CUSTA) in 1989 and then through the North American Free Trade Agreement (NAFTA) in 1994. Because the USA is Canada's largest trading partner, the details of the CUSTA are still being worked out; those of the NAFTA will take several more years before they fully manifest themselves. However, as Manitoba commercial market gardeners have discovered, 'free trade' does not always mean 'fair trade'. In the meantime, Indian policy became embroiled in legal wrangling over land rights and self-government. While Canada's Foreign SAWPs are still intact in anticipation of renewed expansion, the problems of Native farm workers seem to have been lost in the shuffle.

As a result of the CUSTA, the first setback occurred in 1989 when Campbell's Soup Company, an important market for potatoes and vegetables and a major employer in rural Manitoba, announced it would close its Portage la Prairie plant and consolidate its operations in Toronto. In addition to the 168 plant workers who would lose their jobs, an estimated 273 truckers and nine commercial market gardeners would also be affected (Argan, 1989). The second setback occurred when non-tariff barriers were arbitrarily imposed on Manitoba produce entering the USA. Despite assurances by Agriculture Canada (1988) that an open border would mean minimal spot checks, some shipments were being delayed in Minneapolis for three to four times longer than necessary while the US Food and Drug Administration (FDA) checked not the recommended 1–2% but 15% of all shipments for pesticide residue. By the time they were released, the shipments had spoiled (Friesen, 1991). To the frustration of growers, the practice continued into the mid-1990s (Zielinski, 1995).

The only good news that Manitoba commercial market gardeners received as a result of the CUSTA was a recommendation by the Canada Employment and Immigration Commission (CEIC) in 1992 that, due to the ongoing problem of attracting and retaining local labour, the Canada–Commonwealth Caribbean and the Canada–Mexico SAWPs would not be curtailed in those provinces where they were most needed.

As a result of the NAFTA, the two remaining processing plants in Manitoba – McCain in Portage la Prairie and Midwest Food Products (formerly Carnation) in Carberry – announced plans for multi-million dollar expansions to supply the growing global market for potato products. Details of the McCain expansion show that it did not proceed without some cost to the public (Werier, 1968). The federal, provincial and municipal governments financed a Can$15 million upgrade of Portage la Prairie's water-treatment plant to handle an increase in effluent, and the federal and provincial governments contributed another Can$3 million toward the installation of an irrigation system to accommodate McCain's need for an additional 17,000 acres (6880 ha) of potatoes. Even the Manitoba Vegetable Producers Marketing Board took a small risk and called for three or four more growers to supply it with 15 new summer crops (Friesen, 1995).

While Mexican Mennonites continue to supply the labour needs of the capital-intensive potato sector, one can only speculate as to the future of the Canada–Mexico SAWP. Except for occasional criticism in the Manitoba Legislature from the NDP, the issue has seldom been broached since 1977 even though the number of Mexican workers in Manitoba has now reached the programme's imposed cap of 100. A study carried out by the Canadian International Trade Tribunal (CITT, 1991) anticipated changes to both of Canada's Foreign SAWPs. Without specifying what those changes might entail, it hinted that: 'It is a matter of pride to Canadians that migrant workers are treated very well here, but modest *concessions to competitiveness* could be made without jeopardizing social justice or the supply of visiting workers' (CITT, 1991, p. 25, emphasis added). If it means that the USA might challenge the programmes as a subsidy to Canadian farmers and that the government might have to withdraw from its supervisory role, it does not bode well for migrant workers.

For their part, growers seem to have concluded that hiring foreign workers is not just a temporary solution to labour shortages but their right. The best proof of this, according to a CEIC official, is the 'totally unacceptable' practice of laying off local and out-of-province labourers once the Canada–Mexico SAWP has been approved for another year. Growers may approve of the Mexicans' 'strong work ethic', but they have no choice other than to work a 12-hour day, 7 days a week, or they will not be re-hired the following year. Also, despite the fact that they earn minimum wage, it means that hundreds of thousands of dollars leave Canada every year. In the opinion of the CEIC, the Canada–Mexico SAWP is a 'sweetheart deal' that has gone on for far too long.

Soon after the NAFTA was ratified, some new faces appeared in Manitoba vegetable fields. Laotian women, new immigrants to Canada, are now shuttled from Winnipeg to Portage la Prairie and back by the federal government's Day-Haul Transportation Program.

As to the status of the MFWA, which some claim no longer exists, its membership dwindled from 450 in 1976 to 150 in 1993. Although more housing has been provided for them, farm workers are still not covered under Manitoba's Employment Standards Act. The LEAP failed, according to one former activist in the MFWA, because it did not address the most basic issues of members' welfare. Many Native workers are illiterate, unable even to read chemical containers on the jobsite; their brown-bag lunches of potato chips and soft drinks rob them of energy in the hot afternoon sun; and, for some, fetal alcohol syndrome makes hand–eye coordination difficult. In the opinion of the former president of the MFWA, nothing has really changed. Native workers have always been and will continue to be a reserve army of labour for Manitoba agriculture.

Conclusion

To fulfil the dual functions of accumulation and legitimization in agriculture, Canada has attempted to manage various social relations in the overall interests of agribusiness. Even though various levels of government may disagree as to how this should be accomplished, the interests of wholesalers, retailers, processors and suppliers consistently have been and, under the CUSTA and the NAFTA, increasingly are favoured over those of Manitoba farmers. In its attempt to keep the competitive sector afloat, however, the state belies its concern for a 'surplus of workers at the bottom of the employment ladder'. Variously legitimized to the public as an assimilation process, wartime exigency, revisions to immigration policy, or development aid, the labour-power of Native workers, Japanese evacuees, Mexican Mennonites, Mexican guestworkers and Laotian immigrants has been mobilized at different times for the benefit of the decreasing number of Manitoba farmers who have managed to survive the cost-price squeeze imposed on them by monopoly capital – agribusiness.

References

Adachi, K. (1976) *The Enemy That Never Was: a History of the Japanese Canadians.* McClelland and Stewart, Toronto.

Agriculture and Agri-Food Canada and Manitoba Agriculture (1994) *Growing Manitoba.* Agriculture in the Classroom Manitoba, St Pierre-Jolys, Manitoba.

Agriculture Canada (1988) *The Canada–US Free Trade Agreement: an Assessment.* Internal Trade Policy Directorate, Ottawa.

Argan, G. (1989) Soup company closing factory. *Winnipeg Sun* 25 August, p. 3.

Bogacz, R. and Forsyth, P. (1990) Fruit leads harvesters far afield. *Winnipeg Free Press* 22 October, p. 22.

Burawoy, M. (1976) The functions and reproduction of migrant labour: comparative material from Southern Africa and the United States. *American Journal of Sociology* 81, 1050–1087.

Carter, S. (1990) *Lost Harvests: Prairie Indian Reserve Farmers and Government Policy.* McGill-Queen's, Montreal and Kingston.

CITT (1991) *An Inquiry into the Competitiveness of the Canadian Fresh and Processed Fruit and Vegetables Industry.* Department of Supply and Services Research Branch, Canadian International Trade Tribunal, Ottawa.

Dion, L. (1991) The resettlement of Japanese Canadians in Manitoba, 1942–1948. MA thesis, University of Manitoba, Winnipeg, Manitoba.

Drummond, W.M., Anderson, W.J. and Kerr, T.C. (1966) *A Review of Agricultural Policy in Canada.* Agricultural Economics Research Council of Canada, Ottawa.

Elias, P.D. (1988) *The Dakota of the Canadian Northwest: Lessons for Survival.* University of Manitoba Press, Winnipeg, Manitoba.

Francis, J. (1978) Vegetable grower says Mexicans can increase job opportunities. *Winnipeg Tribune* 2 December, p. 53.

Frideres, J. (1999) Altered states: federal policy and aboriginal peoples. In: Li, P.S. (ed.) *Race and Ethnic Relations in Canada*, 2nd edn. Oxford University Press, Don Mills, Ontario, pp. 116–147.

Friesen, R. (1991) Vegetable board calling for new growers for '95. *Manitoba Co-operator* 30 March, p. 10.

Friesen, R. (1995) US inspection unfair. *Manitoba Co-operator* 12 September, pp. 1, 10.

Hawkins, F. (1977) Canadian immigration: a new law and a new approach to management. *International Migration Review* 11, 77–93.

Howard, J.H. (1984) *The Canadian Sioux*. University of Nebraska Press, Lincoln, Nebraska.

Hunter, N. (1975) The other side of the coin: these workers comfortable. *Winnipeg Tribune* 11 July, pp. 1, 5.

La Violette, F.E. (1948) *The Canadian Japanese and World War II: a Sociological and Psychological Account*. University of Toronto Press, Toronto.

Manitoba Co-operator (1960) Trends in vegetable production reviewed by department official. 14 January, p. 11.

Manitoba Co-operator (1976) Farm labor confrontation. 20 May, pp. 1, 14.

Manitoba Department of Agriculture (1982) *100 Years of Agriculture in Manitoba, 1881–1980*. Winnipeg, Manitoba.

Manitoba Department of Agriculture (1989) *Manitoba Agriculture Yearbook*. Winnipeg, Manitoba.

Manitoba Department of Agriculture (1990) *Manitoba Agriculture Yearbook*. Winnipeg, Manitoba.

Manitoba Department of Agriculture (1991) *Manitoba Agriculture Yearbook*. Winnipeg, Manitoba.

Manitoba Department of Agriculture (1992) *Manitoba Agriculture Yearbook*. Winnipeg, Manitoba.

Manitoba Department of Agriculture and Immigration (1959) *A Study of the Population of Indian Ancestry Living in Manitoba, Main Report*. Department of Agriculture and Immigration, Winnipeg, Manitoba.

McCook, S. (1977) Mexican workers 'beautiful'. *Winnipeg Tribune* 16 June, p. 15.

Morton, W.L. (1985) A century of plain and parkland. In: Francis, R.D. and Palmer, H. (eds) *The Prairie West: Historical Readings*. Pica Pica Press, Edmonton, Alberta.

O'Connor, J. (1973) *The Fiscal Crisis of the State*. St Martin's Press, New York.

Panitch, L. (1980) The role and nature of the Canadian state. In: Harp, J. and Hoffey, J.R. (eds) *Structured Inequality in Canada*. Prentice-Hall, Scarborough, Ontario.

Parai, L. (1975) Canada's immigration policy, 1962–74. *International Migration Review* 9, 449–472.

Redekop, C.W. (1969) *The Old Colony Mennonites: Dilemmas of Ethnic Minority Life*. Routledge and Kegan Paul, London.

Rosner, C. (1976) Farm hands want Mexicans banned. *Winnipeg Free Press*, 12 May, pp. 1, 4.

Roy, P.E., Granatstein, J.L., Iino, M. and Takamura, H. (1990) *Mutual Hostages: Canadians and Japanese During the Second World War*. University of Toronto Press, Toronto.

Sawatzky, H.L. (1971) *They Sought a Country: Mennonite Colonization in Mexico*. University of California, Berkeley.

Statistics Canada (1982) 1981 Census of Canada, *Agriculture Manitoba*, Catalogue pp. 96–908.

Statistics Canada (1987) 1986 Census of Canada, *Agriculture Manitoba*, Catalogue pp. 96–109.

Statistics Canada (1992) 1991 Census of Canada, *Agriculture Manitoba (Pt. 1)*, Catalogue pp. 95–363.

Statistics Canada (1997) 1996 Census of Canada, *Agricultural Profile of Manitoba*, Catalogue pp. 95–178-XPB.

Stone, G.E. (1968) Potential of potato production in Manitoba. *Manitoba Co-operator* 7 March, p. 47.

Takata, T. (1983) *Nikkei Legacy: the Story of Japanese Canadians from Settlement to Today*. NC Press, Toronto.

Wall, L. (1977) Vegetable men want Mexicans. *Winnipeg Tribune* 10 August, p. 4.

Weinfeld, M. and Wilkinson, L.A. (1999) Immigration, diversity, and minority communities. In: Li, P.S. (ed.) *Race and Ethnic Relations in Canada*, 2nd edn. Oxford University Press, Don Mills, Ontario.

Werier, V. (1968) The amazing tax deal at Portage. *Winnipeg Tribune* 25 May, p. 6.

Winnipeg Free Press (1978) Working in Canadian fields golden chance for Mexicans. 14 June, p. 4.

Winnipeg Tribune (1943) 2,000 Indians may help next harvest. 12 March, p. 11.

Winnipeg Tribune (1958) Revolution looms in special crops. 8 November, p. 5.

Winnipeg Tribune (1966) Gov't told farm laborers trucked in here like cattle. 8 November, p. 1.

Winnipeg Tribune (1973) Ontario growers seek imported help. 10 August, p. 2.

Winnipeg Tribune (1974) 'Sweated-labor policies' cited. 26 January, p. 2.

Winnipeg Tribune (1976a) Farmers, laborers at odds in Portage la Prairie fields. 12 May, p. 3.

Winnipeg Tribune (1976b) Mexicans leave families for farm work in Manitoba. 16 June, p. 15.

Winnipeg Tribune (1977) Mexicans not needed: manpower. 9 September, p. 4.

Zielinski, S. (1995) Carrot seizure by US infuriates vegetable grower. *Central Manitoba Farmer* October, p. 9.

14 Cycles of Deepening Poverty in Rural California: the San Joaquin Valley Towns of McFarland and Farmersville

Fred Krissman
Center for US–Mexican Studies, University of California at San Diego, La Jolla, California, USA

Abstract

Labour-intensive crop industries have together comprised one of the most profitable segments of the American economy for more than a half century. However, the groups that supply most of the 2–3 million workers for the nation's farm labour markets are among the most impoverished of its workers. California's San Joaquin Valley is an example of such regional wealth disparities. Crop industries within its three top farm counties generate more than US$7 billion in revenues annually. Nevertheless, these same counties contain seven out of the ten poorest communities in California, with average annual per capita incomes of US$5500. The seven communities are populated primarily by Mexican immigrants, and farm jobs are the principal source of employment. This chapter discusses the situation and prospects for two of these farm worker towns – McFarland and Farmersville – using ethnographic, archival and secondary data. These cases are used to address three questions. What historical and structural factors have contributed to the unequal distribution of sectoral wealth in rural California? Is the Californian case an aberration or a model for changes occurring across rural communities? Can public policies promote changes in the cycles of deepening poverty for every generation of farm workers?

Introduction

America's farm sector is the single largest revenue producer, export earner and private sector employer in the nation. Labour-intensive crop industries have long comprised one of US agriculture's most profitable segments (Martin, 1985). However, the 2–3 million wage workers needed to produce the nation's US$150 billion in farm revenue have always been the most impoverished of the 'working poor' (Griffith and Kissam, 1995). This paradox of poverty in the midst of prosperity has been the focus of numerous studies and diverse social programmes during the past century. Nevertheless, rural living and working conditions in California, the top farm state in the USA, have deteriorated significantly in recent decades (Taylor *et al*, 1997; Mines and Alarcon, 1999).

This chapter examines two communities in the San Joaquin Valley, California. By the

 The Dynamics of Hired Farm Labour (eds J.L. Findeis, A.M. Vandeman, J.M. Larson and J.L. Runyan)

mid-1990s agricultural industries within its three top farm counties (the adjacent municipalities of Fresno, Tulare and Kern) generated more than US$7 billion in revenues annually, making this area the most productive farming region in the world. Nevertheless, these same counties contain seven of the ten poorest incorporated cities in all of California, each a farm worker town where agricultural jobs are the principal source of income. In these communities, predominantly populated by Mexicans and their families, annual per capita incomes average only about US$5500. The endemic poverty in towns such as McFarland and Farmersville has economic, demographic and political aspects that mirror those in scores of other rural Californian settlements. An extensive literature suggests that these cases are also representative of hundreds of other rural communities across the USA.

A combination of ethnographic, archival and secondary data are used in this chapter to examine the current situation and future prospects for McFarland and Farmersville. The chapter outlines the rise and contemporary effects of agribusiness control over rural California's towns, illustrates the trend toward cycles of deepening poverty by looking at the experiences of two groups of farm worker families, and reviews proposals that could help to revitalize the state's rural communities. Historical and structural factors contributing to the unequal distribution of regional wealth are identified, the likelihood of the California model for expanding areas across rural America is considered, and public policy changes that could reverse the spread of rural poverty are addressed.

Corporate Agriculture in Rural California

While scholars may debate the costs and benefits derived from the industrialization of agriculture to society at large, the negative consequences on the communities scattered across the hinterland were demonstrated conclusively more than 50 years ago. During the waning days of the great depression, Walter Goldschmidt conducted a set of studies in the San Joaquin Valley. The politically explosive findings were suppressed by the federal Department of Agriculture until a senator had the reports published directly into the *Congressional Record*. A summary of the findings in *As You Sow* (Goldschmidt, 1978, first published in 1947) provides the historical context for discussing contemporary rural California (see McWilliams, 1939; Steinbeck, 1939).

The situation in the three towns studied by Goldschmidt represented alternative futures for rural California. Goldschmidt conducted fieldwork in Arvin and Wasco (in Kern County) and in Dinuba (in neighbouring Tulare County). The data collected by Krissman for communities in the same counties document the continuing socioeconomic decay of rural California (for specific points of comparison and contrast, see Krissman, 1997a). One path would support the maintenance of vibrant regional trading centres, comprising a citizenry with diverse occupations and a sense of civic duty. The other route would lead to the devolution of these communities into 'neo-feudal' villages, with few amenities and little more than two classes – a small number of wealthy landlords and a majority population of indigent farm worker families. For rural California to take the first path would have required fundamental changes in key public policies, which Goldschmidt argued were propelling rural communities down the second. As important as these findings were, his most significant prediction was that much of the rest of rural America would eventually follow California's lead down one road or the other.

In Part I of Goldschmidt's *As You Sow,* the case of the town of Wasco demonstrated that the industrialization of farming was already well entrenched in California and increasingly common across the USA. The industrial mode of farming was shown to have a number of characteristics that would alter rural social organization in California. While many of the changes were deemed harmful to the continuation of a vibrant community structure, Goldschmidt saw industrialization even at this early date as a *fait accompli.* At that point his aim was to suggest ways to ameliorate some of its negative effects on rural communities.

Part II of Goldschmidt's work examined one particularly disturbing feature of industrialized farming – the concentration of control over farmland, typically a result of the rise of corporate farming. Large-scale farm production has several forms, including concentration of control over landholdings (either through purchase or lease, or under contract) and vertical integration of several types (with the goal to increase control over the total supply of the commodity). Regardless of the form, Goldschmidt (1978) found that rural communities located in regions with more concentrated control over farm production suffered a number of negative consequences.

The negative effects were demonstrated in a comparison of the California towns of Arvin and Dinuba. Arvin, surrounded by far fewer, much larger farms, had the following undesirable characteristics despite higher average crop revenues: twice the number of low-wage farm workers; per capita incomes one-fifth lower; one-third fewer local purchases; half the local businesses; 50% fewer self-employed breadwinners; half the civic organizations; inferior public institutions; and fewer social amenities (including more limited basic social services). In sum, Arvin was a much more socioeconomically impoverished rural community, with only two highly differentiated classes. Today, poverty experts describe this as an 'hourglass' economy. The benefits of economic expansion go disproportionately to a few, largely through the increased productivity of a great number of low-wage workers, but the workforce bears the brunt of downturns in the economic cycle.

Goldschmidt (1978) argued for public policies that would benefit the bulk of the rural population, not just the relatively few already privileged landowners. He believed that the traditional pro-grower policies of government could be altered due to a historic transformation in the composition of the workforce. After a succession of waves of largely non-white, foreign-born, disenfranchised migrant men (mainly from east Asia and Mexico) between the 1870s and 1920s, white citizen families had become a major portion of California's farm labour force in the 1930s. As the larger society became aware of the dismal conditions suffered by white families in rural California, popular pressure increased to ensure decent living conditions for farm workers. Picardo (1995) documents how agents representing a number of major industries (rail, banking, insurance, agricultural supply and processing) with interests in highly extractive farming helped to fund a statewide vigilante movement to suppress violently any efforts at social reform (see Krissman, 1996a, for documentation and description). However, growers soon realized that the continued abuse of white Americans was not politically tenable.

It can be argued that corporate business strategies undermined a growing national consensus for better conditions in rural California by displacing white citizen farm workers with an 'army' of new migrant workers recruited from Mexico. While the immediate result was to maintain low wages, increase profits and continue the expansion of corporate farming, the long-term consequences included the socioeconomic transformation of California; Piore (1979) noted these benefits for America's urban industries that recruit new migrant workers. A profile of two groups of foreign-origin farm work families from Mexico is presented later in this chapter to illustrate working and living conditions in rural California today. Firstly, important trends are outlined in the post-Second World War era to update Goldschmidt's research.

Important Trends in the Post-Second World War Era

The State did not revise its pro-grower bias in the contemporary period. Therefore, concentration of farm land in California has continued since the end of the Second World War. This situation has been well documented by Liebman (1983), though land concentration has dramatically accelerated since her study. Effective control over production has been reduced in many farm a small handful of land- land tenure has largely shifted perators (such as the 'family [illegible]') to corporate firms run by hired managers.

Ethnographic data collected for the study reported here demonstrate that a few dozen corporations dominate two of the San Joaquin Valley's billion-dollar-plus crop industries (citrus and table grapes) which today occupy broad swathes of farmland in much of adjacent Fresno, Tulare and Kern counties. These production belts run through the irrigation districts that surround both McFarland and Farmersville, while a few types of industry-related businesses – especially farm labour contractor firms – have operations within the towns.

The levels of corporate control in these industries are illustrated by three firms operating in the San Joaquin Valley in 1990 (Krissman, 1996a). Firstly, Paramount Citrus Association (PCA) owned more than 27,000 ha of regional citrus orchards and two multi-million-dollar packinghouses, yet was only one subsidiary of the American Protection Industries company (based on the East Coast). PCA, in turn, was one of the most important corporate influences within Sunkist Growers. Secondly, Superior Farming Company was a wholly-owned subsidiary of Texas-based Superior Oil Company until sold to the even larger Sun World in 1990. Superior Farming planted over 81,000 ha of labour-intensive crops, including about 9600 ha of vineyards. Thirdly, Dole Foods, Inc. in the mid-1990s generated more than US$3.5 billion annually and employed more than 28,000 seasonal farm workers in California alone. Each of these corporate firms has received federal and state support, including ongoing access to a foreign-origin workforce.

The Bracero Program recruited about 5 million Mexican men to the western USA between 1942 and 1964, effectively displacing the white Americans who were subsequently absorbed into wartime industries such as the military, and then the post-war economy. As a result of the recruitment programme, Mexico became the source for more than 90% of the 700,000 farm workers employed in California and now provides the majority of farm labourers across the USA. The process by which many Mexican farm workers shift from international migration to settled immigration led to the demographic transformation of the southwestern USA. Mexican families came to constitute a major portion of the population in at least 70 incorporated rural California towns by 1980. A decade later, Mexico-origin residents had become the majority in almost all of these towns due to their settlement and the consequent 'flight' of white residents. A representative sample from the San Joaquin Valley, where over a third of all such towns are located, illustrates the overall trend (Table 14.1).

As the population has shifted in ethnic composition, the once notable economic differences among the state's rural communities have diminished. A general economic decline has settled on most rural California towns, even as agricultural revenues have soared in recent decades (*Los Angeles Times*, 1992b, 1993). Officials in both McFarland and Farmersville have seriously discussed *dis*-incorporation due to insufficient local tax bases to support essential services such as police and fire departments (Table 14.2).

In a moment of candor, one of McFarland's mayors actually referred to his town as an 'incorporated farm labour camp' (Bakersfield *Californian*, 1979). Economic dependence upon low-paying, seasonal agricultural jobs ensures that eight of the ten poorest cities in all of California are located in rural areas (Table 14.3), not the urban *barrios* and ghettoes usually associated with chronic poverty. Only the two cities in Los Angeles County are not farm worker towns. Note, too, that the five poorest communities in the state are all located in Fresno County, the number one agricultural county in the US self-promoted as 'Agribusiness Capital of the World'.

Along with ethnic transformation and economic deterioration, there are other contemporary changes of note in rural California's towns. A particularly troubling issue is the lack of representative local institutions. The remaining minority of white residents generally continue to enjoy political and economic dominance in these towns until their portion of the total population falls below about 10%. Many of the Mexico-origin

Table 14.1. The Mexico-origin portion of residents in selected farm worker towns of the San Joaquin Valley, California. (Source: Rochin and Castillo, 1993.)

Town	County	Percentage Latino		Percentage increase[c]
		1980	1990	1980–1990
Arvin[a]	Kern	57.9	75.0	75.1
Dinuba[a]	Tulare	48.6	60.4	59.8
Farmersville[b]	Tulare	41.7	59.3	54.6
Huron	Fresno	91.3	96.5	81.9
McFarland[b]	Kern	75.8	82.9	48.8
Mendota	Fresno	84.7	93.9	50.1
Orange Cove	Fresno	72.2	86.0	65.8
Parlier	Fresno	93.0	97.1	191.8
San Joaquin	Fresno	60.2	75.4	50.0
Wasco[a]	Kern	48.0	63.3	70.3

[a]Towns studied by Goldschmidt.
[b]Towns studied by Krissman.
[c]The dramatic increase in the Latino proportion of the population in most of these towns during one decade is due to: (i) the increase in Latino residents; (ii) the large growth in the population size of the towns; and (iii) the rapid decline in the number of non-Latino white residents (so-called 'white flight').

Table 14.2. Local tax base for Farmersville versus all California cities, 1982. (Source: *California Journal*, 1983.)

Base	Farmersville (%)	All California cities (%)
State revenues	38.9	15.5
Local fees	29.4	38.4
Local taxes	15.0	31.3
Interest/rentals	7.5	7.6
Licences/fines	5.1	2.8
Other	4.1	4.4

Table 14.3. California's ten poorest incorporated cities. (Source: *Los Angles Times*, 1992a.)

City	County	Per capita income 1989 (US$)
Orange Cove	Fresno	4385
Parlier	Fresno	4784
Mendota	Fresno	4920
San Joaquin	Fresno	5356
Huron	Fresno	5501
Coachella	Riverside	5760
Farmersville	**Tulare**	**5857**
Cudahy	Los Angeles	5935
McFarland	**Kern**	**6056**
Bell Gardens	Los Angeles	6125

majority are relatively new immigrants, or still engage in international migration. Typically these residents are not American citizens and therefore are ineligible to participate in most political forums. In fact, major corporate farms employ an unknown but large number of workers in the USA without legal documents, and are usually in violation of immigration and labour laws.

Unrepresentative government is bolstered by the prevalence of anti-democratic election procedures that favour white voters and wealthy property owners. Still common non-representative political practices include: the exclusion of non-citizen and undocumented residents from local participation; appointed offices; voting systems based on property valuation (especially on 'water boards' that often control a variety of additional vital social services); and at-large elections (Krissman, 2001). As a result, most officials on the city councils on school, utility, and water boards, and within the police and fire departments are white. Few of these local officials are bilingual, even though the majority of adult residents speak Spanish as their principal language. Everyday official practices often have the flavour of a system of apartheid (Burawoy, 1981; Krissman, 2001).

Due to their political power, the white minority retain economic control in these towns. Public expenditures generally favour the neighbourhoods, businesses and interests of the minority over that of the majority. The minority also generally support the interests of corporate agribusiness at the expense of the majority of residents in struggles over issues such as worker organization, pesticide use, immigration laws and welfare policies (Krissman, 1996a). The alliance of the white minority and the region's growers is particularly disturbing in light of the fact that the latter seldom reside locally. Indeed, most of the land is controlled by corporate firms, whose boards of directors and major investors live far from the rural regions where farming occurs.

Absentee landlords and corporate managers hire supervisory staff to oversee the day-to-day work in California's rural areas. Although most skilled and supervisory positions were long reserved for white men, longtime residents and children of Mexican immigrants raised in the USA have experienced a gradual process of upward mobility into better-paying jobs. The use of Mexicans and Mexican Americans as supervisory staff accelerated in the aftermath of the rise of the United Farm Workers (UFW) union in the 1970s. Agribusiness owners seeking to displace union workers hired those in closest contact with Mexican workers to recruit new workers. The proliferation of ranch foremen, field supervisors, pesticide applicators, custom harvesters and farm labour contractors (FLCs) has given rise to internal differentiation of the Mexico-origin majority. Labour recruiters and field supervisors comprise much of an emerging middle-class in many of rural California's farm worker towns.

As Menchaca (1995) documented in *The Mexican Outsiders*, a small Mexico-origin *bourgeoisie* has been at the forefront of struggles against ethnic segregation and discrimination. This group also provides much of the promise for a nascent revitalization in the rural towns, starting and expanding businesses, purchasing and remodelling homes, and growing particularly labour-intensive crops under contract to processing firms. While the rise of some farm worker families into a rural middle class has led to some improvements in conditions in these towns, this group is typically aligned with the white minority on issues related to the farm labour system because of their roles as supervisory and intermediary personnel. Mexico-origin residents have become divided over basic socioeconomic equity, with ideological rationales providing for mutual rancour between union and environmental activists on the one hand, and FLC owners and farm supervisors on the other (see Kwong, 1997, for the strikingly similar situation in New York City's contemporary Chinatown).

Solidarity has also declined among the Mexico-origin majority due to increased intra-ethnic differentiation. Most long-time international migrant networks link employers to workers from central Mexico, where the majority of the population is of mixed (*mes-*

tizo) ancestry. However, labour recruiters began soliciting indigenous workers from southern Mexico in the 1970s to displace the *mestizos* who were members of the UFW. During the 1990s, tensions between *mestizo* and indigenous Mexicans were often more pronounced than those between Mexicans and whites.

The Mexican majority is also differentiated by the socioeconomic ties that bind international migrants to their native hometowns. Informal tensions, as well as ritualized conflicts, often occur between groups from different regions of Mexico. The formation of cohorts often crystallizes around groups of *paisanos* (fellow countrymen) from the same hometown or region of origin (Martinez, 1997). Formal self-help associations in California's rural towns usually arise based on the same criteria, with a principal goal to raise funds for small development projects in a common Mexican community of origin (Krissman, 1996a). While there are many positive aspects of this international phenomenon, these activities separate – and may divide – the Mexico-origin population in the USA. White community officials often escalate these tensions by labelling such groups as 'gangs' and using non-Spanish-speaking police to crack down on gatherings in parks, schools and other public places.

In sum, rural California has undergone a major and broad-based socioeconomic transformation as a result of the corporatization of its agricultural sector. The demands of corporate growers for low-cost labour have led to the recruitment of foreign-origin workers as well as increased differentiation of the Mexico-origin workforce, which is today the majority of the residents in scores of California's rural towns. Both the increasing power of fewer and larger agribusiness firms and a heterogeneous Mexican migrant labour force present new challenges for officials, activists and workers grappling with the problems related to the cycles of deepening poverty in rural California. Two groups of farm workers, one *mestizo* and the other indigenous, are presented as examples of this pattern of progressive rural impoverishment.

Networks of *Mestizo* and Indigenous Mexican Farm Workers

Mexican nationals have been an important part of California's farm labour force since 1900. However, Native Americans, Chinese, Japanese, Filipinos and even white Americans have comprised a large, and sometimes a majority, share of California's farm workers between 1769 and 1942 (Krissman, 1997b). It is only since the 1950s, as a direct result of the US-sponsored Bracero Program, that agribusiness has come to rely almost exclusively upon Mexico to supply a vast migrant workforce.

Krissman (1996a) traced the development of international migration from two regions of rural Mexico to the San Joaquin Valley. Examining a representative sample of the workforce in two major regional crop industries, the research demonstrates that international migrant networks are initiated and sustained by continuous recruitment activities by agribusinesses in the USA. Growers recruit new migrant workers to maintain access to low-cost labour in the face of pressure to improve conditions for resident farm workers. These recruitment strategies exacerbate already profound rural poverty. A flow of *mestizo* Mexicans from central Mexico was initiated by US companies, including railways, mines and growers, as early as the 1880s. This flow was reinvigorated after the depression era. A new wave of indigenous migrants from southern Mexico was recruited by California's growers to help to break farm labour strikes in the 1970s. These different migrant flows to the San Joaquin Valley are briefly described in the discussion below.

Mestizo migrants from Zacatecas

McFarland, California, is a principal destination for members of a migrant network from the municipality of Huanusco, Zacatecas in central Mexico. A full third of the town's residents are from this one Zacatecan municipality. Children from Huanusco constitute almost two-thirds of the participants in McFarland's migrant education programme. Several sources (including municipal authorities in Huanusco and immigrant leaders of the municipality's international migrant association 'Club Huanusco') agree that at least 12,000 people born in Huanusco now reside in north Kern County, while only about 10,000 still live in Huanusco itself. Indeed, Huanusco suffers a negative population growth rate in spite of a relatively high birth rate and a lengthening average life expectancy (Krissman, 1996a).

It has been reported that at least 500,000 Zacatecans reside in the state of California (INEGI, 1991) and that more than a million Zacatecans live in the USA (*Los Angeles Times*, 1999), with Zacatecas having the highest rate of emigration per capita of any Mexican state. The literature documents the migration of Zacatecans to work on railways in the USA as early as the 1880s (Cardoso, 1980). Informants' life and work histories collected by Krissman demonstrate that international migration between Huanusco and California goes back at least to the 1920s. A number of factors led to the current concentration in north Kern County. During the Bracero era, Zacatecas was one of the principal sources of farm workers to California (Galarza, 1964) and many thousands worked in labour-intensive crop industries in the San Joaquin Valley. Zacatecans were among the early supporters of the UFW union during the 1960s, while many more became members when the union won a contract from the table-grape industry in the 1970s.

Zacatecan informants recall with nostalgia the recent past, when it seemed that upward mobility was within their grasp. Zacatecans used to work almost year-round by combining harvest work in the region's citrus industry during the winter/spring with a variety of tasks in the summer/autumn grape industry. UFW contracts recognized seniority and provided a variety of benefits as well. Scores of Zacatecan families also took advantage of federally subsidized housing built in the town in the early 1980s, anchoring hundreds to McFarland. When the table-grape industry nullified union contracts, some Zacatecans began farm labour contractor firms. This international community is still affected by the continuing labour strife in California since the 1970s. Now the farm labour contractors headed by men from Huanusco offer several thousand seasonal jobs, but primarily to new migrant workers, not the long-time immigrants settled in Kern County. Meanwhile, citrus growers obtained a new workforce unaffected by the labour activism of the 1970s.

Indigenous Mixtec migrants from Oaxaca

Farmersville is the new home for several hundred indigenous migrants from the Mixteca region of the southern Mexican state of Oaxaca. Indeed, most San Joaquin Valley towns (including McFarland) have growing Mixtec settlements within them (Runsten and Kearney, 1994). Although the Mixtec have migrated from their hometowns since the end of the 19th century, most Mixtec migration was internal to Mexico until the 1970s. At first, Mixtecs flocked to seasonal farm jobs on plantations of the Gulf and Pacific coasts. Later, they took advantage of sustained economic growth in Mexico City. Finally, especially after the collapse of the Mexican economy in the 1960s, Mixtecs headed north to plantations just south of the border with the USA. Vegetable growers in north Mexico provided buses to transport entire families more than a 1000 miles to work in tomato, pepper and cucumber fields (Runsten *et al.*, 1993). These plantations provide US consumers with two-thirds of their winter vegetables.

When growers in nearby California began searching for a non-union alternative to the *mestizo* members of the UFW in the 1970s, these indigenous labourers were an ideal replacement workforce (Krissman, 1994). The citrus industry provides a classic case study of labour displacement, as Zacatecan union activists were replaced by new workers, such as the Mixtecs brought up from north Mexico (see Mines and Anzaldua, 1982, on the displacement of long-time citrus harvesters in nearby Ventura County by Mixtec workers recruited from the border area).

The Mixtec in Farmersville work in a number of regional crop industries, including raisin and wine grapes, stone fruits, tomatoes and olives. However, they are concentrated in citrus as a result of some small upward mobility. Eight Mixtec women have obtained better seasonal jobs inside a regional citrus packinghouse through a fellow villager who is the new line foreman, while another Mixtec foreman recruits for a citrus farm labour contractor that picks the fruit.

Cycles of deepening poverty

The migration and farm labour histories of two international migrant networks that link disparate regions of Mexico to two San Joaquin Valley communities were provided above. Their histories document the successive integration of Zacatecans and Oaxacans into the region's farm labour market. The earlier integration of Zacatecans permitted their settlement in McFarland (and in the region's better-paid table-grape industry and more secure non-farm jobs), whereas the newly integrated Mixtecs, while growing rapidly in number, are still at the very bottom of the region's farm labour market. This comparison can be seen in the relatively better conditions of life and work among the Zacatecan subsample (Table 14.4), discussed below.

Due to their very mature and dense network, Zacatecans in McFarland are in an advantaged position. The median per capita income for the Zacatecan sample is US$5971, roughly equal to the overall average in McFarland. While the average household size is large (9.4 members), the majority of the households' wage earners are no longer farm workers. For every 0.9 farm workers per household, there are 1.5 non-farm workers. Less than one-third of Zacatecans work outside the San Joaquin Valley; almost all of these migrants work in Mexico, typically in their hometown during extended return visits each winter.

The relatively privileged position of Zacatecans within the region's farm labour market is attributed to two factors: the high value of the very fragile table-grape crop and the union-organizing campaigns of the UFW during the 1970s. However, the majority of the Zacatecan sample report that living and working conditions for farm workers have been deteriorating since the 1980s. By 1992 hourly wages had declined by as much as 20% (in the grape harvest), total hours

Table 14.4. Comparative data for California-based households, 1990[a]. (Source: Krissman, 1996a.)

	Zacatecans (n=34)	Mixtecs (n =12)
Income (US$)		
Per household	53,300	31,300
Per worker	11,677	4,334
Per capita	5,971	2,887
Household composition (no. of persons)		
Total size	9.4	10.6
Worker/dependent ratio	2.4/7.0	6.5/4.1
Farm/non-farm worker ratios	0.9/1.5	6.3/0.2
Migrancy for work (%)		
Within the USA	1.6	72.4
Within Mexico	29.8	67.8

[a]Data for only subsample of Mexicans who have resided continuously in California for at least the past 5 years. Total study sample included another 61 households of international migrants, who remain based in their rural Mexican communities (Krissman, 1996a). See Krissman (forthcoming) for a comparative discussion of migrant and immigrant farm workers from Mexico.

worked declined an average 15% and almost all job benefits were eliminated. Most blame the demise of the union and the continued influx of new migrant workers from throughout Mexico willing to take any job at almost any wage.

Due to their relatively new and diffuse international network contacts, the Mixtecs in the case study sample are much poorer than the average resident in Farmersville. Based on life and work history data, the exacerbated poverty of the members of this new network can be gauged. The average household contains 10.6 people. These households have a very high ratio of workers to dependents. This is because relatively few non-workers migrate from southern Mexico, due to much higher costs and greater poverty, while the Mixtec immigrant community is so new that relatively few children have been born in the USA. Instead, the typical Mixtec household consists mainly of related adult male workers. The importance of farm work is clear – 6.3 adults worked in agriculture for every 0.2 adults who worked in a non-farm position. Mixtec workers averaged only US$4334 annually, while household per capita income of US$2877 was only half the average for Farmersville (US$5857 annually). Mixtecs had a high incidence of migration for work, both within the USA and between the USA and Mexico. Of the sample, almost three-quarters worked for wages outside the San Joaquin Valley, while over two-thirds also worked in Mexico, typically in north Mexico and in Mexico City, not in the Mixteca itself.

A comparison of these two international migrant networks documents a relationship between ongoing migration and the deterioration of living and working conditions for all farm workers through the processes of labour market 'oversupply' and displacement. Nevertheless, the onus of rural impoverishment should not be placed on the victims of these cycles of deepening rural poverty. Rather, historical and contemporary evidence links the poverty endemic in farm worker towns to the practices of agribusiness firms. Californian agribusiness continues to generate the highest revenues in those counties with the most impoverished rural communities. In the current period, growers have used ethnic intermediaries such as field supervisory personnel and farm labour contractor firms to continue to draw upon a largely undocumented workforce in violation of federal immigration law, with the aim of undermining the organization of farm workers in violation of state labour law.

For a Better Future Rural California

Two community case studies have been provided here to illustrate the conditions within scores of rural Californian towns. Goldschmidt's 1947 classic, *As You Sow,* concluded with a warning that government policies that promoted the concentration of farm production into ever fewer hands had to change. More than half a century since his work was published, both the corporatization and widespread impoverishment of rural California are indisputable. California's poorest communities are rural farm worker towns, while 'agribusiness' is the preferred and more accurate term for the contemporary farm enterprise, not the 'family farm'.

While *As You Sow* is still critical today to understand the basic causes and consequences of the impoverishment of rural Californian communities, recent research shows that the transformation of rural life has gone far beyond even the significant concerns of that period. Subsequent technological innovations, including the development of rural infrastructure such as freeways and irrigation works, have helped to greatly increase the scale of industrialized agricultural production. The implementation and institutionalization of the federal Bracero Program, followed by the resurrection of the farm labour contract system, have also been crucial in permitting agribusiness to maintain access to an ample and low-cost migrant workforce for the last half century. The constant recruitment of new immigrants into California's farm labour markets has transformed the demography and economy of scores of rural towns beyond even Goldschmidt's warnings of the 1940s.

Agribusiness strategies to maintain a low-wage migrant labour force have had a number of ramifications for the migrant communities on both sides of the US–Mexican border as well (Krissman, 2000). Firstly, continual recruitment has fostered the continued influx of new immigrants, and the spread of labour-sending communities beyond the traditional confines of central Mexico and into new regions of indigenous southern Mexico. This spreading of out-migration has exacerbated anti-immigrant tendencies in California and across the USA. Continuous recruitment also pits the members of different migrant networks against one another within California's rural towns for scarce jobs, housing and other services. Secondly, agribusiness restructuring of labour relations has proceeded with the aid of 'oversupplied' labour markets, decline of the unions, and increased worker competition for seasonal jobs. Thirdly, the Mexican majority within farm worker towns has been fragmented among those whose good jobs are threatened by increased subcontracting, those who gain upward mobility within the FLC system, and the new migrants recruited by FLC supervisory personnel.

Finally, as conditions in farm labour markets deteriorate, so do the local services that are provided within rural communities. An impoverished local population cannot provide an adequate economic base for a thriving local commercial sector. Without viable local commerce, the local tax bases for city governments have stagnated or contracted. Increasing rural impoverishment increases white flight from rural communities, further homogenizing these towns into farm labour camps. The cultural, racial and class gulf between the local elite and the majority of the populace has resulted in increased repression within these communities. The local elite aims to develop the local economy, not through demands for regional socioeconomic equity, but through real estate speculation and government subsidies (Krissman, 1996a). This elite reflects concerns alien to the general populace, and represents outside interests in opposition to the community on issues ranging from health to policing. Such dissonance between the policies of the minority elite and the needs of the majority of residents can escalate to crisis dimensions, such as in the cases of the protests over McFarland's cancer cluster in 1984 and the US Immigration and Naturalization Service raids in Farmersville in 1994 (Krissman, 2001).

Policy recommendations

Although most of Goldschmidt's recommendations are unlikely to be implemented at this late date, those that would empower the workforce could still belatedly improve conditions in rural California. Such policies would permit towns that have devolved into farm labour camps to recover some of their pre-Second World War socioeconomic vibrancy. These recommendations fall within the general arenas of domestic politics, bilateral relations and domestic policy planning and implementation.

Firstly, public officials must recognize the impossibility of stopping undocumented influxes at the long and permeable US–Mexico border. The long-stated public policy of border 'control' has always been problematic because of the well-established linkages between many labour-intensive industries and immigrant workers – linkages forged by government to foster the growth of the nation's most important economic sectors, especially agriculture (Krissman, 1996b). If the USA suddenly determines that undocumented migration is counterproductive, out-migration from rural Mexican communities could be reduced significantly. American and international organizations could implement meaningful and long-term development projects within Mexico's labour-sending regions, especially inside the rural communities that have become centres for the reproduction of a huge migrant workforce in the USA.

Unfortunately, to date the actions of these entities have coincided with ongoing Mexican economic policies that actually exacerbate rural out-migration. Urban development through the extraction of rural resources is a longtime Mexican government strategy (Hansen, 1971). In the most recent Mexican crisis, these policies are now being supplemented by programmes of privatization, restructuring and austerity mandated by international organizations. The Mexican government recognizes that these policies exacerbate both rural poverty and outmigration, yet only a few demonstration programmes are aimed at softening their impacts within rural regions.

Meanwhile, within the USA there are domestic policies that can help to slow the 'revolving door' caused by continued migrant labour recruitment. The four policies suggested here are grounded in the need for increased socioeconomic equity – the same recommendation proposed a half century ago by Goldschmidt (1978, pp. 262–273).

1. Government can apply pressure to reform farm labour markets by hiring enough regulators to enforce labour, workplace and sanctions laws. Currently, the number of State regulators is shrinking even faster than labour-intensive crop industries in the agricultural sector are expanding.
2. The organization of farm labour into unions should be supported by the government to prevent labour market deterioration. Instead, Mexico recently launched a protest that existing labour laws in the USA are not adequately enforced in the case of farm workers.
3. The minimum wage must be raised substantially so that all work pays a living wage. The current substandard minimum has created a growing working class fraction – the 'working poor' – made up largely of disadvantaged minorities, women and immigrants.
4. The ultimate employers of labour must be responsible for the conditions of all of their workers, including those hired via subcontract arrangements with ethnic intermediaries such as those within FLC firms. Labour contractor firms arose out of efforts by agribusinesses to evade all responsibility for their workers.

Within the scores of farm worker towns a number of policies within two general areas could promote better conditions. Firstly, nonrepresentative local governments should be reformed in a systematic way, outlawing the systems that still exist in rural communities. Currently such practices are only undermined on a costly and time-consuming case-by-case basis. Goldschmidt (1978, p. xliii) underscored the crucial importance of farm worker participation in local politics.

Secondly, government must renew (in a comprehensive fashion) support for subsidized housing, medical care and other basic services for seasonal farm workers. Of course, the amount of direct government intervention required within rural communities would be determined by the level of State regulation in the farm labour markets. Government should decide whether it will continue to subsidize large agribusiness corporations by allowing them to maintain substandard conditions for the workforce. However, government seems to prefer to greatly expand its guestworker (e.g. *bracero* programme) policies.

Is there a need for such remedies beyond California? About US$150 billion in farm revenues are produced annually by between 2 and 3 million workers, an increasing majority of which are new migrants from Mexico. Krissman (1999) has documented increasing corporate control over the billion-dollar apple industry, as well as the deepening impoverishment of at least half a dozen Mexican majority farm worker towns in Washington State. The Californian corporate farm model has spread not only among the nation's crop industries but also to previously privileged labour markets such as meat packing, maintenance, construction, manufacturing and even 'high-tech' firms (Krissman, 2000). It is time for citizens to take a stand against the spread of the Californian agricultural model of prosperity for a few via a workforce made up of impoverished immigrants.

References and Further Reading

Bakersfield *Californian* (1979) McFarland's past, present, and future tied to farming. 6 August, p. 2.

Burawoy, M. (1981) The functions and reproduction of migrant labour: comparative material from southern Africa and the US. *American Journal of Sociology* 81, 1012–1050.

California Journal (1983) Farmersville: pride and uncertainty in California's rural cities, p. 10.

Cardoso, L.A. (1980) *Mexican Emigration to the US, 1897–1931*. The University of Arizona Press, Tucson, Arizona.

Galarza, E. (1964) *Merchants of Labour: the Bracero story.* McNally and Loftin, Santa Barbara, California.

Goldschmidt, W. (1978[1947]) *As You Sow: Three Studies in the Social Consequences of Agribusiness*. Allanheld, Osmun, and Co., Montclair, New Jersey.

Griffith, D. and Kissam, E. (1995) *Working Poor: Farm Workers in the US*. Temple University Press, Philadelphia.

Hansen, R. (1971) *The Politics of Mexican Development*. Johns Hopkins University Press, New York.

INEGI (1991) *La Situacion Actual de la Migracion Zacatecana*. Instituto Nacional de Estadistica Geografia e Informatica, Zacatecas City, Mexico.

Krissman, F. (1994) The transnationalization of the fruit, vegetable, and horticultural agricultural industries of North America. Paper presented at the *Latin American Studies Association* congress, Atlanta, Georgia.

Krissman, F. (1996a) Californian agribusiness and Mexican farm workers (1942–1992): a binational agricultural system of production/reproduction. PhD dissertation, University of California, Santa Barbara, California.

Krissman, F. (1996b) Border crackdown is a political trick: economic problems collide at the US-Mexican Border. *San Diego Union-Tribune* 30 January, p. B7.

Krissman, F. (1997a) Fifty years after: an ethnographic update of rural California's alternative futures. Paper presented in the session 'Culture and Agriculture Honors Walter Goldschmidt', at the *American Anthropological Association*, Washington, DC.

Krissman, F. (1997b) California's agricultural labour markets: historical variations on the use of unfree labour, c. 1769–1994. In: Brass, T. and van der Linden, M. (eds) *Free and Unfree Labour: the Debate Continues*. Peter Lang, AG, Bern, Switzerland.

Krissman, F. (1999) Agribusiness strategies to divide the workforce by class, ethnicity, and legal status in California and Washington. In: Wong, P. (ed.) *Race, Ethnicity, and Nationality in the United States: Toward the Twenty- first Century.* Westview Press, Boulder, Colorado.

Krissman, F. (2000) Immigrant labour recruitment: the personnel practices of US agribusiness and undocumented migration from Mexico. In: Foner, N., Rumbaut, R. and Gold, S. (eds) *Transformations: immigration and immigration research in the US*. Russell Sage Foundation, New York.

Krissman, F. (2001) Undocumented Mexicans in California: disenfranchising some of our best and brightest 21st century citizens. *Center for Southern Californian Studies' Working Papers Series*.

Krissman, F. (forthcoming) *Coming or Going? Mexican Farm Workers in California*. Forthcoming in the New Immigrants Series. Allyn and Bacon, Needham Heights, Massachusetts.

Kwong, P. (1997) *Forbidden Workers: Illegal Chinese Immigrants and American Labour*. The New Press, New York.

Liebman, E. (1983) *California Farmland: a History of Large Agricultural Landholdings*. Rowman and Allanheld, New York.

Mahler, S.J. (1995) *American Dreaming: Immigrant Life on the Margins*. Princeton University Press, Princeton, New Jersey.

Martin, P.L. (1985) *Seasonal Workers in American Agriculture: Background and Issues*. National Commission for Employment Policy, Washington, DC.

Martinez, R. (1997) Beyond borders: culture, movement, and bedlam on both sides of the Rio Grande. In: *North American Congress on Latin America*, January, 36–39.

McWilliams, C. (1939) *Factories in the Fields: the Story of Migratory Farm Labour in California*. Little, Brown, and Company, New York.

Menchaca, M. (1995) *The Mexican Outsiders: a Community History of Marginalization and Discrimination in California*. University of Texas Press, Austin, Texas.

Mines, R. and Alarcon, R. (1999) Family settlement and technological change in labour-intensive US agriculture. Paper presented at the conference, *Dynamics of Hired Farm Labor: Constraints and Community Response*, Philadelphia, October.

Mines, R. and Anzaldua, R. (1982) *New Migrants vs. Old Migrants: Alternative Labour Market Structures in the California Citrus Industry.* Center for US-Mexican Studies, La Jolla, California.

Picardo, N.A. (1995) The power elite and elite-driven countermovements: the associated farmers of California in the 1930s. *Sociological Forum* 10, 21–49.

Piore, M.J. (1979) *Birds of Passage: Migrant Labour and Industrial Societies*. Cambridge University Press, Cambridge, UK.

Rochin, R. and Castillo, M. (1993) *Immigration, Colonia Formation, and Latino Poor in Rural California: Evolving 'Immiseration'*. The Tomas Rivera Center, Claremont, California.

Runsten, D. and Kearney, M. (1994) *A Survey of Oaxacan Village Networks in California Agriculture*. California Institute of Rural Studies, Davis, California.

Runsten, D., Cook, R., Garcia, A. and Villarejo, D. (1993) The tomato industry in California and Baja California. In: *Case Studies and Research Reports, Appendix 1*. A Report for the Commission on Agricultural Workers, Washington, DC.

Steinbeck, J. (1967 [1939]) *The Grapes of Wrath*. The Viking Press, New York.

Taylor, J.E., Martin, P.L. and Fix, M. (1997) *Poverty Amid Prosperity: Immigration and the Changing Face of Rural California*. The Urban Institute, Washington, DC.

Los Angeles Times (1992a) California's cities: rich and poor. 6 July, p. A3.

Los Angeles Times (1992b) Farming: California's green economic oasis ... sector may be the most prosperous. 9 November, p. D1.

Los Angeles Times (1993) A growth industry: state's agribusiness rides out recession, insects and bad weather. 26 July, p. D1.

Los Angeles Times (1999) A Mexican state's crowning glory. 20 November, p. A1.

Section III

Farm Worker Health and Safety

Farm worker health and safety and access to health care remain important issues. Farm production is hazardous, both for farm family members engaged in farming and for farm workers. Farm equipment and tools pose safety issues, and the improper use of pesticides and herbicides can have major impacts on health status. In addition, musculoskeletal problems are common and long-lasting. Unfortunately, most farm workers are without health insurance and may lack access to health care facilities able to meet their needs.

There is a large and growing literature on this important issue. The three chapters in this section provide additional background for understanding some of the health and safety issues that farm workers face. Chapter 15 (Seiz and Downey) provides the detailed results of a pilot study undertaken in Colorado to better understand health issues faced by farm workers. Based on interviews with farm employees, the chapter describes in great detail the views of farm workers regarding their safety and the safety of their families engaged in farm work. The chapter underscores the hazardous nature of farm work, which can be made even more hazardous by lack of language skills and cultural differences.

Chapter 16 (McNamara and Ranney) examines farm worker and farmer participation in health insurance and shows that, despite the safety issues inherent in farm work, many involved in farm work in the USA lack health insurance. Chapter 17 (Hennessy) assesses the problem of access to health care among workers in the Alaska fishing and seafood industry. Hennessy notes that workers in these industries in Alaska ('fish workers') are very much like migrant and seasonal farm workers in the contiguous USA, and in fact may be one and the same. That is, the seasonal nature of farm work in the USA encourages migration to other seasonal 'off-season' jobs. The problem is that these workers are engaged in high-risk work in Alaska but have very limited access to health care – their access is limited by remoteness and by issues of definition in the law, as well as by their poverty.

Returning to the issue of technology development raised in Section I, an issue that is also raised in Section III is the extent to which new technology can reduce the health problems attributable to farm work. Technological change has provided capital (farm machinery and equipment) and other inputs to agriculture, including pesticides and herbicides, that have reduced the need for labour. However, the labour that continues to be needed – both hired farm worker and farm family labour – still encounters significant occupational health risks. The challenge now is to develop technologies that reduce this risk.

(eds J.L. Findeis, A.M. Vandeman, J.M. Larson and J.L. Runyan)

15 Safety and Health Attitudes and Practices in Migrant Farm Labour Families

Robert C. Seiz and Eleanor Pepi Downey
School of Social Work, Colorado State University, Fort Collins, Colorado, USA

Abstract

A pilot qualitative research study in Weld County, Colorado, was designed to obtain the perspectives of migrant farm labourer families on the occupational health and safety risks they face and to gain their insights into extant and potential preventive measures. Through the use of a structured interview schedule, over 700 pages of transcription data were collected from lengthy, in-depth interviews with migrant farm labourers and their families, conducted in migrant farm labour camps. An iterative process of qualitative data analysis identified four themes: risks, options/obstacles, strengths and supports. Considerations relevant to educators for the development, design and delivery of prevention programmes were extracted from the data and presented. Future directions in research with this and other migrant farm labour populations are offered.

Introduction

Since the 1960s, determined efforts have been mounted to lower the unacceptably high rates of agricultural workplace-related injuries, illnesses, disabilities and deaths (US Department of Health and Human Services, 1990, 1998). Federal and state governments have sought to enhance agricultural workplace safety by: (i) supporting research into the causes of farm accidents; (ii) facilitating the engineering of enhanced safety features into farm equipment; (iii) regulating chemicals, behaviours and practices; and (iv) developing education and training programmes for farmers and farm labourers in the handling of dangerous situations, in the use of hazardous substances and in the operation of powerful machinery.

Despite continuing efforts, farming remains one of the three most dangerous occupations in the USA (National Safety Council, 1997). The current disappointing statistics on farm-related injuries and fatalities indicate that more remains to be done to protect the safety and health of the agricultural population (US Department of Health and Human Services, 1998). There is an increasing awareness of a need for additional approaches, including behavioural approaches (Gunderson, 1995). Along with the important life-and-limb-saving safety devices and mechanisms that are being continually engineered into farming systems and equipment, an equally important safety con-

(eds J.L. Findeis, A.M. Vandeman, J.M. Larson and J.L. Runyan)

stant must also be considered: the person behind the wheel, using the tool, working on and under the machinery, interacting with the animal, handling the chemical, and climbing on the structure.

Organized educational and training programmes undoubtedly have an important role to play in providing farmers and farm workers with the types and kinds of information they need to maintain their safety in the workplace. Educational efforts, however, have not been proven to be as effective as originally hoped in reducing the elevated rates of agriculturally related accidents, diseases, illnesses and fatalities, and have come under increased scrutiny and criticism (Murphy, 1981; Edwards *et al.*, 1990; Elkind, 1992; Shutske, 1994; Rodriquez *et al.*, 1997; Quandt *et al.*, 1998; Thu *et al.*, 1998).

In some cases, governmental regulations have been shown to be effective in limiting the effects of accidents (Margolis *et al.*, 1996), but the vast majority of American farmers vehemently oppose any more governmental regulations and restrictions in the farm arena (Thu *et al.*, 1990) and would strenuously resist any regulation that they perceive as being manipulative, coercive, politically or commercially directed, paternalistic, or financially and/or culturally unrealistic (Green and Kreuter, 1991). In the current climate, it is 'highly unlikely that significant additional state or federal legislation could be passed that would effectively regulate the farm work environment' (Thu *et al.*, 1998, p. 162).

Perry and Bloom (1998) noted that for prevention programmes to be effective they must be responsive to the concerns and values expressed by the population to whom they are directed. Green and Krueter (1991, p. 21) warned that if professionals fail to consult adequately with the target population to determine their needs, problems and aspirations, they risk having their programmes and policies remain 'sterile, technocratic solutions to problems that may not exist or that hold a low priority in the minds of the [target] people'. Thu *et al.* (1998, p 162) urged that closer consideration be given to the 'financial and cultural factors that impinge on day-to-day farming conditions that inhibit the effectiveness of educational efforts'.

The pilot study reported here assumed that a comprehensive understanding of agricultural safety must include the human quotient in the equation of safety by discovering and taking into account the thinking, attitudes, beliefs, perceptions and practices of migrant farm labourers in matters affecting their own safety and the safety of their loved ones. The objectives of the study were to discover the thinking and opinions of migrant farm labourers and their families in five safety-related areas: (i) the nature and extent of health and safety risks; (ii) the accessibility and validity of safety information; (iii) the nature of behavioural changes required to implement safety information; (iv) the viability of implementing safety options; and (v) the existence and usefulness of safety-related support systems. The study used qualitative research methods in keeping with a growing trend in agricultural health and safety studies (Baker, 1995; Garkovich *et al.*, 1995; Perry and Bloom, 1998; Quandt *et al.*, 1998; Thu, 1998; Thu *et al.*, 1998).

Study Location

The study was conducted in Weld County, Colorado, in the northeastern section of the state and approximately a 1h drive north of the state's capital (Denver) along I-25. Weld County ranks in the top two counties in Colorado in the number of farms, with approximately 2909 farms and an average of 717 acres (290 ha) per farm (US Government, 1992). Over 2.08 million acres (0.8 million ha) of land in Weld County are farmland. Most farms operate on between 50 and 500 acres (20–200 ha) of land. Some farmers have over 5000 acres (2000 ha), while others farm as few as 9 acres (3.7 ha). Family histories of area farmers indicate that many have been farming in Weld County for two, three and more generations.

According to the Colorado Migrant Health Program, 97% of migrant farm labourers in Colorado are Hispanic, of which 67% come from Mexico and 21% from Texas. One-third (33%) speak English, another 18% speak limited English, and the remaining 48% speak Spanish only.

Seasonal cycles keep these migrant farm labourers and their families constantly on the move, and constantly at work. Local Colorado growers depend heavily on migrant farm labourers from May to October to plant, tend, harvest and ship their crops. Work in the fields is strenuous, the pay generally low, the working conditions arduous, and the living conditions basic. Although migrant and seasonal farm labourers are employed statewide in Colorado, most work along the US Highway 85 corridor, which runs through the centre of Weld County in northeastern Colorado.

Greeley is the largest city in the county and the home of the University of Northern Colorado. According to the *Greeley Tribune* (1997), approximately 4000 migrant labourers come to Weld County annually to work in the sugarbeet, pickle, onion and potato fields. There are four migrant labour camps in the area with beds for 400 workers. The Peckham camp has 12 units, the Gilcreast camp (the largest) has 30 units, the Hudson facility has 20 units and the Ault camp, which uses two Second World War Quonset huts, has 14 units. On average, five-plus people live in each unit.

Selection of Sample

The study reported here was part of a larger supported study by the National Institute of Occupational Safety and Health (NIOSH) that included in the sample not only migrant farm labour families but also farm operator families. In keeping with the guidelines for the grant, the study targeted only families with children between the ages of 10 and 18 years. This distinctive stipulation significantly limited the research pool from which the sample was drawn. Children are generally viewed as a population at particular risk for farm injuries. As such, they fit a priority research area established by the National Occupational Research Agenda (US Department of Health and Human Services, 1998). According to the US Department of Labor's National Agricultural Workers Survey for 1994–1995, approximately 7% of migrant farm labourers are 10–17-year-olds and are more likely to come from farm backgrounds. Also, 53% of migrant farm labourers 17 years of age and under live with at least one parent.

As a qualitative pilot study, no statistical formulation was employed to determine either the number or location of the participants recruited. The purpose of the pilot study, as well as logistical considerations, directed the conceptualization of the sampling frame, the data collected and the area canvassed. The study population was not intended to be reflective of views from a population-based representative sample; therefore, the study's findings must be interpreted with caution and any efforts to generalize ought to be resisted. Nevertheless, the study's data do provide a basis for further studies on the attitudes and practices of migrant farm labourer families.

Survey Methods

Sample: migrant farm-labour families

Seven of the ten migrant farm labour families were recruited for the study with the assistance of a social worker at a migrant job placement social service agency and an outreach worker at a rural migrant health clinic. The remaining three families were recruited using a snowball sampling methodology (Grinnell, 1997). The agency and clinic workers initially approached potential participants for the researchers and vouched for their authenticity and trustworthiness. Social relationships in the Mexican-American culture rely more on personal relationships than on formal social structures (Green, 1995). It was anticipated that members of the migrant community in Weld County would be more likely to accept someone recommended by a trusted person, rather than because they carried a formal credential.

Families willing to participate provided a time and location for meeting with a researcher for the interview. The ten migrant families identified themselves as Hispanic with roots in both Texas and Mexico. The Texas-based migrants came from the towns of

Eagle Pass, Pharr and McAllen near the Texas–Mexico border. They migrate because they are unable to find full-time employment in their own area. Most travel a 'circuit' through Arizona and California to Oregon, Nevada, Colorado and Wyoming. The participating families were two-parent families, with one to seven children living with them. Many of the families were represented by three generations. The ages of the fathers ranged from 25 to 48 years, with a mean age of 36.6 years; the mothers ranged in ages from 25 to 48 years, with a mean age of 33.3 years.

The ten families were all involved in weeding, thinning, topping, picking and loading produce such as onions, pickles, beets and grain. The work is often backbreaking. For example, to top onions, the worker must bend over all day pulling onions from the ground and cutting off the dried tops. This work must be interrupted during periods of sustained intense heat, since high temperatures will scald the exposed onions. Likewise, thinning and weeding crops cannot be done during prolonged periods of rain. None of the migrants reported using heavy equipment or machinery during the last 12 months.

Interview schedule

An in-depth, open-ended question interview schedule was used to elicit safety and health attitudes and practices from the perspective of the major stakeholders, who had the opportunity to present their own concepts, constructs, insights and ideas. Data collection involved face-to-face interviews using the pre-tested interview schedule (Gorden, 1992). The structured interview guide permitted comparison of responses between respondents (Patton, 1990), reduced interviewer bias, and increased consistency between multiple interviewers (Rubin and Babbie, 1993). The open-ended interview format provided respondents with opportunities to explain the rationale for their views fully, thereby overcoming limitations usually associated with surveys, polls and questionnaires that are formatted using limited-range response options. Participants were asked about their agricultural experience, history of injuries, knowledge of safety information and availability of safety resources. Questions were designed to solicit their opinions on the major cause(s) of farming accidents and about the decision-making components they used for determining when and how their children would participate in work on the farm.

Each participant signed the consent form and was provided with a copy. A confidentiality agreement was established to protect the privacy of the children and parents who voluntarily participated in the research. Permission was obtained from all participants to audiotape the interviews. A time frame was fixed when all primary source data (i.e. audiotape recordings and identifying information) would be destroyed.

Definition of terms

The definitions in Table 15.1 were employed in the development of the research design and interview schedule. These definitions also served as specifiers when the study's participants requested a clarification of meaning.

Data collection

The interviews were conducted by two researchers from Colorado State University, each with over 20 years of experience in interviewing. The ten migrant families were interviewed either in their residences in the migrant labour camp or in individual houses owned and provided by the farmers employing them. Parents and children were interviewed in separate groups, but parents were frequently either present or close by during the children's interviews. Prior to beginning the interviews, all participants received a US$50 cash honorarium per family for their participation. The families were told that they could terminate the interviews at any point without forfeiting the honorarium. The interviews were all tape-recorded. The tapes were valuable for three reasons: they

Table 15.1. Definitions.

Term	Definition	Reference
Accident	An unintentional event that leads to injuries which require professional medical care, or result in one day or more time lost from usual activities	Murphy, 1981 (p. 332)
Attitudes	A predisposition to react in a certain patterned way when confronted with a particular stimulus	Murphy, 1981 (p. 332)
	A relatively constant feeling that is directed toward an object, person or situation and entails an evaluative component	Green and Kreuter, 1991 (p. 158)
Perceived barriers	The potential negative aspects of a particular health/safety action that may act as impediments to undertaking the recommended behaviour	Janz and Becker, 1984 (p. 2)
Perceived susceptibility	One's subjective perceptions of the risk of contracting a condition or suffering a negative effect	Janz and Becker, 1984 (p. 2)
Risk factor	A biological or environmental variable that contributes to a problem occurring, makes the problem worse, and/or makes the problem last longer	Johnson, 1999 (p. 3)
Protective factor	A biological or environmental variable that contributes to preventing a problem from occurring, lessens its severity, and/or helps it get better faster	Johnson, 1999 (p. 3)
Values	The cultural, intergenerational perspectives on matters of consequence. Ultimately, the basis for justifying one's actions in moral or ethical terms	Green and Kreuter, 1991 (p. 157)
Strategy	A plan of action that anticipates barriers and resources in relation to achieving a specific objective	Green and Kreuter, 1991 (p. 435)

allowed the use of exact quotes of pertinent comments by participants; they assured the accuracy of the data collected; and, when necessary, they provided a check to clarify initial impressions and interpretations.

Seven of the interviews were conducted with the assistance of a Spanish-speaking interpreter. Family members and a migrant outreach worker assumed this important role. To ascertain the accuracy of each translation, a bilingual student was recruited to review the tapes and compare them for accuracy with the typed transcriptions. She provided the researchers with little additional information that had not been translated by the interpreters and certified that the interviews had all been accurately translated and transcribed.

All interviews were conducted between June and August 1998, a period when migrant farm workers were available and a period of intense activity called 'push time' by local farmers. Despite attempts to limit the length of the interviews, they averaged 2.5 hours.

Data Analysis

As an exploratory study, primary emphasis was placed on gathering impressions as well as facts. Toward this end, open-ended questions were used to capture qualitative information on the research questions. The analysis of the data is primarily descriptive and thematic. A descriptive account is valuable in itself for unravelling the complexities of group perspectives and attitudes regarding quality of life and safety issues.

The audiotapes were transcribed by the graduate research assistant and the transcriptions were reviewed by the researcher conducting the interview. Transcriptions were reviewed to ensure their accuracy and to fill in difficult-to-understand passages. There were 11 quantitative questions on the interview guide that requested respondents to answer using a 10-point scale. These data were extricated and coded separately for quantitative analysis.

The reliability and validity of qualitative data hinge on the ability to convince the consumer of the research that 'the findings of an inquiry are worth paying attention to, worth taking account of' (Lincoln and Guba, 1985, p. 290). The ability to convince is strengthened by the qualitative techniques used, including triangulation, multiple informants and cross-checking the accuracy of transcriptions. To assure the accuracy of the data, transcriptions were completed by the research assistant within a week of the interview and then compared with the audio recording for congruence by the researcher conducting that interview. Corrections were made prior to data analysis. In the case of interviews conducted in Spanish, the bilingual student independently listened to the tapes while simultaneously reviewing the English transcripts.

An iterative process of qualitative data analysis was conducted to draw out major themes from the participants' perspectives and attitudes regarding the quality of life and the role of safety in the lives of their farm families. For the purpose of qualitative data analysis, 'a theme is a pattern found in the information that at a minimum describes and organizes the possible observations and at a maximum interprets aspects of the phenomenon' (Boyatzis, 1998, p. 4). During the first phase of analysis, the researchers individually sorted the interview content contained in over 700 pages of transcriptions into the study's five objectives. Following the independent sorting procedure, the researchers worked together to synthesize and integrate their findings. Subsequently, through a process of comparison, reflection and lengthy discussion, four major themes were identified and agreed upon by the researchers: (i) risks; (ii) options/obstacles; (iii) strengths; and (iv) supports.

Risks and options as reported by migrant farm labourers

Working for minimum wages, most migrant labourers are employed without many of the amenities the average American worker views as entitlements: paid sick leave, paid vacations, public holidays and medical insurance (Baker, 1995). The ability to work safely in such an environment speaks to the strength and resiliency of the ten families that were interviewed. Saleebey (1997, p. 13) may well have been describing this population when he wrote: 'Many people who struggle to find their daily bread, a job, or shelter are already resilient, resourceful and though in pain, motivated for achievement on their terms'. It is these and other inherent strengths that contribute to migrant labourers' attempts to create and maintain a safe and healthy work environment for their families.

Risks

EXPERIENCES WITH INJURIES

Almost half (40%) of the migrant farm labour families interviewed admitted to having experienced a farm-related injury that was serious enough to warrant medical attention and/or resulted in lost days of work. In addition, 70% had knowledge of specific individuals who were injured to this degree while working on a farm. The injuries reported included broken bones, severed or lacerated fingers, back pain and spasms, chemical inhalation, impact injuries, and, in one case, the loss of a limb. All but one of the stories involved either themselves or family members.

A few respondents reported suffering significant injuries. For example, a 32-year-old Hispanic father of five showed the interviewer three badly scarred fingers on his left hand. The fingernails appeared darkly discoloured and cracked. He related that his fingers were broken 'picking up tracks ... had to move all the tracks off the rollers to fix them ... one of the bolts broke and hit my hand'. This same migrant farm labourer reported suffering from severe back spasms. As he tells the story:

> It can happen here any time, because we're working the pickles, and by the end of the day each family has over a 100 bags of pickles and we have to pick them up and throw them in the truck and if you have 15 families that's 1000 pounds ... you can easily throw out your back, pull a muscle, herniate a disk, you know ... you think twice about it ... Currently I'm on a TIMS unit ... back spasms because of all the heavy lifting and stuff ... they [doctors at the migrant health clinic] recommended a TIMS unit ... actually I'm using it now. [*Interviewer*:] How frequently do you use the TIMS? [*Migrant*:] I have it right here, as soon as I take a shower, I'll leave it on all afternoon, all night and take it off in the morning and go to work.

This same worker reported another injury that resulted in 2 to 3 days lost work when he cut his left thumb.

> I didn't lose my finger, [but] it was a pretty good cut. I was cutting spinach and you cut it with a knife you know, and suddenly I just was cutting and hurt my finger by myself and I went to the doctor and [he] gave me some stitches on it and got to wait 2 to 3 days 'til it heals and then I start working again.

The majority of the migrant respondents reported knowing about others who were injured seriously while working on the farm. The 32-year-old father of five related a particularly graphic story about his brother in Texas, who still suffers significant symptoms from an accident that happened 10 years ago.

> See, that happened to my brother, he got a bad case of poisoning. We were out doing pesticides and all of a sudden I saw him and the machine going round and round in the middle of the fields, and I thought, what is he trying to do, get us fired? When I went to approach him I saw he had blood coming out of his nose and he was passed out and I guess he got a bad dose of the poison in him ... he's had ... for a while we thought he had muscular dystrophy because his muscles just haven't been the same ever since ... and this was like 10 years ago ... involuntary contractions in his hands, shaking so bad ... I guess his symptoms are not as bad as they were but you can still tell, his hands are still shaking a little bit, and he was wearing masks and everything, but they weren't filtered ... he was down wind from me.

This same migrant told of another accident that was, for him, an object lesson in the dangers of working around powerful machinery. He related:

> I work with cotton pickers back home with my father and another farmer ... they had shown us a film about the new cotton pickers ... the first thing they said was 'don't reach in and try to pick the cotton out of the spindles' ... spindles going around and round that pick up the cotton ... I seen this guy I was working with who didn't see the film and my dad said, 'Didn't you see the film? Don't be reaching in and try to clean up the spindles while the machine is running cause it might pull you in' ... when the guy acted distracted and my dad said, 'Alright, I told you, I warned you' ... my dad turned around to work in the field and I saw the machine pull the guy in and yank all the meat off his arm ... you know, four drums of spindles with 800 spindles, you know, it just literally tore the meat off his arm and there was nothing but bone.

The 21-year-old son of a migrant family told of the injury his younger sister sustained while cutting onions.

> Working on contract we go real fast; like last year my sister cut her finger working on the onions, working real fast, and the months are still cold, she couldn't even feel her finger, she chopped off her finger ... at the knuckle [*points to knuckle of his right index finger*].

A 24-year-old Hispanic mother of three told of her younger sister's injury working in the fields.

> [*Mother*:] We were picking cantaloupe, melons ... and it's big van and in order to put it on it ... It's a big thing that the tractor has on the back, and the thing fell down on her back and she

> still like has some pain sometimes ... I felt sad because she went to a hospital and she was OK for what is it ... less than a year, then it started again, and I feel sorry for her because they put shots in her back and that hurt. [*Interviewer*:] Does she still have pain? [*Mother*:] Yes, she has pain. [*Interviewer*:] Does she still work? [*Mother*:] Yes. [*Interviewer*:] How old was she when it happened? [*Mother*:] Like 16, she's going to be 17 ... Now she can't walk as much because she feels tired, she can't swim because it hurts, her back hurts ... she can't run, because of the same thing ... From what happened, she just wants to be asleep ... she doesn't want to do nothing. Right now she's sleeping.

The 26-year-old Hispanic male told of his father's strong influence on his whole family in teaching them about safety. The father used his own incapacitating injury as a lesson for the family.

> And he tells us all the time be careful, it all depends where we're at you know, like if we were right here, just tell us be careful when you carry a sack or something, my dad had an accident like that, carrying a sack over in Oregon. He got a disk moved by carrying a sack of 80 pounds. And he always told us, 'You see what happened, you've got to be careful, you've got to have your belt on,' and all that you know. He always told us a lot of things you know, carry water, a gallon of water in the field, always be drinking and everything like that.

CHANCES OF BEING INJURED

The migrant farm labourers were asked to make a subjective prediction, using a scale of 1 to 10, of how likely they thought it was for them to be injured on any given day while working on the farm. Eight of the ten who responded to this question predicted moderate chances (4–7) of being injured on any given day. No one estimated a high probability, while two predicted a low probability.

Those predicting a low chance of being injured tended to see fewer dangers. A 27-year-old mother of six felt her only danger was in 'handling the tools'. She worked with hoes, knives and scissors. A 41-year-old mother of three and grandmother of four did not think she would hurt herself because, as she stated, she always consciously tried to 'work in a safe way'.

Other respondents were not so certain. The 32-year-old father of five related:

> Everyday I get up and put my pants on and think there is a 70% chance I am going to get injured that day because of the job that I do ... but if I take the necessary precautions and the time to do my job the way it's supposed to be done and not the way I want it to be done I figure I might get through without an injury ... but the farmers I work for and the employers I have are demanding and they push you to the point where what is logical is illogical and you just want to get through ... I believe that when I wake up and pull my pants on that morning that I have a 70% chance I'm going to get injured that day ... whether it be a small or a big injury due to people rushing us.

Others attributed their greater chance of injury to the uncertainty of not knowing what others might be doing, or what in general will happen. A 24-year-old mother of four was concerned about the presence of unknown chemicals in the fields and gave herself a 70% chance of injury 'because we don't know what [chemical] they are throwing, if they are dangerous to us and if we are going to get sick'.

NATURE OF THE RISKS

Respondents identified pesticides, non-powered cutting tools, pests, wild animals, weather, lifting heavy objects and sanitation conditions as constituting the primary risks to their health and safety. Most of the respondents cited only one or two – at most, three – hazards from the above list. No one cited all of them. Eight families, a substantial majority, specifically identified pesticides, while four cited cutting tools, three the hazard from snakes, one mentioned parasites, three identified the weather conditions (lightning and the hot sun) and one pointed to heavy lifting.

Notably absent were concerns about tractors, farm machinery, ponds, farm structures, silos and farm animals. While respondents failed to mention these latter risks specifically, it cannot be concluded that they went unrecognized by the respondents. During visits to the labour camps by the interviewers, parents were observed warning their children to avoid going into a field next to the camp where cows could be seen grazing. In two separate instances, children were cautioned about going close to a shallow pond approximately 300 yards behind the camp where unaccompanied migrant farm labourers would typically fish for an evening meal of carp and catfish. Also, among the injury stories told by the migrants were stories of relatives and acquaintances being struck or grabbed by farm machinery. It is reasonable to infer that, at least for these workers, the dangers involved in being proximal to farm animals, working with farm machinery and playing near or on unguarded natural hazards are real, known and respected.

As stated, a majority of respondents identified chemicals, especially pesticides, as the primary risk facing them. The 27-year-old migrant mother of six told of her family's reliance on the farmer to know when this danger is present.

> Well we have to be ... we have to know, like for the pesticides. That they are using pesticides we know that we have to get out and not to go in ... some are 24 hours, some are 48. It all depends on which kind they use and we know the farmer, the owner tells us when to go in.

The 32-year-old Hispanic father of five echoed this concern about the dangers of poisonous pesticides, saying that 'the poisons they use to destroy bugs in the crops and stuff ... We have to be aware of which poisons are harmful and which ones aren't'. This father also identified dangers from prolonged exposure to the hot sun and from wild animals hidden in the fields ready to strike the unsuspecting field labourer: 'Mother Nature ... wild animals in the fields, snakes and stuff like that ... Heat stroke and stuff like that. Sweating so much ... have to drink a lot of water to not get a heat stroke.'

The 34-year-old mother of three, however, seemed to downplay the dangers, philosophically placing them in a comparative context with other occupations. She did not deny that the dangers in the fields are real and present, but simply refused to grant them extraordinary status: 'Well ... I guess if you grow up doing this you won't find it so different from what other people do ... because, you know, everybody's got their own things they do ... I don't know ... to me it seems like not anything weird or dangerous.'

A 41-year-old mother of three spoke poignantly about the head-in-the-sand attitudes of some migrant labourers. 'A lot of them don't believe about dangers, there is no need to for safety out in the fields ... They don't believe that there is danger out there.' Others commented on the dangers posed by the sharp hand tools used in harvesting the crops. The tools mentioned included hoes, knives and scissors and several specifically commented on the sharpness of these tools. An older brother insisted that his younger sister, who was just starting out working in the fields, should only use scissors, never a knife, for cutting the onions, 'because [scissors] are safe[er]'.

CAUSES OF ACCIDENTS

Most respondents avoided attributing a single cause to any accident. Instead, they either tended to see multiple causes, or recognized a difficulty in trying to discern one cause easily from among the complexity of possible causes. Taken in the aggregate, however, the causes cited for accidents can be divided into two categories: those that lay within the control of the individual and those that were beyond individual control. The causes most cited by respondents were those within the control of individuals to some degree. Yet it is always difficult to divorce the decisions made by individuals from the environmental conditions and circumstances that influence, and may at times even control, those choices.

Causes deemed within individual control included such things as a lack of focus, improper use of tools, alcohol impairment, negative attitudes, excessive speed and fatigue. The causes cited as being beyond individual control were unwitting exposure to pesticides and fields that were improperly prepared prior to harvesting.

The following excerpts provide a cross-section of opinions by the migrant farm labourers in the study about the causes of farm accidents.

- 'Stupidity ... trying to beat the clock, not taking the time to do it right.' [*28-year-old male*]
- 'Carelessness ... That's the main one right there ... if you don't pay attention to what you're doing ... Not giving a hoot.' [*37-year-old male and 34-year-old female*]
- 'She [his sister] chopped [her finger] off ... it happens basically because we were working too fast and in the sun, we get tired, not enough water.' [*21-year-old male*]
- 'Not paying attention to what they're doing while they're working ... Like there are people that are going to be drinking at the same time, if you're using the tools that can be dangerous because you're drunk and you don't know what you're doing. Or either you're talking to someone else and you're not paying attention to what you're doing ... Not using the tool you're supposed to use, like to work or stuff like that.' [*45-year old male*]

It was clear throughout the interviews that almost all of the respondents viewed pesticides as posing a distinct danger to their health. Yet curiously few, when directly asked what causes farm accidents, cited pesticides in their responses. This was not the case, however, for a 24-year-old mother of three who, when asked by the interviewer for her opinion of what causes farm accidents, used a crisp, one-word response: 'Herbicides.' Her 26-year-old husband was quick to enlarge on her answer: 'Herbicides and pesticides.'

Options

Options were defined as recognizing alternatives, identifying choices and feeling personally empowered to choose freely between viable alternatives. Having and exercising options entails, in part, a subjective determination of how much control one believes one has over one's behaviours and choices. Respondents were asked to estimate, using a scale from 1 to 10, the level of control they felt they had over their own safety on the farm. Parents were also asked to estimate the level of control they felt they had over the safety of family members.

LEVEL OF CONTROL OVER SELF

A significant majority (77.7%) felt they had high levels of control over their own safety on the farm. High levels were defined as scores at 8 or higher on the scale of 10, moderate levels were scores between 4 and 7 and low levels were scores of 1 to 3. Two respondents (22.3%) were identified as believing they had moderate control. Significantly, no one saw themselves as having low control.

> [I] take responsibility for making sure the tools are handled properly so everyone can stay safe ... 90%. [*27-year-old female*]

> Yeah, because if it starts raining and I see lightning I know that I'm coming home, I don't want to get hit by lightning, because I think I have control over my safety out there. [*37-year-old male*]

> It depends what kind of work we are doing. Like picking the vegetables I need to wear gloves because I use a knife, when we are doing onions it's the same thing ... and right now we are picking up the weeds, we need to wear like hats and all the stuff, so it depends on what kind of work we are doing. [*Interviewer*:] So, just in general, how much control do you think you have? [*Migrant*:] Like 8. [*24-year-old female*]

LEVEL OF CONTROL OVER THE FAMILY

A majority of respondents felt they had more control over the safety of members of their families than they did over their own safety. Four respondents gave a 10 rating. The lowest rating was a 5, and that was by only one respondent. This would be a curious phenomenon except when viewed within the cultural perspective of the dominant Mexican-American family structure where parents tend to exercise tighter control and influence over their children's behaviours (Green, 1995). The following excerpts provide a flavour of the respondents' reasoning.

> [*Interviewer*:] How about your family's safety on the farm? [*Migrant*:] 100% ... I decide whether they do it or not do it because it's my

> family ... about me, I'll slide a little bit, but when it comes to my family I got 100% control. [*32-year-old male*]

> A 7 ... If [I] see something that's wrong with one of the kids [I] stop them right away. [*45-year-old male*]

> I go for an 8 to a 10 ... because I always be looking at my sisters and my brother-in-law, be careful there, take care there. [*26-year-old male*]

> [*Migrant*:] Yeah, 10 ... Because every day before they go out to work [I] explain how they should take care of themselves and what needs to be done to prevent accidents. [*Interviewer*:] Is this a ritual every day? [*Migrant*:] Yeah. [*41-year-old female*]

Obstacles

Despite their claims of being in control of their own safety, many migrants expressed feelings of not being empowered to take alternatives that could directly affect their safety. The migrant farm labourers in this study appear to distinguish between situations over which they can exercise control without jeopardizing their means of livelihood and those that may place that livelihood at risk. Living continually on the edges of poverty, they have an understandably strong motivation to do all that needs to be done to maintain their jobs and earn that next paycheque, even if that means cutting corners and taking risks they otherwise might not take. A 48-year-old migrant mother of four put it succinctly: 'Everybody has to do that around here ... in the fields ... take our chances.' Other migrants echoed similar sentiments.

> [*Interviewer*:] Is there a reason you wouldn't say to them [the farmers] 'I don't want to go in there because you just sprayed?' [*Migrant*:] No, we can't, we need money. [*Interviewer*:] You wouldn't feel free to say that? [*Migrant*:] No, because we need some money, like right now we just got to work, we need to work. [*24-year-old mother of three*]

Other migrants identified youthful exuberance, the desire to please, and excessive competition as obstacles to making good judgments. One migrant opined that perhaps one only learns to exercise good safety options through experiences that come with making mistakes and gaining maturity and mastery, provided one is fortunate in surviving long enough without a serious injury to learn those lessons.

> You bend down and you just work because it hurts your back ... because the more you lift, they pay like 10 cents a sack ... they got about 5 guys and whoever picks up the most gets a bonus at the end ... Lately, yes, at first no ... I would have to say that throughout the years I got more [safe] ... Through the years with experience it's kind of grown on the inside ... at first no, I didn't. I'd do anything, try to impress your boss, take risks, that's when injuries occur ... takes time to figure out what's going wrong ... maybe get by injury free. [*32-year old male*]

This migrant labourer spoke soberly about his youthful days working in the fields and about his skewed feelings of being invulnerable, of being able to 'beat the machine'. He sobered from his youthful hubris only after he saw at first hand the unforgiving and ghastly consequences of losing to the machine.

> I was younger, faster, and I could beat the machine ... but what happened to him scared me to death ... he lost his arm because he thought he was faster than the machine ... it could have been me, you know ... never again ... I decided to turn the machine off and take more time ... it's not worth it to lose an arm because of negligence.

Paradoxically, this same migrant labourer admitted to the likelihood of his taking unsafe risks just to get the job done sooner so he could go home early: 'I probably would ... yeah ... probably take a risk ... know I'm not supposed to but, if I really wanted to get out of there and go home and I know I needed to pick 2 more, I probably would, ignorance would probably take over.'

Other obstacles to exercising good safety practices came from the nature of the job coupled with the dictates and vagaries of the weather. Several migrants saw themselves at the mercy of the weather in terms of having to do what has to be done when the weather conditions permit, and even sometimes when they do not permit.

> Well, in the spinach ... if it rains we can't go in because the spinach has to be dry in order to cut it, and, if it's raining, a lot of the spinach grows to a certain point that it's not good. And for the onions they have to be dry in order to cut them. The pickles ... cause we're going ... we're taking a certain amount of the money [on contract] ... so we have to go even if it's raining. [*27-year-old mother of six*]

PREVENTION OPTIONS

Some of the migrant participants did feel freer to implement safety options in areas they viewed as being part of their zone of responsibility. With a mindset of self-reliance, they viewed themselves, not someone else, as being essentially responsible for their safety in the fields. These migrants customarily assumed responsibility for themselves and their families, and did not rely unduly or unrealistically on others to watch out for them. It was in the arenas where they took responsibility that they felt especially empowered and in control of their own safety. To a degree, and in some matters, they knew they had to rely on the candour and probity of others, but they also recognized there were things they could do for themselves with a little forethought and knowledge. There were steps and actions they could and would take as precautions.

To deal with the hot sun, migrants counselled the use of 'a large hat' and carrying jugs of water into the fields with them. One family regularly carried 25 gallons (114 litres) of water with them into the fields and would send out to get more if the water ran out.

> Whenever we ran out of water someone comes in to get more ... we're out and we're far from the house, we're just out, the first farm we get to we ask for some water ... we never drink from ponds or from the ditches or whatever. [*27-year-old mother of six*]

The 32-year-old father of five asserted the argument that it may actually be better for the safety consciousness of migrant labourers if employers were to encourage and expect labourers to take more care of their own safety needs in the fields. The thrust of his argument is that when labourers are expected to assume responsibility in those areas of safety over which they have some control, this helps to sensitize them to potential dangers and motivates them to stay alert and exercise due diligence. Using himself as an example, he continued later in the interview:

> Oh yeah, we got our own gloves and water jugs, we go prepared. There's no need for us to be dehydrated all day long and waiting for somebody to do you a favour, you can do yourself a favour and supply it yourself.

A 26-year-old male and his wife added the significant qualifier of 'experience' to the notion of being able to care for oneself in the fields. Left unanswered was how much control and responsibility one can expect from young migrant workers who lack knowledge, experience and, often, maturity. And what, if anything, should be done to help them gain the necessary levels of knowledge, experience and maturity?

> [*Male*:] Well, like ... a lot of people go in [to the fields] and they [farmers] just tell you just work and that's it ... they don't tell you hey, you gotta have your safety, you gotta do this and be careful or do that or something like that, and take your safety. But most of the time they don't tell you, [they] just [say] this is the field and that's the price and that's it ... and ourselves we got to take our safety.
> [*Interviewer*:] Do you feel [farmers] assume you know how to take care of yourself and that's why they don't tell you? [*Male*:] Well, probably, because, like they say, we're migrants and they know that we already working in the fields you know, that's why they know that we know our safety already, take care of it and everything.

Besides citing the general prevention options of wearing large-brimmed hats, carrying water with you into the fields, and having and wearing gloves, several migrants focused forcefully on the necessity of forethought and concentration when working in the fields – of concentrating 'on what you are doing ... and just take care of looking and be sure [of] what you're doing, and don't get distracted ... turn the radio off and everything, just concentrate on what we're doing'. And regularly seeking to 'be more careful, work slow, and see how it is, everything, before you start ... see before [you] do, think about what [you're] going to do, and be serious, it's gotta be right'. One migrant summed it up well when he said:

> Yeah, there is always something going through the back of my mind especially working with equipment ... try to foresee what might go wrong before hand ... especially when you are underneath heavy equipment ... can almost visualize something going wrong before it happens and try to take precautions ... There's not much you can do except take all the necessary precautions ... If something is going to go wrong, it's going to go wrong no matter what you do but at least you know in your mind that you did everything that you could for it not to happen ... if it happens it happens, regardless ... I've seen jacks go out and a 20 ton machine just squash a man and that's with air bags all around the machine ... and jack stands. [*32-year-old male*]

The migrants in the study overwhelmingly favoured working under a contract arrangement, rather than working for an hourly wage. They saw the contract option as giving them greater control and flexibility. With a contract, they felt better enabled to make judgements for themselves under changing situations affecting safety, without fear of penalty or retribution.

> When we're working by the hour ... not real fast ... Just doing the nice job but not to make it sloppy, and when we're working by contract we can stop whenever we want ... when we get tired we can go like fast and then stop for 30 minutes and have a snack or whatever and then go back to work. [*32-year-old male*]

MAKING GOOD SAFETY DECISIONS

One way the study sought to gauge migrants' perceptions about options was to ask them if, in general, they believed they made good decisions relating to safety. Once again, respondents were instructed to use a scale of 1 to 10.

Two respondents (22.2%) believed they usually do not make good safety decisions and scored themselves at 1. One wonders whether these migrants fully understood the question, or whether its meaning was lost in translation from English to Spanish and back again. When asked to elaborate on their answer, they could not. The remaining 77.8% distributed themselves between the mid to high ranges, with 44.4% scoring between 6 and 7, and the remaining 33.4% between 8 and 10. Those who scored high tended to have a comprehensive attitude about safety – an attitude summarized by the 34-year-old mother of three who said, 'Yeah, we don't take too many chance out there.' Those scoring in the mid range admitted to being influenced by circumstances that pushed them to bend the boundaries of safety.

> I give myself a 6 because I do, sometimes I don't ... Like sometimes we need the money, so we have to work, it doesn't matter if it's been sprayed or anything ... we have to work if we need the money ... sometimes there's not a lot of work and sometimes there is, like yesterday we really didn't have to work. [*Note: it rained.*] [*45-year-old father of six*]

The migrant labourers were next asked whether they were generally satisfied with the safety decisions they made. Respondents generally tended to score levels of satisfaction at the same levels they scored themselves on the previous question. Those who believed they generally do not make good safety decisions tended to be dissatisfied with the safety decisions they made. Those who scored themselves higher tended to be more satisfied with their safety decisions.

Remarkably, many migrant workers in this study believed they had choices available to them that could lessen their risks and decrease their exposure to danger while working in the fields. A statement of this belief can be found in the comments of a 24-year-old mother of three. After defining choices as 'to either do something, not to do it, or to do it a different way', she was asked if she thought she had any choices available that would lessen the dangers of working on a farm. She replied, 'We do whatever we want ... if we like the work we do it, if we don't, we don't do it. Like, we 3 weeks ago quit a farmer. If we don't like the work we quit it. We have too much choices ... We do whatever we want.'

Strengths and supports reported by migrant farm labourers

Strengths

THE MIGRANT FARM LABOUR FAMILY AS A WORKING UNIT

Families serve as the primary group of self-identification for most individuals and it is within the context of the family that most human beings conduct the major components of their lives (Schriver, 1998). The same may be said of migrant farm labourers who not only live together as a family but share the same work environment. Family members, who were interviewed for this study, reported that often the entire family goes into the fields; some children work alongside their parents, and for younger children the fields are their day-care setting.

Several fathers recalled their own experiences as children growing up in a migrant family and the ways the family worked together in the fields.

> Every time, he [his father] always kept us together, all my family, all my brothers and my sisters and my mom ... We were always together, if we came over here we were together, if we go to Florida, we'd work together, all the time together. [*26-year-old male*]

A 16-year-old man, who worked 7 h a day in the fields alongside his parents, sisters, brother-in-law, nieces and nephews was asked about his early experiences working in the fields. He replied, 'My dad said I could [work] because I was sitting down in the truck and he told me that I could go into the fields. He [father] would then go in, and I would be just in the middle of my dad and mom, so they could watch me.'

FATHERS AND GRANDFATHERS AS SAFETY EDUCATORS

The first and continuing lessons on agricultural safety come from within the structure of the family and are delivered as on-the-job-training. Several of the interviewed parents reminisced about working in the fields as children and the role that their grandfathers (as well as their fathers) had in providing them with safety education and direction.

> He [father] tells us all the time to be careful, it all depends where we're at ... if we were right here, just tell us to be careful when we carry a sack or something. My dad had an accident like that ... He always told us, 'You see what happened, you've got to be careful, you've got to have your belt on and all that.' He always told us a lot of things ... 'Carry water, a gallon of water in the field, always be drinking.' [*32-year-old male*]

> My dad always tells us, 'Be safe, don't talk in the fields, you can talk when you're resting, but take care of yourself.' [*37-year old male*]

Pesticides and other agricultural chemicals found on farms are seen by most migrant labourers as the most dangerous aspect of their work. Children were taught about their dangers and were instructed to avoid them. It appears to be the father's role as head of the family to obtain new health and safety information about the chemicals and to communicate this information to his family.

> We grew up on a farm and there's always fertilizers and all kinds of pesticides lying around. My dad taught me to stay away from them. General knowledge came from our parents ... any particular item we didn't understand ... learn from the boss ... if my father didn't know I'd get the information from someone else. [*32-year-old male*]

When three generations of a family are working together it was not unusual for health and safety information to be dispensed by a grandfather. A father of five talked about his own experience as a child and the role that his grandfather played in keeping him safe in the fields.

> [*Migrant*:] He [grandfather] was extremely knowledgeable because he took precautions with everything he did ... took time to learn the good and bad things about everything he did, what to do and what not to do. [*Interviewer*:] You credit him with a lot of knowledge? [*Migrant*:] Oh, yeah, my grandfather's knowledge ... my knowledge came from them ... he set the standards and if he didn't do it, you couldn't do it ... if he didn't do it there was a reason and if you didn't know why he would tell you why. [*Interviewer*:] It sounds like he was the type of man that explained things to you. [*Migrant*:] He wouldn't want you to do anything he wouldn't do himself ... and he wanted you to know why he wouldn't do it ... he would teach us why we couldn't do it ... what could happen and what couldn't happen ... basically he teaches us. [*32-year-old male*]

The role of the multi-generational family in farm safety education was mentioned by the current generation of adolescents.

> [*Interviewer*:] Who are the best people to teach you about farm safety? [*Adolescent*:] My grandfather. He is always with us, telling us about the fields and everything. [*Sibling*:] When he was a little boy, a young man, he used to work with his father and mother in the fields.

It appears that information and direction came through the existing Hispanic patriarchal system to the younger generations and, therefore, children might well seek out advice, guidance, or information from either a father or a grandfather. Mothers, too, play a significant but somewhat different role in monitoring their family's health and safety.

MOTHERS AS SAFETY EDUCATORS

Mothers keep an ever-watchful eye on their children in the fields and also seek out spiritual support to protect their children. When asked who could do the best job of teaching her children good safety practices, a 41-year-old mother of six replied that she was in the best position to teach them. When asked why, she replied, 'Because I am their mother and I care what happens to them and nobody is going to care like me!'

A 33-year-old mother commented on her role in keeping her children safe:

> I keep an eye on the smaller kids because there's an irrigation ditch right there, and I don't want them to go there. Because I am afraid they might fall in there, and I tell my daughter, don't let her [sibling] go out there because she might fall in the ditch and never get out because the dirt is kind of loose, so that why we always keep an eye on them.

In some families, mothers are responsible for the spiritual aspects of keeping the family safe. According to one adolescent daughter, 'My mom, every time one of us gets sick, she ... asks God or a promise that she makes or something like that.'

The 43-year-old mother of three, living and working with an extended family of eleven, mentioned a similar source of help and support against the dangers on the farm. She was speaking through an interpreter:

> [*Interviewer*:] Does she have any philosophy of life or any religious beliefs that help her deal with some of the uncertainties of working on a farm? [*Interviewer*:] Yes. [*Interviewer*:] Could she explain? [*Interpreter*:] She has a lot of faith in God. Every day she prays before we go out. [*Interviewer*:] For what? [*Interpreter*:] For their safety.

SAFETY AS A FAMILY NORM

A mutual concern and responsibility for the safety of each family member appears to be a well-understood family norm. A 16-year-old adolescent male told the story of killing a rattlesnake that his family had encountered while working in the field.

> [*Interviewer*:] Have you ever done anything you knew was not safe? [*Adolescent*:] I killed a rattlesnake that almost bit my sister. [*Interviewer*:] You did something unsafe to protect your family. [*Adolescent*:] Yeah. [*Interviewer*:] Would you do it again? [*Adolescent*:] Yeah, to protect my family, I will do anything.

In a similar vein another young man talked about working too long in the fields when he was tired. Nevertheless, he had stayed on to finish the work. When asked why he did that, the young man responded, 'I was trying to help my family.' The interviewer asked 'Would you do it again?' 'For my family, yeah,' the adolescent replied.

Another young man talked about the need for protecting his siblings when there are chemicals in the fields.

> When the people are working they have to tell us they're going to put pesticides or whatever ... and tell you not to go because it can be dangerous or not to get near the fields because we have little brothers and sister that go. So like don't let the kids over there because there are pesticides. [*19-year-old male*]

A 32-year-old father of five summed up his role and responsibility when asked how much control he had over his family's safety. '100% ... I decide whether they do it or not do it because it's my family ... about me, I'll slide a little but when it comes to my family I got 100% control.'

For the migrant family that lives and works together, the existing family structure and norms provide a powerful mechanism for maintaining safe working conditions. The multi-generational nature of these families, particularly the presence of an experienced grandparent, is a further strength as support in the working environment. The families in the study appear to reflect a strong sense of familism, a significant characteristic of the Hispanic family. According to Schaefer (1998, p. 286), 'By familism is meant pride and closeness in the family, which results in family obligations and loyalty coming before individual needs. The family, therefore, is the primary source of both social interaction and care giving.'

From conversations with this group of migrant families, it appears that the existing family structure provides a great deal of education, supervision, support and protection. In contrast to the knowledge and concern about safety mentioned by most of the families, a 33-year-old husband and father described the careless and devil-may-care attitude of young single males who worked in the area: 'They don't have brains of what they do ... They're ignorant ... too fast and just want more and more money, they cut themselves ... the man, I think, was fast and cut himself because he wanted more money.' Family structure is a strength that may well be useful in designing programmes to increase the safety and well-being of this population.

SELF-RELIANCE

Learning to be prepared and be responsible for one's own health and safety is another strength demonstrated by this group of migrant farm families. A 32-year-old father of five had strong feelings and a definite opinion about the need to take care of oneself rather than relying on others, particularly the farmers, to do it for them. He had observed that, unlike the practice in other areas of the country, the migrants with whom he currently worked provided their own water and safety equipment.

> I noticed in Colorado they [farm workers] do more taking care of themselves than they do back home [Texas]. Back home they kind of expect the boss to take care of you, take you water, tell you when to take a break. Back home they take it more for granted than they do here. Here everybody supplies themselves, they don't expect anybody to do anything for you. Back home ... they almost demand it from the boss. They bring water and stuff like that, and we supply ourselves, we take our own water ... Yes, they take care of you more. Here because they leave you alone so you, so you tend to yourself. Because there you rely on the boss too much and if the boss don't show up and you take it for granted that he was going to show up and you didn't take water and you were in the field 12 hours with no water that day you'd be dehydrated and collapse out there. Yeah, I feel they take care of you better out here because they expect you to take care of yourself, over there they take everything for granted ... We carry our own 25 gallons of water ... We just take it ourselves ... And you know cause clean water ... I mean we always take our own water. We don't tell the farmer if they're going to provide water. We take it. [*32-year-old male*].

For other migrant labourers, self-reliance means taking an active role in making safety decisions by asking questions. A 35-year-old father of three pointed out that you need to ask questions and rely on your own experience. He believed that the workers' own experiences gave them the basis for making decisions about the safety of a field after chemical spraying. 'I would think as long as you go out there and ask the questions, you can figure things out. It's our judgment, it comes to our judgment.'

A 45-year-old father of four advocated a proactive stance that included being more assertive with farmers, learning from other migrant workers, making choices about where one works and passing information on to other families.

> [*Migrant*:] We feel more safe. [*Interviewer*:] Why? [*Migrant*:] A few years ago the farmers don't pay nothing. Now we talk more to the farmers. [*Interviewer*:] So something changed. [*Migrant*:] Yes, basically we changed because 4 or 5 years ago we didn't know what we were doing ... and the farmers took advantage of us, but now that we've been coming every other year ... [*Interviewer*:] So you've learned how to better talk to the farmers? [*Migrant*:] Yes. [*Interviewer*:] What do you do now that's different in the way you talk to the farmers? [*Migrant*:] We learn how to better defend ourselves, because the first years we just said we just going to pick a little ... we want to get paid a little bit more ... just do it so that we could look good in front of the farmer ... [*Interviewer*:] So really you guys have changed in your own attitudes, so you take more control, is that a fair word? [*Migrant*:] Yes. And if the farmers don't like it, we just changed. [*Interviewer*:] Why did you change? [*Migrant*:] More money and we learned how the crops looked ... [*Interviewer*:] You got smart. [*Migrant*:] Now we can tell the farmers what you think, you know, I don't like this. [*Interviewer*:] When you change like that and got quote/unquote 'smarter', is it because someone told you something, you learned from someone else, or you just came to the conclusion yourself? [*Migrant*:] No. We learned from somebody else. I think some farmers that they could ... take advantage. [*Interviewer*:] So do you guys then tell other people about it? [*Migrant*:] Not to be taken advantage of.

Supports

In addition to the strengths found within these migrant farm labour families, there are professionals and agencies that provide support in the form of services and information on agricultural health and safety. The migrant medical clinics are a major source of support not only because they provide free or low-cost medical treatment but because they also provide safety information. When asked where they would go to obtain additional safety information, many responded that they would talk to a nurse or go to the clinic. According to one migrant labourer, 'The clinic over here in F.C. [a town] they give us a lot of brochures that are for safety tips.' A 39-year-old father of three with 20 years of experience elaborated on the information and supplies provided by the local migrant clinic.

> Yeah, the year before last, in May, we got some papers and pamphlets and a videotape I was telling you about. Showing us about pesticides, about what we had to wear, long sleeves, gloves, whatever. When going to the bathroom washing your hands and all that ... The clinic people would come by here. They would give us glasses for the rays of sun, they gave us gloves, they gave us a first aid kit. Those people gave us information about all that, so I would think we were a little bit educated on that part ... These people around here, the clinics and all that, they do a real good job at that, letting us know the dangers of mostly everything out there ... what we can be allergic to or whatever.

Other labourers talked about the availability of information at the clinic site. Also, based upon the researchers' observations in the migrant camps, the visiting nurses are very well trusted by migrant families and are generally seen as useful and truthful sources of health and safety information.

A number of labourers talked about safety programmes that they had attended in other states and the availability of safety information.

> We went to Wyoming. They got us all the workers ... how to protect yourself from chemicals, from the sun, from the water that we drink, everything, how to protect ourselves from everything that poisons. [*45-year-old male*]

When asked where he would go to obtain safety information, a 32-year-old worker replied:

> To tell you the truth, I've never asked for it up here mainly because back home [Texas] they post them up [safety information] on the walls where we're at about the chemicals we are working with. So we read up on it, this is what we're going to do and how we're gonna do it and hopefully everything will go all right.

The 32-year-old father of six believes that he is safer today than he was in the past and attributes the improvement to the greater availability of safety information. 'The government is getting more safety tips to the farms or to us [migrants] by the brochures they give out.'

Summary

The interviews with the migrant farm labour families painted a picture of self-reliance, resourcefulness, and the availability of options in the face of dangers involved in agricultural work. These migrant families took a proactive rather than passive stance toward their own safety and the safety of other family members. They stressed the importance of family members protecting each other and being responsible for each other. Further, it appears that the role of the father and/or grandfather was to assume primary responsibility for the family's safety. Again, family members – sometimes three generations living and working together – shared the responsibility for the health and safety of each other.

Discussion

The findings of this pilot study should be interpreted within the context of two cautions. The first concerns the nature of a pilot study that uses qualitative methodology. This study, with its limited and non-representative sample, sacrifices generalizability for depth of understanding, for sensitive self-expression and for nuance. As a pilot study it offers direction for future studies that use more traditional research designs that can test the sustainability and generalizability of its findings to larger populations of migrant farm labour families.

Secondly, this study does not directly address the safety attitudes and practices of a very important and rapidly growing population of migrant farm labourers – those who live and work separated from their spouses and children. The study's interviews, however, do suggest vital areas of concern for studies that undertake an assessment of the safety attitudes and practices of unaccompanied migrant farm labourers. For example, future studies of this group ought to include an investigation into the prevalence of drug and alcohol use and the role of drugs and alcohol in accidents involving this population. In addition, since an emphasis of the present study was on the family unit, it would be of interest to discover whether unaccompanied migrant farm labourers' attitudes and beliefs differ – and, if so, by what degree and direction – from those voiced by the family respondents in the present study; and what the implications are of any differences for the design and delivery of safety information and programmes.

With these two cautions in mind, the results of the interviews suggested a variety of perceptions and practices by migrant farm labour families that ought to be considered in order to keep pushing forward with the development of robust farm safety intervention and prevention approaches. When working with this subset of the migrant farm labour population, the important role played by families in consciously motivating migrant farm labourers to be attentive to safety issues, to act in safe manners and to teach and model safe practices for the sake of loved ones is an aspect of their work ethic and *zeitgeist* that should be fully explored and exploited in the design, packaging and delivery of farm safety educational programmes. Also, attention should be directed toward discovering the nature of the information passed down among and between migrant farm family members and assuring its accuracy.

When interacting with migrant farm labour families in efforts to increase safety, it may be of benefit to emphasize the following:

- Farm safety should be viewed as a family issue that requires full family involvement in establishing and developing a *zeitgeist* of safety.
- Migrant parents should be supported in their role as teachers of safety for their children.
- The approach should be one that treats migrant parents as intelligent and competent consumers of safety information and as individuals deeply concerned with their own safety and with the safety of their families.
- Programmes should be designed to provide migrant farm families with support services and pragmatic techniques for dealing with the multitude of family stresses (including economic) inherent in today's farming.

Finally, it appears that migrant farm labourers periodically, but not cavalierly, cut corners on safety in an effort to maintain their economic viability and competitiveness in the labour market. This phenomenon ought to be recognized and openly discussed with them with an eye to developing alternatives within a climate of problem solving.

References and Further Reading

Baker, R. (1995) *Los dos Mundos: Rural Mexican Americans, Another America.* Utah State University Press, Utah.

Boyatzis, R.E. (1998) *Transforming Qualitative Information: Thematic Analysis and Code Development.* Sage Publications, Thousand Oaks, California.

Edwards, J., Tindale, R., Health, L. and Posavac, J. (eds) (1990) *Social Influence Process and Prevention.* Plenum, New York.

Elkind, P. (1992) Attitudes and risk behavior. In: Myers, M.L., Herrick, R.F., Olenchock, S.A., Myers, J.R., Parker, J.E., Hard, D.L. and Wilson, K. (eds) *Papers and Proceedings of the Surgeon General's Conference on Agricultural Safety and Health,* 30 April–3 May 1991, Des Moines, Iowa. DHHS (NIOSH) Publication Number 92-105. US Department of Health and Human Services, Washington, DC, pp. 123–128.

Garkovich, L., Bokemeier, J. and Foote, B. (1995) *Harvest of Hope: Family Farming/Farming Families.* The University Press of Kentucky, Lexington, Kentucky.

Gorden, R. (1992) *Basic Interviewing Skills.* Peacock Publishers, Inc., Itasca, Illinois.

Greeley Tribune (1997) Migrant farmworkers in Weld County, Colorado. 31 August, Section A, pp. 1, 12. Series by Jared Fiel.

Green, J. (1995) *Cultural Awareness in the Human Services: a Multi-ethnic Approach,* 2nd edn. Allyn and Bacon, Boston, Massachusetts.

Green, L. and Kreuter, M. (1991) *Health Promotion Planning: an Educational and Environmental Approach,* 2nd edn. Mayfield Publishing Company, Mountain View, California.

Grinnell, R. (1997) *Social Work Research and Evaluation: Quantitative and Qualitative Approaches,* 5th edn. F.E. Peacock, Itasca, Illinois.

Gunderson, P.D. (1995) JASH Editorial: Health promotion and disease prevention ... revisiting the role of education. *Journal of Agricultural Health and Safety* 1, i–ii.

Janz, N. and Becker, M. (1984) The health belief model: a decade later. *Health Education Quarterly* 11, 1–47.

Johnson, H. (1999) *Psyche, Synapse, and Substances: the Role of Neurobiology in Emotions, Behavior, Thinking and Addiction.* Deerfield Valley Publishing, Greenfield, Massachusetts.

Lincoln, Y.S. and Guba, E.G. (1985) *Naturalistic Inquiry.* Sage Publications, Newbury, California.

Margolis, L., Bracken, J. and Stewart, J. (1996) Effects of North Carolina's mandatory safety belt law on children. *Injury Prevention* 2, 32–35.

Murphy, D. (1981) Farm safety attitudes and accident involvement. *Accident Analysis and Prevention* 13, 331–337.

National Safety Council (1997) *Accident Facts: 1997 edition.* National Safety Council, Itasca, Illinois, p. 48.

Patton, M.Q. (1990) *Qualitative Evaluation and Research Methods.* Sage Publications, Newbury Park, California.

Perry, M. and Bloom, F. (1998) Perceptions of pesticide associated cancer risks among farmers: a qualitative assessment. *Human Organization* 57, 342–349.

Quandt, S., Arcury, T., Austin, C. and Saavedra, R. (1998) Farm worker and farmer perceptions of farm worker agricultural chemical exposure in North Carolina. *Human Organization* 57, 359–368.

Rodriguez, L., Schwab, C., Peterson, J. and Miller, L. (1997) The impact of an Iowa public information campaign. *Journal of Agricultural Safety and Health* 3, 109–123.

Rubin, A. and Babbie, E. (1993) *Research Methods for Social Work*, 2nd edn. Brooks/Cole, Pacific Grove, California.

Saleebey, D. (1997) Power to the people. In: Saleebey, D. (ed.) *The Strengths Perspective in Social Work Practice*, 2nd edn. Longman, New York, pp. 3–19.

Schaefer, R.T. (1998) *Racial and Ethnic Groups*, 7th edn. Longman, New York.

Schriver, J.M. (1998) *Human Behavior in the Social Environment: Shifting Paradigms in Essential Knowledge for Social Work Practice*, 2nd edn. Allyn and Bacon, Boston, Massachusetts.

Shutske, J.M. (1994) An educator's perspective on childhood agricultural injury. *Journal of Agromedicine* 1, 31–46.

Thu, K., Donham, K., Yoder, D. and Ogilvie, L. (1990) The farm family perception of occupational health: a multistate survey of knowledge, attitudes, behaviors, and ideas. *American Journal of Industrial Medicine* 18, 427–431.

Thu, K., Pies, B., Roy, N., Von Essen, S. and Donham, K. (1998) A quantitative assessment of farmer responses to the certified safe farm concept in Iowa and Nebraska. *Journal of Agricultural Safety and Health* 4, 161–171.

Thu, K.M. (1998) The health consequences of industrialized agriculture for farmers in the United States. *Human Organization* 57, 335–341.

US Department of Health and Human Services (1990) *Healthy People 2000: National Health Promotion and Disease Prevention Objectives*. US Printing Office, Washington, DC.

US Department of Health and Human Services. (1998) *National Occupational Research Agenda (NORA): Update July, 1998*. Publication No. 98–134. US Printing Office, Washington, DC.

US Department of Labor (1998) The National Agricultural Workers Survey (NAWS). US Department of Labor, Office of the Assistant Secretary for Policy, Office of Program Economics, Washington, DC. NAWS Report No. 7.

US Government (1992) *Census of Agriculture, Weld County, Colorado*. US Printing Office, Washington, DC.

16 Hired Farm Labour and Health Insurance Coverage

Paul E. McNamara[1] and Christine K. Ranney[2]
[1]Department of Agricultural and Consumer Economics, University of Illinois at Urbana-Champaign, Urbana, Illinois, USA; [2]Department of Applied Economics and Management, Cornell University, Ithaca, New York, USA

Abstract
This chapter examines the health insurance coverage of US hired farm workers with data from the Current Population Survey from 1995 through 1999. Compared with the average US worker, hired farm workers tend to be poorer, have less education, are more likely to be Hispanic and are more likely to work for a small firm. All of these characteristics correlate with the lack of health insurance coverage. Of the 7204 hired farm workers in the sample, 38% lacked health insurance coverage, a level more than twice the national average. The multivariate statistical analysis shows that even after controlling for socioeconomic and demographic variables, hired farm workers are 3 percentage points more likely than other workers to be without health insurance. This effect is in addition to the negative effect of Hispanic ethnicity, low income, low levels of education, and other demographic characteristics associated with hired farm workers. These results may help to target public policy efforts aimed at increasing health insurance coverage rates for working adults.

Introduction

Agricultural workers have long faced difficulties in obtaining and affording health insurance coverage in the US system of employment-based health insurance. Previous research has tended to focus on the health insurance coverage of farm families and has not directly considered health insurance access in the hired agricultural labour population. Although few extant studies quantify the nature of the problem, public health advocates express concern that farm employees and their families may forego needed medical care because of financial constraints and a fear of jeopardizing their health insurance coverage (e.g. Waller, 1992). This study aims to contribute to the understanding of this rural policy issue.

This chapter sheds light on the prevalence and trends of health insurance coverage among hired farm workers and their families. What is the percentage of farm workers who have health insurance coverage? What is their source of health insurance coverage and what types of health insurance coverage do they own? What variables influence their decision to purchase coverage? Are agricultural workers at higher risk to be without health insurance coverage relative to workers in other seasonal industries (e.g. forestry, fishing, tourism)? Additionally, if policy makers intend to increase health insurance coverage in the farm labour population, what might be the most effective avenues? What questions should be addressed by future research in this area?

To answer these questions, data from the Current Population Survey (CPS) March Demographic Supplement from the years 1995–1999 are analysed in this chapter. These data include information on the health insurance coverage of individuals, including persons employed in the agricultural sector. The aim of the statistical model is to measure quantitatively the levels of health insurance coverage and the factors affecting the choice of coverage. Within the scope of the analysis, summary level statistics are estimated pertaining to health insurance coverage of hired farm workers and their families. In addition, a multivariate probability model is estimated that illuminates the distinct impact of policy-relevant variables on health insurance coverage.

The chapter takes the following outline. The literature on health insurance coverage, particularly as it relates to the situation of agricultural labour, is briefly reviewed; this section discusses related literature from the areas of agricultural economics and health services research. Then the data used in the analysis are discussed, and descriptive statistics concerning the characteristics of hired farm workers and their health insurance coverage are presented. The next section presents the results of a probit econometric model that illustrates the distinct impact that variables (such as education and income) have on the likelihood of health insurance coverage for hired farm workers. The chapter concludes with observations on the implications of the analysis for efforts in the policy arena that seek to increase the health insurance coverage of farm workers. In addition, a research agenda for health economic research concerning farm labour and its health insurance coverage is proposed.

Related Literature

At a national level in the USA, access to affordable health insurance coverage remains problematic, with roughly 16% of US residents lacking health insurance coverage during 1997 (Vistnes and Zuvekas, 1999). Public health researchers have documented a link between being uninsured and problems in obtaining and paying for medical care (Donelan *et al.*, 1996). Moreover, in terms of employed persons without health insurance coverage, people working in small firms (establishments with fewer than 25 employees) constitute 50% of the uninsured. Farmers, farm employees and other rural people are disproportionately represented in that group (Davis, 1993). Farm families are at risk for the lack of coverage and pay more for health insurance policies that cover less, even when compared with urban self-employed persons (Jensen and Saupe, 1987; Kralewski *et al.*, 1990; Gripp and Ford, 1997).

In addition to the problems of uninsured persons, many persons in the health insurance market face the situation of being underinsured. Underinsured persons are defined as those for whom out-of-pocket health care expenses exceed some (subjectively defined) cut-off point, say 10% of income (Comer and Mueller, 1992). For farm workers, being underinsured results, at least in part, from their position in the individual health insurance market. The lack of group membership (through an employer) puts many persons in the individual health insurance market at risk of limited coverage and higher than standard rate premiums, or exclusion from health insurance coverage (US GAO, 1996).

Despite this national-level research on health insurance coverage, little research exists on the problems of agricultural workers and their health insurance coverage. While several that deal specifically with the health insurance coverage of farm families have been identified (Jensen and Saupe, 1987; Kralewski *et al.*, 1990; Gripp and Ford, 1997), there exists a dearth of research concerning the problems of coverage faced by hired farm workers. None the less, state legislatures and federal policy makers continue to consider proposals to subsidize or otherwise encourage affordable health insurance coverage for hard-to-insure persons, such as low-income farm employees. The problems that agricultural labourers face in obtaining health insurance coverage often serve as motivation for these legislative efforts. This study aims to provide current research on this rural policy issue.

Data Issues

The words 'hired farm worker' usually bring to mind images of migrant farm workers harvesting a field of melons or seasonal labourers de-tasselling maize under the hot Illinois sun in the middle of July. While these images represent an important part of the roles that hired farm workers play, the variety of positions that they fill extends far beyond seasonal help on farms. For the purposes of this study, hired farm workers are defined as those persons employed in production agriculture, but not in the roles of farmers, farm owners or farm managers. Thus, the term 'hired farm workers' represents a class of workers that includes an accountant, a secretary, or a diesel mechanic on a large farm. It also includes hired help that is year-round in nature, such as a hired hand on a dairy farm, and, of course, migrant and seasonal workers.

In this study the health insurance coverage of hired farm workers and their families is examined, and their coverage is compared with that of other groups of employed persons. One important comparison group is other people working in production agriculture. All persons working in production agriculture are divided in this research into two groups: (i) hired farm workers; and (ii) farmers and farm managers. Additional industry groups for comparison purposes include persons working in the forestry and fishing industries, construction industry workers, restaurant industry workers, and hotel workers. Hired farm workers are compared with all other persons that are not in one of the industry-related comparison groups.

Data description

The March Current Population Survey (CPS), which the Census Bureau fields annually, collects information about labour market participation and health insurance, among other information, from a representative sample of the US population. Roughly 55,000 households containing 145,000 individuals are interviewed each year. The March CPS includes supplemental questions concerning work experience and non-cash benefits (i.e. health insurance and retirement benefits). This analysis of the CPS uses a sample of 7204 hired farm workers and a sample of 30,163 persons in families with a hired farm worker.

As a data source for comparative studies of health insurance coverage by type of industry and employment, the March CPS has a number of strengths. As opposed to survey data focusing on migrant farmers, the March CPS allows comparisons of health insurance coverage across industries as well as within an industry by demographic and economic variables. However, the CPS does not survey firms and obtain detailed information about the firms' offer of health insurance to employees or the price the firms pay for coverage and the employee portion of that cost. Furthermore, since the CPS does not focus on issues of hired farm workers, different types of hired farm workers are grouped together, even though it is expected that health insurance coverage may differ between workers employed at one operation year-round versus those employed temporarily. These data limitations bound this chapter's analysis and they point out possible future research topics.

The health insurance questions in the March CPS refer to health insurance coverage over the previous 12 months. However, some researchers feel that respondents answer the health insurance questions as if they were with respect to a particular point in time or a period of time less than 1 year. These opinions are based on comparisons of the CPS health insurance data with data from the Survey of Income and Program Participation (SIPP) (see, for example, Employee Benefit Research Institute, 1994). If this observation concerning the respondent's interpretation of the questions holds true, then the CPS-based estimates of the number of persons without health insurance for an entire year may be biased upwards.

Description of variables

The dependent variable in the analysis is 'health insurance coverage', where '1' denotes that the individual has coverage and '0' shows that the individual is not covered by health insurance. The independent variables in the analysis control for social and economic variables that are expected to influence a person's decision to obtain health insurance coverage. Table 16.1 shows the variable names and definitions and Table 16.2 presents descriptive statistics concerning the variables. Previous research suggests age, race, region, education, firm size and income as important determinants of coverage. Results from the Medical Expenditure Panel Survey (MEPS) show that younger adults (aged 19 to 24 years), Hispanic males and people in the south were more likely to be without coverage in 1997 (Vistnes and Zuvekas, 1999).

Age (AGE) correlates strongly with a person's health insurance status. To account for the potential nonlinear effects of age on

Table 16.1. Definition of variables.

Variable	Definition[a]
AGE	age in years
SQAGE	age×age
DCHILD	= 1 if years of age <= 17, 0 otherwise
DELDER	= 1 if years of age >= 65, 0 otherwise
DFEMALE	= 1 if female, 0 if male
INCOME	= household income/poverty threshold for household
DHISPANIC	= 1 if Hispanic, 0 otherwise
DBLACK	= 1 if race is black, 0 otherwise
DLTHIGHSCHOOL	= 1 if educational attainment is less than high school graduation, 0 otherwise
DHIGHSCHOOL	= 1 if educational attainment is high school, 0 otherwise
DSOMECOLLEGE	= 1 if educational attainment is some college, 0 otherwise
DHFW	= 1 if hired farm worker, 0 otherwise
DFARMER/MANAGER	= 1 if farmer or farm manager, 0 otherwise
DOFFFARM	= 1 if family with a hired farm worker or farmer has an off-farm worker
DFORESTRY/FISHING	= 1 if works in forestry or fishing industry, 0 otherwise
DCONSTRUCTION	= 1 if works in construction industry, 0 otherwise
DRESTAURANT	= 1 if works in restaurant industry, 0 otherwise
DHOTEL	= 1 if works in hotel industry, 0 otherwise
D1996	= 1 if survey year is 1996
D1997	= 1 if survey year is 1997
D1998	= 1 if survey year is 1998
D1999	= 1 if survey year is 1999
DNORTHEAST	= 1 if lives in Northeast USA
DSOUTH	= 1 if lives in South USA
DWEST	= 1 if lives in West USA
DSIZE1	= 1 if firm has 1–9 employees
DSIZE2	= 1 if firm has 10–99 employees
DSIZE3	= 1 if firm has 100–999 employees

[a]For family regression, the variables referring to occupation or industry (e.g. DHFW or DCONSTRUCTION) mean that the family includes a person working in that industry/occupation if the variable equals 1.

Table 16.2. Characteristics of hired farm labour, farmers and farm managers, and other workers. (Numbers in parentheses are estimated standard errors.)

Characteristic	Hired farm labour	Farmers and farm managers	All other workers
Mean age (years)	35.63 (0.2046)	49.22 (0.3605)	38.75 (0.0835)
Mean household income (US$)	27,747 (357.24)	31,238 (549.55)	39,723 (170.45)
Income/poverty threshold (%)[a]	3.3129 (0.0431)	3.7559 (0.0662)	4.7518 (0.0205)
Minority race/ethnicity (%)[a]			
Hispanic	0.3356 (0.0056)	0.0231 (0.0024)	0.0931 (0.0021)
Black	0.0425 (0.0023)	0.0046 (0.0011)	0.1208 (0.0024)
Education (%)[a]			
Less than high school	0.3949 (0.2046)	0.1537 (0.0090)	0.1129 (0.0021)
High school	0.3155 (0.0071)	0.4478 (0.0105)	0.3252 (0.0031)
Some college	0.1902 (0.0057)	0.2312 (0.0084)	0.2918 (0.0029)
College degree or more	0.0993 (0.0043)	0.1673 (0.0079)	0.2700 (0.0029)
Geographical region (%)[a]			
Northeast	0.1129 (0.0047)	0.0718 (0.0054)	0.1937 (0.0025)
Midwest	0.1863 (0.0060)	0.4046 (0.0114)	0.2377 (0.0028)
South	0.3275 (0.0079)	0.3469 (0.0116)	0.3519 (0.0033)
West	0.3732 (0.0087)	0.1766 (0.0085)	0.2166 (0.0029)
Firm size, by number of employees (%)[a]			
Fewer than 10	0.4300 (0.0080)	0.8061 (0.0087)	0.1840 (0.0025)
10–99	0.2853 (0.0076)	0.0671 (0.0058)	0.2114 (0.0027)
100–999	0.1115 (0.0055)	0.0165 (0.0027)	0.1924 (0.0025)
1000 or more	0.1059 (0.0048)	0.0227 (0.0032)	0.3810 (0.0031)
Self-employed or other	0.0672 (0.0040)	0.0875 (0.0061)	0.0312 (0.0010)

[a]The percentages reported here are divided by 100.

the probability of having health insurance coverage, the square of age (SQAGE) is added to the econometric analyses. Also, since most elderly persons have Medicare health insurance or Medicaid health insurance, a dummy variable equal to 1 for persons aged 65 years and older is added (DELDER). Additionally, in the family equations, a variable (DCHILD) equal to 1 for persons under 18 years of age is added.

There is some evidence that attitudes towards risks vary by gender and that gender may play a role in determining whether or not a person has health insurance. To control for gender-related differences in health insurance coverage, an indicator equal to 1 for females was included. Income is measured at the household level and is expressed as a ratio of family income to the poverty threshold for that family. A squared income term was tested, but it was not statistically significant and was excluded from the analysis.

Indicator variables for Hispanic ethnicity (DHISPANIC) and for black race (DBLACK) are included. Similarly, indicator variables for educational attainment (DLTHIGHSCHOOL, DHIGHSCHOOL, DSOMECOLLEGE) are included, and the omitted category is a college degree or higher. In the family equation, children are assigned their mother's educational attainment unless the mother is not present, in which case the child is assigned the father's or the designated parent's education level.

In the agricultural workers equation, an indicator variable equal to 1 for hired farm workers (DHFW) is included. In the families of agricultural workers model, indicator variables defined to control for the presence of a hired farm worker (DHFW) and an off-farm worker (DOFFFARM) are included. Because the ability to pool risks allows larger firms to obtain health insurance less expensively than smaller firms, dummy variables (DSIZE1, DSIZE2, DSIZE3) are included in the analysis. DSIZE1 captures firms with fewer than ten employees, DSIZE2 equals 1 for firms with ten to 99 employees, and DSIZE3 equals 1 for firms with 100 to 999 employees. The firm size category of 1000 employees and more is omitted to identify the firm-size effect. To control for potential trends over time, indicator variables for the years 1996 to 1999 are included, with 1995 being omitted. Also, a set of regional dummy variables are defined (DNORTHEAST, DSOUTH, DWEST) and the Midwest category is omitted.

Hired Farm Workers and Health Insurance

Compared with the average worker, hired farm workers tend to be poorer, have less education, are more likely to be Hispanic and are more likely to work for a small firm (Table 16.2). As Table 16.2 shows, the average worker's household income in the USA is US$10,000 more than the household income of the average hired farm worker. While 11% of workers in general have not achieved high school graduation, over 39% of hired farm workers have not completed high school. Furthermore, 9% of non-agricultural workers are Hispanic, but more than 33% of hired farm workers are Hispanic. Lastly, while 18% of all other workers are in firms with fewer than ten employees, these small firms constitute the workplace for 43% of hired farm workers. Overall, these descriptive statistics imply that hired farm workers have unique characteristics, and some of these characteristics, such as low educational attainment and working for small firms, have relevance to the design of policies that aim to increase health insurance coverage.

Who are the uninsured hired farm workers?

Of the 7204 hired farm workers in our sample, 2747 (38%) were without health insurance (Fig. 16.1). The Medical Expenditure Panel Survey data from the first half of 1997 revealed that 18% of all workers were uninsured (Vistnes and Zuvekas, 1999). Thus, based on the simple frequencies, hired farm workers were more than twice as likely as the average worker to lack health insurance coverage. Compared with the average hired farm worker, a hired farm worker without health insurance is more likely to be poor, Hispanic and less educated. These same characteristics have been identified by previous researchers studying health insurance coverage at a national level as important correlates of the lack of health insurance. Therefore, these factors could be important variables in explaining the likelihood that a hired farm worker is uninsured.

In the sample, 53% of low-income hired farm workers (defined as persons whose household income is less than twice the poverty threshold for that household size) were without health insurance. In comparison, 40% of middle-income hired farm workers (household income from two to four times the poverty threshold) were without health insurance, and 18% of high-income hired farm workers (household incomes above four times the poverty threshold) were uninsured.

Race and ethnicity also relate closely to being uninsured for hired farm workers. While 33% of hired farm workers are Hispanic, 56% of uninsured hired farm workers are Hispanic. Looked at differently, of the 2418 Hispanic hired farm workers in our sample, 1539 (64%) did not have health insurance. Compared with the rates for white (non-Hispanic) hired farm workers (24%), for black (non-Hispanic) hired farm workers (45%) or for 'other' (non-Hispanic) hired farm workers (29%), Hispanic hired farm workers have the highest uninsured status rate.

Hired farm workers without health insurance also tend to have lower levels of educational attainment. The frequency data show that of the 2747 hired farm workers without health insurance, 57% had less than a high school degree and 26% had a high school degree. Along with educational achievement, regional differences exist in the rate of uninsurance among hired farm workers. Hired farm workers in the Northeast and

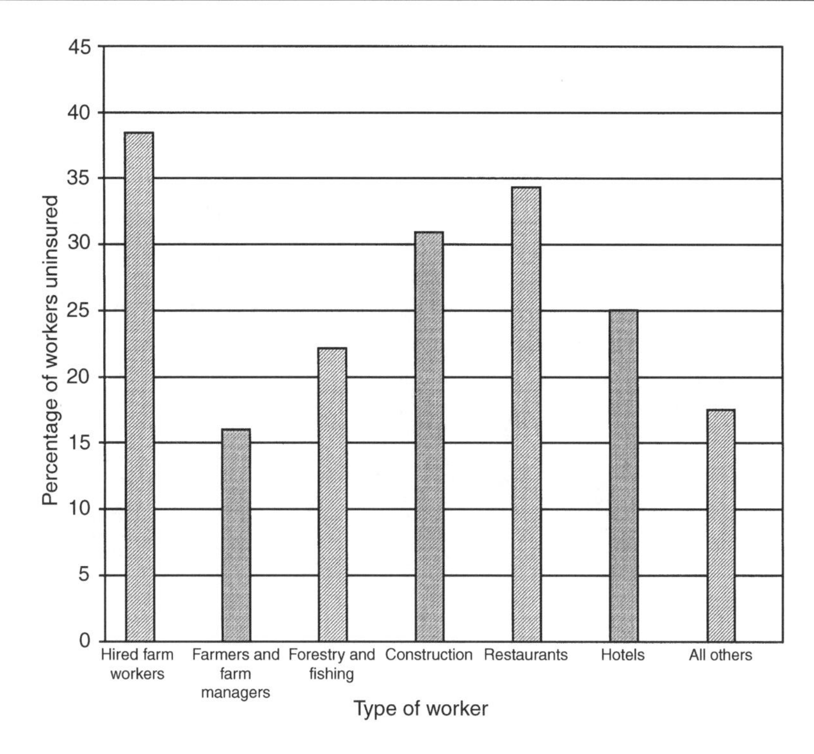

Fig. 16.1. Percentage of uninsured workers: hired farm workers and other types of workers, 1994–1998. (Source: Current Population Survey data.)

Midwest are more likely to have health insurance than are hired farm workers in the South or West. In contrast to the differences across educational attainment or geographical regions, differences in the lack of health insurance do not appear to be strongly related to firm size on the basis of simple frequencies. For example, while firms with fewer than ten employees engage 43% of hired farm workers, 45% of uninsured hired farm workers work for firms in that category.

When all family members are considered (in families with a hired farm worker), the same variables, such as low income, Hispanic ethnicity and low levels of educational attainment, correlate strongly with the lack of health insurance. For instance, 60% of the family sample's low-income persons lack health insurance, while only 33% of hired farm worker family members fall in the low-income group. Similar to the case of workers considered alone, when the health insurance coverage of all family members is examined, the family members of hired farm workers have the highest level of lack of health insurance compared with the family members of workers from all the other industry categories we considered (Fig. 16.2).

Compared with farmers and farm managers, hired farm workers are more likely to be uninsured and to have Medicaid health insurance (Figs 16.3 and 16.4). A total of 38% of hired farm workers are uninsured versus 16% of farmers and farm managers, and 6% of hired farm workers have Medicaid insurance as compared with only 2% of farmers and farm managers. Nearly the same levels of hired farm workers and of farmers and farm managers have group health insurance – from a group such as a cooperative or a spouse's employment – at 39% and 34%, respectively. Fewer hired farm workers (13%) purchase non-group health insurance (i.e. insurance purchased in the individual market) compared with farmers and farm managers (34%).

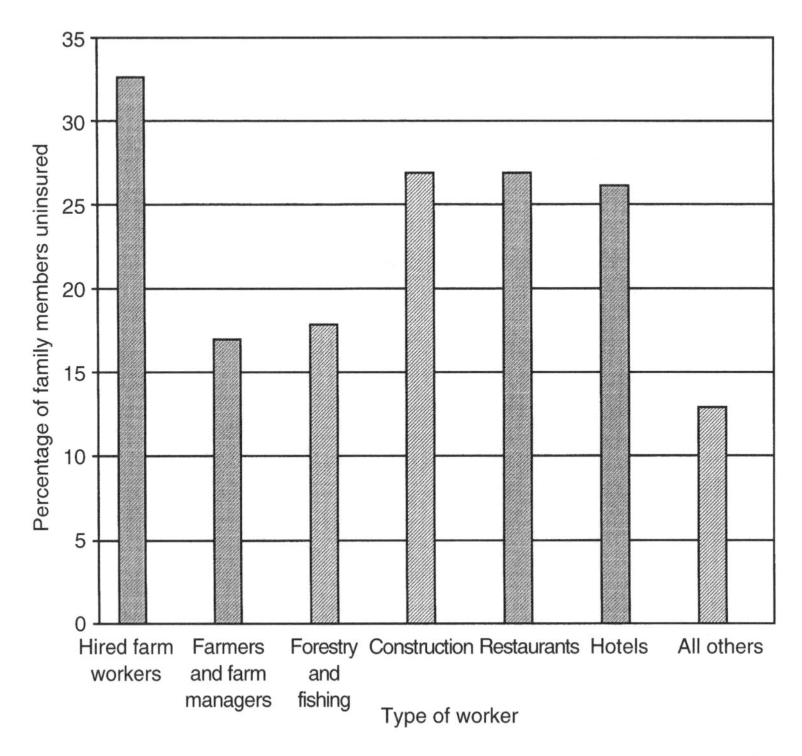

Fig. 16.2. Percentage of uninsured family members: by industry/occupation, 1994–1 998. (Source: Current Population Survey data.)

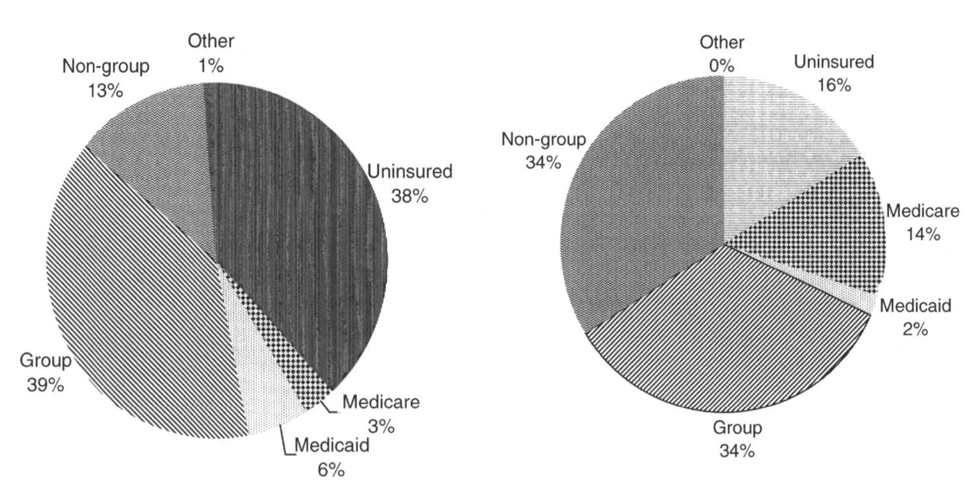

Fig. 16.3. Type of health insurance held by hired farm workers.

Fig. 16.4. Type of health insurance held by farmers and farm managers.

While simple descriptive statistics indicate the basic characteristics of those hired farm workers without health insurance, they may also suffer from the effects of confounding influences. To control for the potential for associations between variables, such as educational attainment and income, the next section presents a probit multivariate regression model of health insurance coverage as a function of a set of explanatory variables.

Probit Regression Results

Probit regression models (for a description see Maddala, 1983, or Greene, 1993) were estimated on two data sets. The first set of data is all workers from the CPS from the years 1995–1999. This data set includes all agricultural workers (hired farm workers, farmers and farm managers) and the sample includes workers from other low-wage and seasonal occupations from the forestry and fisheries sector, the construction industry, the restaurant industry and the hotel industry. In addition to these workers, we drew a 10% subsample of the remaining workers to include in the all-workers sample. The total sample size of the data set of all workers is 110,776. A second data set is also constructed to look at all family members, and this data set consists of all of these workers and their household members. The number of individuals in the family members data set is 188,370.

The 'all workers' model and the 'family members' model are structured similarly. The dependent variable is health insurance coverage. The independent variables consist of demographic variables (AGE, SQAGE, DCHILD, DELDER, DFEMALE, DHISPANIC and DBLACK), a set of education variables (DLTHIGHSCHOOL, DHIGHSCHOOL and DSOMECOLLEGE), a set of industry variables (DHFW, DFARMER/MANAGER, DOFF-FARM, DFORESTRY/FISHING, DCONSTRUCTION, DRESTAURANT and DHOTEL), a set of dummy variables to capture year-specific effects (D1996, D1997, D1998 and D1999), a set of geographic dummy variables, an income variable (INCOME), a set of firm size variables (DSIZE1, DSIZE2 and DSIZE3) and a constant.

The estimated coefficients from the probit regression model cannot be interpreted as simply as the coefficients from a simple linear regression. That is, the coefficients multiplied by their relevant variables are summed to form an index, which is transformed according to the standard cumulative normal distribution to provide the probability of a positive outcome. It should be noted that the marginal effects for the probit model vary over the range of the variable in question. Thus, estimates of the marginal effects (labelled dF/dx) are provided in the regression output. These marginal effects are estimates calculated at the means of the relevant independent variable. In the case of a continuous variable such as age or income, the marginal effect is interpreted as the change in the probability of a positive outcome, given a single unit change in the independent variable. In the case of an indicator variable, such as DHISPANIC, the reported marginal effect gives the change in the probability of a positive outcome, given a switch in the indicator variable from 0 to 1. This model is estimated with Stata software via a maximum likelihood routine. The regression model estimation accounts for the differential probability of selection of observations into the sample. This is achieved by the use of the Current Population Survey's AFNLWGT variable as a weighting variable in the estimation.

All-workers model results

Probit regression results from the all-workers model are presented in Table 16.3. The estimation results reveal the relative importance of variables like income and race/ethnicity, which were described above as strong correlates with the lack of health insurance. By itself, working as a hired farm worker leads to a statistically significant reduction in the likelihood of holding health insurance. Relative to the average US worker, switching to the occupation of a hired farm worker leads to an estimated decrease of 3.1% in the chance of having health insurance. Of course, this direct effect of hired farm worker status does not include the additional impact of (potential) minority race/ethnicity, working for a small firm and living in a low-income household. All of these effects would decrease the chance of having health insurance further for a hired farm worker. The lower likelihood of having health insurance

Table 16.3. Probit regression results: probability of health insurance coverage for all workers (n=110,776; pseudo R^2=0.1745).

Variable	Coefficient	*t*-statistic	D*F*/d*x*[a]	Mean
AGE	−0.0130	−5.001	−0.0036	36.8299
SQAGE	0.0004	9.925	0.0001	1531.5500
DELDER	0.7655	10.052	0.1480	0.0241
DFEMALE	0.1207	10.518	0.0334	0.4336
INCOME	0.1136	56.516	0.0316	4.3079
DHISPANIC	−0.5093	−31.873	−0.1624	0.1304
DBLACK	−0.1935	−10.458	−0.0573	0.0963
DLTHIGHSCHOOL	−0.5200	−26.271	−0.1633	0.1832
DHIGHSCHOOL	−0.2912	−16.834	−0.0835	0.3700
DSOMECOLLEGE	−0.1449	−7.969	−0.0414	0.2765
DHFW	−0.1082	−5.008	−0.0313	0.0655
DFARMER/MANAGER	0.1991	5.862	0.0509	0.0288
DFORESTRY/FISHING	0.0063	0.083	0.0018	0.0034
DCONSTRUCTION	−0.1333	−8.976	−0.0384	0.1926
DRESTAURANT	−0.2794	−18.631	−0.0838	0.1594
DHOTEL	−0.0865	−2.950	−0.0249	0.0322
D1996	−0.0045	−0.275	−0.0012	0.1955
D1997	−0.0298	−1.848	−0.0084	0.2027
D1998	−0.0792	−4.929	−0.0225	0.2034
D1999	−0.0982	−6.053	−0.0280	0.2027
DNORTHEAST	−0.1303	−8.083	−0.0376	0.1707
DSOUTH	−0.2630	−18.381	−0.0756	0.3473
DWEST	−0.2217	−13.985	−0.0646	0.2410
DSIZE1	−0.5650	−39.129	−0.1746	0.2468
DSIZE2	−0.2840	−20.137	−0.0838	0.2434
DSIZE3	−0.0140	−0.841	−0.0039	0.1651
CONSTANT	1.0603	20.448		

[a]D*F*/d*x* is an estimate of the effect of a change in the variable *x* on the probability of health insurance coverage, evaluated at variable means.

among hired farm workers relative to the average US worker is similar to the decrease of 3.8% seen for construction workers. Interestingly, farmers and farm managers are 5% more likely than the average US worker to have health insurance. For the other industries examined in this analysis, working in the restaurant industry leads to the largest decrease (8%) in the likelihood of health insurance coverage.

Other statistically important effects arise from income, race and ethnicity, education and firm size. An increase in household income by one unit of the poverty threshhold leads to a 3 percentage point increase in the likelihood of health insurance coverage, estimated at the mean for income. Figure 16.5 shows how income affects the probability of health insurance coverage for a 30-year-old Hispanic hired farm worker. With a household income equal to the poverty threshold, he will have health insurance coverage with a 12.5% probability. As household income increases to twice the poverty threshold, the chance of coverage for this worker increases to about 15.5%. While we tested a model specification with a squared-income term to help control for nonlinear effects, that term proved statistically insignificant and further research about the shape of the income effect over the range of observed income in the sample is required.

As seen in the descriptive statistics, Hispanic and black workers in the USA lack health insurance with greater frequency than other workers. Relative to non-Hispanic workers, Hispanic workers have an estimated

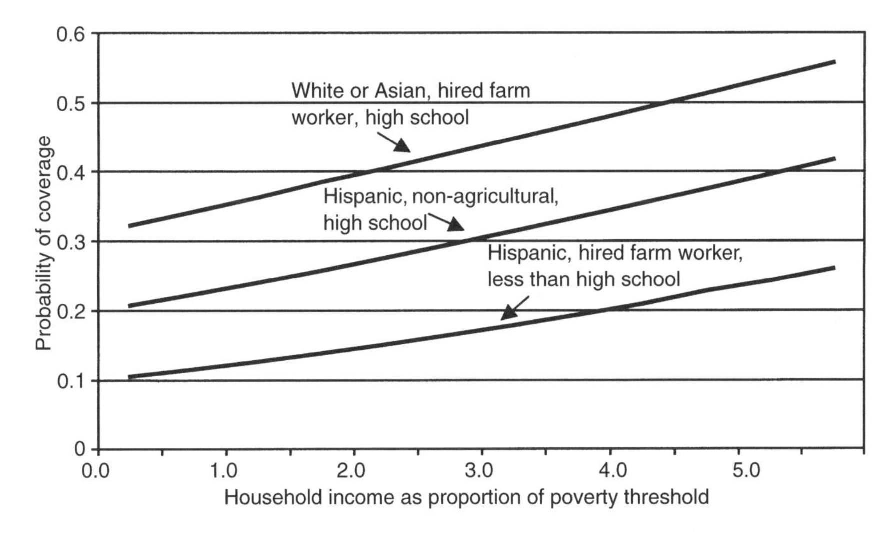

Fig. 16.5. Marginal effects for income changes from probit for all-workers model.

decrease in the likelihood of holding health insurance of 16 percentage points. The estimated effect of being black on health insurance coverage is also negative (–0.057). Other large negative effects on the likelihood of health insurance coverage result from having less than a high school education (–0.16), being in the South (–0.08) and working in a small firm with fewer than ten employees (–0.17).

Family members model results

In the family members model (reported in Table 16.4), similar estimates of the effects of variables on the likelihood of health insurance coverage are seen. The same strong effects of income, race/ethnicity, industry and occupation group and firm size appear. Children are more likely than adults to have coverage, which may be due to Medicaid and the Childrens' Health Insurance Programs around the country.

The time trend effects for health insurance coverage are also significant in both the family members model and in the all-workers model. In the all-family-members model, dummy variables for 1997, 1998 and 1999 are statistically significant. The estimated effects imply that in these years individuals were less likely to have health insurance coverage than in 1995, and that effect increased over these 3 years. Thus, in 1999 an individual was about 3 percentage points less likely to have health insurance coverage than a similar individual in 1995, all other variables held constant. This effect might arise for at least two reasons. Firstly, a trend in the marketplace for health insurance coverage might be occurring, where over time firms are less likely to offer employees health insurance due to cost or some other labour market factor. Secondly, an underlying trend may be occurring that is unobserved but that is correlated or related to health insurance coverage. An example of such a trend might be undocumented workers. If their numbers increased over the period and if they are less likely to have health insurance coverage, it may result in the time trend variables showing significance.

Implications for the Research Agenda and for Public Policy

Future research in the area of health insurance coverage for hired farm workers has the potential to contribute much to the under-

Table 16.4. Probit regression results: probability of health insurance coverage for all persons (n=188,360; pseudo R^2=0.1551).

Variable	Coefficient	*t*-statistic	D*F*/d*x*[a]	Mean
AGE	−0.0011	−0.759	−0.0003	30.9710
SQAGE	0.0002	11.010	0.0001	1344.0200
DCHILD	0.5645	30.918	0.1355	0.3143
DELDER	0.8373	19.120	0.1513	0.0599
DFEMALE	0.1221	15.329	0.0323	0.4853
INCOME	0.0974	60.834	0.0258	4.0177
DHISPANIC	−0.4296	−36.314	−0.1285	0.1517
DBLACK	−0.0903	−6.591	−0.0247	0.1078
DLTHIGHSCHOOL	−0.5064	−32.598	−0.1500	0.2289
DHIGHSCHOOL	−0.2670	−19.128	−0.0731	0.3598
DSOMECOLLEGE	−0.1230	−8.420	−0.0335	0.2588
DHFW	−0.1690	−7.056	−0.0474	0.1181
DFARMER/MANAGER	−0.0403	−1.600	−0.0109	0.1428
DOFFFARM	−0.0299	−1.226	−0.0080	0.1328
DFORESTRY/FISHING	−0.0227	−0.458	−0.0061	0.0052
DCONSTRUCTION	−0.2633	−27.243	−0.0726	0.3288
DRESTAURANT	−0.2567	−25.897	−0.0716	0.2758
DHOTEL	−0.1577	−9.001	−0.0444	0.0545
D1996	−0.0157	−1.252	−0.0042	0.1968
D1997	−0.0568	−4.557	−0.0153	0.2030
D1998	−0.0841	−6.764	−0.0228	0.2037
D1999	−0.1098	−8.772	−0.0300	0.2011
DNORTHEAST	−0.0974	−7.728	−0.0266	0.1691
DSOUTH	−0.2550	−22.924	−0.0699	0.3499
DWEST	−0.1956	−15.861	−0.0542	0.2484
DSIZE1	−0.4539	−30.945	−0.1402	0.0732
DSIZE2	−0.2114	−17.972	−0.0599	0.1316
DSIZE3	−0.1715	−6.443	−0.0194	0.1585
CONSTANT	−0.7679	23.254		

[a]D*F*/d*x* is an estimate of the effect of a change in the variable *x* on the probability of health insurance coverage, evaluated at variable means.

standing of the role of economic factors in the health of hired farm workers and their families, as well as a broader understanding of the barriers faced by low-income working people in the US's employment-based system of health care insurance. Future health economic research in the area might focus on developing economic models that help in understanding the scope and limits of market-based policies intended to increase health insurance among the working poor. Examples of such policies include tax credits or other tax policies and the formation of health insurance purchasing pools that include small businesses. How effective will expansion of tax subsidies on health insurance premiums be in covering low-income people?

Another research question regards the relative importance of risk aversion versus budget constraint in explaining the lack of health insurance coverage. Some researchers have suggested that it is heterogeneity in risk aversion that explains the observed variation in health insurance coverage. Our research advances the explanation that people may choose not to purchase health insurance because of simple lack of funds and the budget constraint. Which explanation is more important?

Another area where health economics and agricultural economics can contribute to the research agenda on hired farm workers concerns the decision of a firm to offer fringe benefits to its employees. What agricultural businesses are more likely to offer health insurance benefits to their employees? Which hired employees receive the benefits? Would unionization significantly help hired farm workers to receive health insurance?

Public policy implications

This research yields several implications for public policies intended to increase the health insurance coverage of hired farm workers and, by generalization, other low-income working people. These findings demonstrate the quantitative importance of the relationship between income and health insurance coverage. In our family sample, more than 60% of the persons in a low-income household with a hired farm worker lack health insurance. Given that health insurance premiums can range from US$3000 to more than US$6000 per year for a family, it is not surprising that people in households below or near the poverty threshold do not purchase health insurance. Given the low-income levels at which some of these households function, the additional expense of even a partial contribution towards a health insurance premium may be too high. Secondly, some of these households suffer significant health shocks or income shocks, making the payment of a steady stream of premium payments difficult. Researchers and public programme managers should investigate whether current premium contributions for the Children's Health Insurance Programs (CHIP) that states are now implementing are too high for these low-income hired farm worker households to afford. Additionally, researchers should investigate whether hired farm workers who are eligible for these state-level efforts at children's health insurance coverage are aware of the programmes and can successfully apply for them. Does the migratory nature of some of the hired farm worker population generate barriers that make it difficult to apply for and receive CHIP benefits? How many of the uninsured hired farm workers and their family members are undocumented or in the country illegally and how does immigration status affect health insurance coverage? Given the relatively low educational attainment levels of the hired farm workers without health coverage, do the current outreach and application support programmes offer sufficient assistance in filling out CHIP applications for low-literacy persons? What are the gaps in the safety net of migrant health centres and community health centres that serve uninsured hired farm workers and their families?

Another implication for public policy from this research arises from the concentration of hired farm workers in small firms. Small firms, like most agricultural firms, might benefit from well-targeted reforms dealing with the tax treatment of health insurance premiums and from membership in small-group programmes that pool small firms at the state level to purchase health insurance. However, care must be taken to ensure that firms receive strong incentives to offer health insurance to workers, if policy makers would like to see low-wage workers covered.

We distil a third implication for the policy arena from the significant link between hired farm workers – especially those who are low-income, black or Hispanic, working in small firms, and less educated – and the lack of health insurance. This link argues for continued support of public funding for community health efforts aimed at hired farm workers, assuming that equitable access to health care is a public policy goal. Such efforts include migrant health centres and other rural community health centres and clinics, and occupational health and safety programmes aimed at hired farm workers. In addition, the analysis suggests that efforts that change the education levels or add skills so that hired farm workers can shift to other industries may also increase health insurance coverage of this population.

References and Further Reading

Comer, J. and Mueller, K. (1992) Correlates of health insurance coverage: evidence from the midwest. *Journal of Health Care for the Poor and Underserved* 3, 305–320.

Davis, K. (1993) Health insurance and access to health care. *Research in Human Capital and Development* 7, 287–306.

Donelan, K., Blendon, R.J., Hill, C.A., Hoffman, C., Rowland, D., Frankel, M. and Altman, D. (1996) Whatever happened to the health insurance crisis in the United States? *Journal of the American Medical Association* 276, 1346–1350.

Employee Benefit Research Institute (1994) *Sources of Health Insurance and Characteristics of the Uninsured.* EBRI Special Report SR-20, Washington, DC.

Greene, W.H. (1993) *Econometric Analysis*, 2nd edn. McMillan, New York.

Gripp, S.I. and Ford, S.A. (1997) Health insurance coverage for Pennsylvania dairy farm managers. *Agricultural and Resource Economics Review* 27, 176–183.

Jensen, H.H. and Saupe, W.E. (1987) Determinants of health insurance coverage for farm family households: a midwestern study. *North Central Journal of Agricultural Economics* 9, 145–155.

Kralewski, J.E., Shapiro, J., Chan, H., Edwards, K. and Liu, Y. (1990) Health insurance coverage of Minnesota farm families. *Minnesota Medicine* 73, 35–38.

Maddala, G.S. (1983) *Limited-Dependent and Qualitative Variables in Econometrics.* Cambridge University Press, New York.

US Census Bureau (1995) *Health Insurance Coverage: 1994 Current Population Reports, Consumer Income P60-190.* US Government Printing Office, Washington, DC.

US GAO (1996) *Private Health Insurance: Millions Relying on Individual Market Face Cost and Coverage Trade-Offs.* GAO/HEHS-97-8, US General Accounting Office, Washington, DC.

Vistnes, J.P. and Zuvekas, S.H. (1999) *Health Insurance Status of the Civilian Noninstitutionalized Population: 1997. MEPS Findings No. 8.* AHCPR Publication No. 99-0030. Agency for Health Care and Policy Research, Rockville, Maryland.

Waller, J.A. (1992) Injuries to farmers and farm families in a dairy state. *Journal of Occupational Medicine* 34, 414–421.

17 Seasonal Migration: Farm Workers in the Alaska Fishing and Seafood Industry and the Impact on Health Care Systems in Communities Where They Work

Patricia M. Hennessy
Alaska Primary Care Association, Anchorage, Alaska, USA

Abstract

Section 330 of the Public Health Services Act, which provides funding to community health centres and special programmes to ensure access to health care in the USA, specifically defines the eligible programmes and target population. Under the Act, the definition of migrant and seasonal labour is specific to the agricultural industry. This is problematic for fishing communities in Alaska because agricultural workers migrate to the fishing industry. This chapter highlights this issue and proposes four alternatives to deal effectively with the influx of seasonal and migrant workers into fishing communities throughout Alaska.

Introduction

In 1996, the Migrant Health Act was folded into Section 330 of the Public Health Services Act in the USA. This legislation provides funding to community health centres and special programmes to ensure access to basic health care among underserved US populations such as the homeless and farm workers. Section 330 of the Act specifically defines the eligible programmes and target population, and definitions of migrant and seasonal labour are specific to the agricultural industry.

This is problematic for fishing communities in Alaska because agricultural workers migrate to the fishing industry. Affected populations are small, geographically isolated and have limited health care resources. During a fishing season the population in a community can swell to three or four times its year-round size. This occurrence taxes a variety of local resources and limits access to health care for both residents and non-residents in the community. Alaska lacks census data that could verify that the state is part of the western migrant stream.

This chapter proposes four alternatives to deal effectively with the influx of seasonal and migratory workers in fishing communities throughout Alaska. The most appropriate solution will improve access to care for both the resident and non-resident populations of a community. As well, continuity in

 The Dynamics of Hired Farm Labour (eds J.L. Findeis, A.M. Vandeman, J.M. Larson and J.L. Runyan)

care will improve for farm workers. This will result in reducing the economic burden that migrant workers place on a community. Additionally, financial resources legislated through the Public Health Services Act will be reallocated in a manner that recognizes the needs of migrant and seasonal labour, not a specific industry. Finally, administrative support and key leadership will endorse the development of collaborative relationships that encourage information sharing and combining of resources throughout the region.

Overview of the Issue

President John F. Kennedy signed the Migrant Health Act on 25 September, 1962. The original law was funded and administered by the US Department of Health and Human Services. Since that Act was signed, money legislated through it has made primary and supplemental health care services available to migrant and seasonal farm workers and their families. The purpose of this money, then as now, is to help the communities in which migrant and seasonal labourers work adapt existing services to meet the unique needs of this transient population.

The definition of migratory and seasonal labour is as follows:

- 'Migratory' means an individual whose principal employment is in agriculture on a seasonal basis, who has been employed within the last 24 months, and who establishes for the purposes of such employment a temporary abode.
- 'Seasonal' means an individual whose principal employment is in agriculture on a seasonal basis and who is not migratory.

The migrant labourer, as defined, travels and has a distinct pattern to when and where he or she is employed. This individual may be in Washington State for the apple harvest and then travel to California to pick strawberries. The same individual can be a farm worker in California and then work in Alaska during salmon season. The worker moves from one 'harvest' to the next to earn an income. This pattern of employment, known as the 'migrant stream', is a problem for fishing communities in Alaska. Since many migrant workers experience generally poor health, they are more susceptible to chronic illness such as diabetes or heart disease and these conditions require ongoing care. Left unmanaged, the chronic condition can precipitate into an expensive and life-threatening situation that stresses the care system even further. A person can receive health care in both locations where they work, but only the clinic treating the 'agricultural worker' can receive subsidies from the Public Health Services Act.

Bridging the Gap

According to A. Roth, a public health nurse for Alaska who works on Kodiak Island, Alaska (1999, personal communication), many seafood workers can only afford substandard, overcrowded housing and have poor nutrition. She goes on to add that the canneries allow their employees to work long hours. 'Processing is physically demanding but the workers are eager to earn and the employer willing to exploit .... This exponentially increases the risk for occupational injury.' Some of the clinics in fishing communities are federally recognized rural health centres and receive US$59 payment for each patient visit, regardless of the type or level of care provided. Clinics could also be reimbursed if the patient is Medicaid eligible. Farm workers, however, spend a limited amount of time in one community, therefore conditions of their employment generate the most significant barrier to eligibility.

Another barrier to health care for farm workers in Alaska's fishing industry is lack of data. Available data only indicate non-resident employment trends and say nothing of any migratory pattern. *Losing Ground: the Condition of the Farmworker in America* (US Department of Health and Human Services, 1995) raises several questions. Who are migrant farm workers? How many of them are there? Where do they come from? What is their state of health? What are their living conditions? There are sufficient data to answer each of these questions independently of the other as they relate to the agricultural workforce, but not for the fishing and seafood processing industry.

Data that are available support the need to collect even more specific information on this population. The *Nonresidents Working in Alaska 1995* report (Alaska State Department of Labor, 1995) indicates that 77% of all workers in Alaska were non-resident, and 66% of all earnings paid by Alaska's seasonal seafood industry went to non-resident workers. Compared with resident workers, who earned an average of US$24,657 annually, non-resident workers earned only US$11,734. Compared with the average wage of US$6.20 per hour in the agricultural industries in the western USA, the fishing industry paid an average of US$8.84 per hour in 1998. Despite this difference in pay, these are low-paid and temporary positions and most do not offer health insurance or benefits of any kind. Finally, as with other migrant labour trends throughout the USA and according to industry experts, Hispanic workers have a significant presence in the fishing industry and the number of Southeast Asians, Africans and Eastern Europeans is increasing. By collecting data that include information about previous place of residence, occupation and plans for future employment, the first steps can be taken toward building an information system that can be used to support migrant workers regardless of where or how they earn an income.

Current Conditions

The definition of migrant and seasonal labour in the Public Health Services Act is specific to the agricultural industry and correlates with the way in which society now views the migrant and seasonal labourer or, more commonly, the 'farm worker'. Many believe that this individual is single, male, of Hispanic origin, poorly educated and in the USA illegally. Often this image lacks an awareness of the conditions under which farm workers live, their poor health status or their lack of access to health care. Additionally, because the federal definition of migrant and seasonal labour is specific to agriculture, most individuals, including policy makers, associate this group of labourers with only certain regions of the USA where agriculture is easily identified, such as California, Idaho, Oregon and Washington – but not Alaska.

When discussing the farm worker population, it is also important to recognize not only the wage earner but also spouses, partners and dependants. Because of the physically demanding work, substandard living conditions and minimal ties to a community, this sector of the population has become increasingly 'disenfranchised' from even the most basic health care and social services that society offers. Additionally, due to immigration status, language or cultural barriers and transience, these individuals can become even more isolated from the community around them.

Many farm workers in the USA are of Hispanic origin but, in recent years, due to civil unrest and faltering national economies, Southeast Asians, Africans and Eastern Europeans are also found in this work force. Industries that employ migrant and seasonal labour are usually driven by a quota system and frequently involve food production. Often, due to market demands, employers tend to look the other way and do not enforce standard work hours: they exploit labour eager to earn the extra money that an 8 h shift would not allow.

Production quotas also apply to the fishing and seafood industry. In Alaska, fishing stocks are plentiful but seasons are short and, like agriculture, fishing requires a labour force that can quickly deliver the product to the consumer. According to the *Alaska Population Overview, 1997 Estimates* (Alaska State Department of Labor, 1997, p. 13), 'employment in Alaska is highly seasonal in ... fishing and seafood processing'. The report goes on to state that 'when employment in Washington, California and Oregon becomes relatively weak, there is a greater tendency to look to Alaska for opportunity and vice versa'. Alaska state officials who have studied this labour force and the health care providers in fishing communities draw similar comparisons to the individuals who work in the fishing industry and farm workers in the contiguous USA. Employees of the fishing industry live in a physically demanding environment, have little or no access to health care, experience generally poor health and are isolated from their community due to language, cultural or immigration issues.

Clinical administrators and clinicians who work in migrant health centres in Washington, Oregon and Idaho have heard patients comment that in the off-season they fish in Alaska. Patient information from the Eastern Aleutian Tribes, Inc. clinic service area in western Alaska indicates that nearly all of the non-resident workers who fish are from western states in the USA, and some records indicate that patients are from agricultural communities such as Stockton, California. Recruiters with seafood companies based in Seattle, Washington, note that they like to hire farm workers because they are hard workers and are accustomed to working long hours. Three of the largest seafood companies in Alaska have even run employment advertisements written in Spanish in the eastern Washington State farming communities of Othello and Yakima. Since 1990, the Hispanic population in Alaska has grown by 62%. The above examples provide strong evidence that Alaska is part of the western migrant stream.

Since the passage of the 1996 Act and particularly since managed care has become the focal point for health care in the USA, research has been conducted in an attempt to understand fully the needs and demographics of isolated and underserved populations.

Evaluation Criteria

The possibility that Alaska is part of the western migrant stream first came to light in 1998. As part of an effort to educate policymakers and health care administrators on health delivery systems in rural and frontier communities, a group sponsored by the Alaska Primary Care Association visited some of the most isolated communities in western Alaska along the archipelago of islands known as the Aleutian Chain (M. Kasmar, Executive Director APCA, 1999, personal communication). Informal surveys of the health care professionals in these communities revealed a significant influx of seasonal workers employed by the fishing and seafood processing companies in the region. In one such community, the only clinic covers 60% of its annual patient visits in a 3-month period that coincides with the pollock season. In others, where the populations range in size from 100 to 1000, it is not uncommon for seasonal workers to outnumber year-round residents three to one. Further discussions revealed similarities in terms of health status and living conditions to migrant workers in the contiguous USA; this was a common theme in all of the fishing communities that the delegation visited (M. Kasmar, 1999, personal communication).

Concurrent to the site visits in western Alaska, fact finding also took place on Kodiak Island in south-central Alaska. Observations were similar to those of the team visiting the Aleutian Chain. Fishing fosters a boom/bust economy in the region. Availability of work has caused families to move into and out of the community, and fishing families have members of their family absent for months at a time. Planning in order to care for this community is challenging, due to the transient nature of the population.

With similar findings across the state, there is the desire by state and local health care officials to verify that Alaska is part of the western migrant stream. Certain evaluation criteria focusing on access to care, financial aspects, administrative support and political leadership may help to determine the optimal policy to minimize the impact that migrant workers have on the health care systems in the rural and frontier communities where they work.

The first of such criteria is effectiveness. Migrant and seasonal workers experience difficulty accessing health care regardless of where they are employed. This inability to gain entry into the health care system extends to the spouse or partner, as well as other dependants. Facilities in the western migrant stream and other key areas that receive funding through the Public Health Services Act are known as migrant health centres. These health centres have a number of programmes that address the specific needs of farm workers and, in some cases, have programmes that allow health care providers to go directly to the consumer – which often means going to the fields to provide care. This model of delivery may be impractical in targeted Alaskan communities, in part because of the occupational safety risks associated with fishing. Incidents of trauma are very high and injuries usually require stabilization and transport to a tertiary medical care facility. In the event of a trauma injury, the entire medical team of the community may need to respond to the needs of the patient and, therefore, would be unavailable to the rest of the community. Effectiveness is based on improved access to and continuity in care for migrant and seasonal workers and the year-round members of fishing communities in Alaska.

Secondly, farm workers have an economic impact on the communities where they work. On the one hand, this impact is positive: they are consumers and producers. But they may negatively affect a community, specifically in health care resources, because of their numbers. Industries that employ migrant and seasonal workers only hire for a limited time and require a sizeable workforce. Unless a community is prepared for such an increase in its population, it is difficult to allocate resources even under the best of circumstances. Health care workers who work with migrant populations in Washington, Oregon and Idaho, even with the funding support that is available, express difficulty in caring for this population. Part of their concern is due to the increase in patient encounters they experience but, more importantly, it is the severity of the illness or health conditions that they see. Few will disagree that health care for the migrant and seasonal labourer costs more than the general population. The alternative for Alaskan communities with high volumes of migrant labourers should also be evaluated with regard to the cost of implementation. A programme designed to promote access should eventually minimize the economic impact that the seasonal workforce has on the health care system in that community.

Additionally, costs of administering a change designed to improve access to and continuity in health care for both migrant labourers and communities must be taken into consideration. Conversations with administrators and clinicians across the western USA demonstrate a genuine concern for the farm worker population. This concern should be built upon, not duplicated. By approaching the logistical aspect to any alternative in this manner, those who advocate for this population will realize that the farm worker can be employed in other industries, not just agriculture.

Finally, the most critical element of any alternative is its political legitimacy. Key leadership at local, state and national levels must be in agreement with a proposed alternative for it to be viable. Funding available through the Public Health Services Act is finite and currently divided among many competing interests. Fiscal trends in health care indicate that less federal aid will be available in the coming years to care for disenfranchised and underserved populations. Already, there is a strong interest in the notion that farm workers seek employment outside of the agricultural industry and improved data will in the future support the anecdotal information already in circulation. Data will also help decision makers to decide where financial resources should be allocated to help communities and ensure the success of any programme adopted to promote the health and well-being of farm workers.

Proposed Alternatives

Farm workers who migrate to Alaska and work in the fishing and seafood industry warrant our attention. Fishing, as the largest private employer in the state, contributes significantly to the economy of Alaska and to the local communities where companies have a presence. For example, the 1998 pollock harvest generated US$350 million in revenue

(Alaska Seafood Marketing Institute http://www.alaskaseafood.org/). Based on what is known so far about migrant and seasonal workers in Alaska, there are four possible alternatives to address what has been recognized as a problem for fishing communities across the state. To choose the best alternative, it must be understood that the problem in Alaska is not singular; rather, it has three components. The first is a recognition that a significant number of workers come to a community only for a limited amount of time. Secondly, and because of the volume of seasonal workers, the health care system in that community is stressed and residents as well as non-residents have difficulty accessing appropriate care. Finally, health centres in these communities are obligated to care for the public, but additional resources are limited in part because of how the Public Health Services Act defines migrant and seasonal labour as specific to agriculture.

One solution to this problem is to challenge the language of the Act that is specific to the migrant labourer. The definition clearly states that a migrant labourer is an individual whose principal employment is in agriculture on a seasonal basis, who has been employed within the last 24 months. Within a 24-month period, it is possible that an individual can work in agriculture in California and Washington, migrate to Alaska to fish and then return to California. There is already anecdotal evidence to support this theory. However, this option may not be effective in achieving access to primary health care and assuring continuity in care for the farm worker; what is required is simply an expansion of the migrant labour definition.

It is possible that more inclusive language in Section 330 may extend the funding opportunities to health centres caring for this transient population. However, the network of migrant health centres, situated in farming communities throughout the western migrant stream, is well established. Many of the programmes in place to care for the seasonal influx of workers are at capacity and any reallocation of funding may affect the ability to provide care and outreach. Further, farm workers have come to rely on the services many of these health centres provide. It is not uncommon for farm workers to return to a community specifically because of the health care services that they have access to even if the service is only available once a year. To discontinue or alter such a programme would have a significant impact on well-being.

Expanding the definition of migrant and seasonal labour has already occurred on an informal level among those who work directly with this population. It is common knowledge that the farm worker does not work exclusively in the agricultural industry to earn an income. However, because data to support this acknowledgement are lacking particularly in Alaska, key leadership seem unlikely to be willing to endorse expanding the definition of migrant labour to including industries other than agriculture. It is important to note that the definition of agriculture is somewhat open-ended. The US Department of Agriculture considers the pork, poultry and dairy industries as well as fish farming to be 'agri-business'.

Another possible way to address the situation in Alaska is to gain the support from fishing and seafood processing companies that are licensed to operate in Alaska. Suggested forms of support include better health screening on the part of industry during the hiring process, subsidies to health centres in communities where they operate, or providing health insurance to workers so that health centres can receive some sort of payment for services rendered.

Migrant labourers suffer from generally poor health, and screenings would simply shed more light on a problem many already know exists. If workers had to meet a minimum health standard, many would likely be ineligible for employment. For a number of reasons, including immigration status, the migrant worker is often unwilling to give more than the most basic information to an employer. Employers, on the other hand, protect themselves by only asking for minimum data on workers.

Another suggestion is for companies to subsidize health centres in the communities where their employees work. One such example of this model can be found in Dutch Harbor or Unalaska, which has been called

Alaska's most prosperous stretch of coastline, with access to three billion fish (North Pacific Fishery Management Council, 1998). The clinic in this community of 3200 is a showpiece for the region because of the subsidies it receives from the fishing industry. An ironic note to this subsidization is that it is in the form of a trauma surgeon, brought in for each season, because the incidence of occupational injury in this industry is so extreme. Given this fact and that migrant workers experience chronic illness more frequently than the general public, as well as the overall significant cost of health care in the USA, providing health insurance for fishing employees would most likely be opposed by the industry.

Communities could work with industry on a case-by-case basis to gain their support on a local level, but this suggestion, along with health screening and insurance coverage, would be very difficult to implement universally. Many companies have licences to fish in Alaskan waters but have an address outside the state or even overseas. The administrative agencies are not in place to regulate such alternatives, and to establish them would be costly for the industry and for state and local agencies required to enforce such a policy. At this level, leadership may not be in support of this approach. If the industry contributed to the cost of caring for these workers, federal funding would likely remain unaltered.

The third alternative to the situation in Alaskan fishing communities can be built from the recognition that farm workers do not work exclusively in agriculture. Throughout the western USA there is a network of health centres that care for underserved and disenfranchised populations. Unlike other regions in the country, health centres here have strong support at the state and regional levels, resulting in many collaborative relationships that encourage information sharing and the combining of resources. One such example of this type of support comes from the Northwest Regional Primary Care Association based in Seattle, Washington. This organization provides management and technical assistance to a number of health centres in Washington, Oregon, Idaho and Alaska and one of their areas of expertise is farm worker issues. This agency has the capacity to foster relationships between migrant health centres and develop a model health care system that could ensure continuity in care for migrant workers throughout the region. The executive directors of the Eastern Aleutian Tribes, Inc. organization and a health centre in Stockton, California, have already embraced this idea (C. Tevlin, Eastern Aleutian Tribes Inc., 2000, personal communication). Through a series of conversations, they have discovered that their clinics probably 'share' patients and are exploring ways to exchange information on their clientele, as appropriate. If this approach was implemented on a regional basis, it would certainly promote continuity of care. For example, if a patient in Akutan (Alaska) was to be treated for hypertension, the provider could contact the clinic in Stockton (California) where this person was seen last and discuss the best course of continued treatment and management. As promising as this model may seem, it may not resolve the access issue for the migrant population or the residents of the affected communities in Alaska. However, with the continuity issue resolved, access may no longer be as critical as it once was, since treatment plans are not disrupted due to migration. The number of acute clinic visits may decrease, leaving health centre staff more available to the resident population.

Another positive aspect of collaborative relationships is that few (if any) costs would be involved, because many have already been established. Given these relationships, the cost of care may also diminish due to timely management of acute concerns and sustained care for chronic conditions in the migrant population. Finally, through this coordinated effort, critical data can be collected on farm workers and 'fish workers'. In turn these data can be used to gain the support of additional leadership, particularly at the national level, and help to determine the best allocation of monetary resources for caring for migrant workers.

The final approach to address the situation in Alaskan fishing communities is to do nothing. The available information that links these workers to the agricultural industry is only speculative and anecdotal at this time. The influx of seasonal fish workers and the impact they have in Alaska may appear extreme because communities are small and geographically isolated.

Conclusion

The size and location of fishing communities in Alaska is perhaps the most compelling reason to address the issue of migrant and seasonal labour and the impacts of this workforce on health care delivery. Statistics indicating that 80–90% of the seasonal workforce in Alaska is from out-of-state in regions only accessible by small plane or boat suggests a significant burden on local infrastructure.

There are several potential alternative approaches to dealing effectively with this situation. While two possibilities are changing the language of the Public Services Health Act to be more inclusive of migrant workers or working with industry to gain support for initiatives designed to improve health care, perhaps the most viable alternative at this time is to foster collaborative arrangements among health care providers caring for this transient or highly mobile workforce. Communities with economies that employ these workers can learn from each other. Together communities in Alaska, Washington, Oregon and Washington should develop a health care delivery model that encourages collaborative relationships. Development of this model will require good communication and collection of census data, to understand further the health care needs of a workforce that is seasonal and migrant, but not entirely 'agricultural' by strict definitions.

References and Further Reading

Alaska State Department of Labor (1995) *Nonresidents Working in Alaska 1995*. Alaska State Department of Labor, Juneau, Alaska.

Alaska State Department of Labor (1997) *Alaska Population Overview, 1997 Estimates*. Alaska State Department of Labor, Juneau, Alaska.

North Pacific Fishery Management Council (1998) *Faces of Fisheries Community Profiles, Alaska*. North Pacific Fishery Management Council, Anchorage, Alaska.

US Department of Health and Human Services (1995) *Losing Ground: The Condition of the Farmworker in America, 1995*. National Advisory on Migrant Health, Bureau of Primary Health Care.

Section IV

Data Comparisons

The final chapter of this volume (by Larson, Findeis, Swaminathan and Wang) provides a discussion of two data sets that are referred to throughout this book: the Current Population Survey (CPS) and the National Agricultural Workers Survey (NAWS). Both data sets have advantages and both have weaknesses. A description of the data is provided here to help readers better understand the samples that are used for the two surveys and the types of questions asked. A comparison of the two data sets shows that they cover very different populations and, as a result, *who* is a farm worker and the well-being of these workers differs depending on the data source that is used. Regardless, both data sets serve to confirm that hired farm workers are, indeed, among those most in need in the USA.

18 A Comparison of Data Sources for Hired Farm Labour Research: the NAWS and the CPS

Janelle M. Larson,[1] Jill L. Findeis,[2] Hema Swaminathan[1] and Qiuyan Wang[1]

[1]*Department of Agricultural Economics and Rural Sociology, The Pennsylvania State University, University Park, Pennsylvania, USA;* [2]*Department of Agricultural Economics and Rural Sociology and the Population Research Institute, The Pennsylvania State University, University Park, Pennsylvania, USA*

Abstract

This chapter compares two national data sets commonly used in the USA for empirical research on farm workers: the Current Population Survey (CPS) and the National Agricultural Workers Survey (NAWS). As the major national survey of employment in the USA, the CPS covers farm workers as well as employees in other industries in the USA. These data are compared with the NAWS, a major survey of crop workers employed on US farms.

The samples for the CPS and NAWS are shown to be very different, even if crop workers (only) from the CPS Annual Earnings File are compared with the NAWS data. Unlike the CPS sample, the NAWS sample includes the (large and significant) segment of the farm worker population that shows a marked degree of mobility and is very likely to be unauthorized. Farm workers in the NAWS are more likely to be Hispanic, have very low levels of education and are younger than in the CPS. The NAWS also specifically shows that, compared with the past, US farm workers are more likely to be solo males working on farms in the USA but spending at least part of the year 'at home' in Mexico. They are also not likely to access public support systems in the USA, despite their low incomes. Finally, their earnings and incomes are difficult to compare with those of other low-income workers in the USA because only US earnings are included in the NAWS data. Inter-industry comparisons with the CPS are possible but, again, a portion of the farm worker population is missed.

Introduction

Accurate data are critical to assessing the well-being of farm workers and their families. For the USA, several sources of farm worker data are available and each has its own strengths and limitations. Taken together, these data sources provide a reasonable pic-

(eds J.L. Findeis, A.M. Vandeman, J.M. Larson and J.L. Runyan)

ture of farm workers and provide answers to a number of questions that are raised in research and policy formulation.

To better understand farm worker data available in the USA and to provide a broader assessment of the adequacy and usefulness of survey data on farm workers, this chapter assesses two of the most widely used data sources: the Current Population Survey (CPS) and the National Agricultural Workers Survey (NAWS). Two separate data sets from the CPS are assessed: the Current Population Survey March Annual Demographic Supplement (March CPS) and the CPS Annual Earnings File. These two CPS data sets were selected because the March CPS is commonly used to assess employment status, earnings and income of workers in the USA and the CPS Annual Earnings File is used to analyse data on US farm workers. For additional background on several sources for US farm worker data including the (US) Census of Agriculture, the CPS-HFWF data, ES-202 and ES-223 data, the Quarterly Agricultural Labor Survey (QALS), and the NAWS, see Martin and Martin (1994). See also the Economic Research Service (ERS/USDA) Briefing Room farm labour data source comparisons by Runyan (2000a). The data sets include the NAWS, CPS, Farm Labor Survey (FLS) and the Census of Agriculture.

The NAWS provides a wealth of information on farm workers in the USA, but is limited to crop workers, since the survey was initially designed to better understand migrant and seasonal farm workers. The significant advantage of the NAWS is the detailed data that this source provides. The CPS was not designed with farm workers specifically in mind but, instead, is a survey of employment in all US industries. The advantage of the CPS is that, conceptually, it should be possible to compare farm worker data from the CPS with data for workers employed in other industries. This would allow useful comparisons between the employment characteristics of those working on US farms and those working in other industries or occupations employing low-skill labour. Questions of interest include the following. Are real wages comparable across industries? Are total real earnings in a year of work comparable? How much time is allocated to the work? Are benefits received, and at what level? To what extent do workers rely on transfer payments?

This chapter focuses specifically on the CPS and NAWS data. A discussion is initially presented on the methodology used to collect both data sets, and differences between the March CPS and CPS Annual Earnings File. This is followed by a comparison of the demographic characteristics of the farm worker population in the NAWS and those workers in agriculture covered by the CPS. This comparison is important because the two surveys have very different coverages that are clearly reflected in the data comparisons. Following this, a discussion is presented that focuses on the wage, earnings and income data available in the NAWS and CPS. Since comparability of these measures is very limited, the discussion focuses on what can be learned from each of the data sets. Finally, questions of benefit receipt and access to public service programmes are raised, with an assessment of the usefulness of each data set for providing answers to these questions.

Methodology

Current Population Survey

The Current Population Survey (CPS) is sponsored jointly by the US Census Bureau and the US Bureau of Labor Statistics (BLS). The CPS is the primary source of labour statistics for the USA, and it collects extensive data on demographic, economic and social characteristics of the employed, unemployed and persons not in the labour force. The CPS is based on a probability sample of households, intended to represent the civilian, non-institutionalized working-age population (i.e. 16 years or older).

The survey is carried out monthly by the US Census Bureau, which surveys roughly 50,000 households from all 50 states and the District of Columbia. The survey is carried out during the week of the month that includes the 19th, and questions refer to the previous week. Selected households are surveyed for 4 consecutive months, omitted for 8 months, then surveyed again for 4 additional months. This provides the sample with continuity, without being excessively burdensome for the respondents. This structure also means that roughly three-quarters of the sample at any time will be interviewed in the following month, and one-half in the same month the following year.

The current sample design, developed in 1996, includes 754 sample areas and maintains a 1.9% coefficient of variation (CV) on the national monthly estimates of unemployment, assuming a 6% unemployment rate. For each of the 50 states and the District of Columbia, the design maintains a CV of no more than 8% on the annual average estimate of unemployment. To meet the national criterion for reliability, estimates for several large states are substantially more reliable than the design requirements. The sample is state-based, and is designed to ensure that most housing units in a given state have the same probability of selection. To maintain this probability, the sample areas are redesigned following each census. There is generally a 6–7% non-response rate.

The CPS questionnaire was redesigned in 1994. Prior to this, the survey questionnaire had been essentially unchanged for nearly three decades. The redesign of the questionnaire had four main objectives: (i) to measure the official labour force concepts more accurately; (ii) to expand the amount of data available; (iii) to implement changes in concepts and definitions; and (iv) to take advantage of computer-assisted interviewing.

Trained enumerators complete a questionnaire for each household member aged 15 years and older. Because those aged under 16 years are limited in their labour market activity, the US Bureau of Labor Statistics generally publishes labour force data only for those persons 16 years old and older. The initial and fifth monthly interviews are usually conducted in person, while other interviews are generally conducted by telephone. The computerized questionnaire introduced in 1994 allows interviewers to confirm data obtained in previous interviews and facilitates the collection of more detailed and accurate information. Questions focus on each household member's paid (or for-profit) activity during the survey week.

Hired farm workers are employed persons who, during the reference week for the survey, did farm work for cash wages or salary, or did not work but had farm jobs from which they were temporarily absent. According to Huffman (2000), hired farm workers include hired farm managers, supervisors of farm workers and farm and nursery workers. Hired farm workers are classified according to the industry of the establishment for which they work. 'Crop production' refers to employment in establishments primarily engaged in producing crops, plants, vines and trees, excluding forestry operations. 'Livestock production' refers to employment in establishments primarily engaged in the keeping, grazing or feeding of livestock. 'Other agricultural establishments' refers to establishments primarily engaged in agricultural services, forestry, fishing, hunting, trapping, landscape and horticultural services, and other agriculture-related establishments. There are two CPS data sets that are often used for research, including research on hired farm workers, and these data sources are described below. In addition, the US Census Bureau website lists a number of current CPS special topic supplements that provide data on various topics relevant for a better understanding of the US workforce.

CPS Annual Demographic Supplement (March CPS)

The CPS Annual Demographic Supplement (March CPS) collects data on family characteristics, household composition, marital status, migration, income from all sources, weeks worked, occupation and industry classification of the job held longest during the year, health insurance coverage and receipt of non-cash benefits. The sample consists of the March CPS sample plus households interviewed in the regular CPS the previous November containing at least one person of Hispanic origin. As there is no overlap in the sample from November in March, the combined sample essentially doubles the Hispanic sample and improves the reliability of the March CPS estimates for the Hispanic population. The March CPS data are widely available for research.

CPS Annual Earnings File

The CPS Annual Earnings File (CPS Merged Outgoing Rotation Groups) is available from the National Bureau of Economic Research (NBER) and includes a subset of the CPS questions asked *throughout the year*. Information on what workers earn in their primary jobs is collected only for those who are receiving their 4th or 8th monthly interview, about one-fourth of the respondents. Prior to 1994, the hours worked were also only collected in the 4th and 8th months; they are now collected in each interview.

Respondents are asked to report their usual earnings before taxes and to include any overtime pay, tips or commissions they usually receive. Data are collected for wage and salary workers, and these earnings data are used to generate estimates of the distribution of usual weekly earnings and median earnings. The earnings of those who are paid at an hourly rate are used to analyse the characteristics of hourly workers.

For the majority of multiple job-holders, occupation, industry and class-of-worker data for their second jobs are collected only in the 4th or 8th monthly interview. For those classified as 'self-employed unincorporated' in their main jobs, class-of-worker of the second job is collected monthly.

Limitations of the data

Four limitations of the CPS data should be noted. Firstly, because the CPS is a household survey, it is likely to undercount workers who live in non-traditional quarters or who are migratory, many of whom are Hispanic. The implications of this limitation of the CPS are very important, as shown later in this chapter.

Secondly, the CPS classifies workers according to the job at which they worked the greatest number of hours during the reference week. As a result, hired farm workers who spent more time during the reference week at their non-farm job than at their farm job would not be included in the primary employment count as hired farm workers (Runyan, 2000b). Instead, they would be counted as having hired farm work as their second employment, but data on earnings from farm work would not be collected.

Thirdly, since only a small percentage of households in the survey supply data on hired farm workers, reliable state and regional estimates are possible but state-level estimates are not (Huffman, 2000).

Finally, the March CPS basically misses a very sizeable proportion of farm workers. By including data from across the entire year, the CPS Annual Earnings File avoids this particular problem. Relatively small sample sizes may also result when a single year of data is used to examine particular subsamples of interest (e.g. by race/ethnicity or by gender).

National Agricultural Workers Survey (NAWS)

The US Department of Labor (DOL) commissioned the NAWS in response to the Immigration Reform and Control Act of 1986 (IRCA). IRCA required the US Secretaries of Labor and Agriculture to monitor seasonal agricultural wages and working conditions and to determine annually, between 1990 and 1993, whether a shortage of seasonal agricultural workers was to be expected. After the IRCA mandate was complete, the NAWS was revised in 1993 to provide more complete demographic information on crop workers.

The NAWS is an employer-based survey that collects extensive data from crop workers about basic demographic characteristics, legal status, education, family size and composition, wages and working conditions in farm jobs and participation in the US labour force. The NAWS also gathers a limited amount of information about household members associated with the crop worker who is being surveyed.

To overcome the difficulty of reaching migratory farm workers in unconventional living quarters, the NAWS samples worksites rather than residences. Interviews are conducted three times a year, to enable data collection to match seasonal fluctuations in the agricultural workforce. More than 2500 randomly selected crop workers have been interviewed each year since 1989. Starting in 1999, NAWS has collaborated with the National Institute of Occupational Safety and Health (NIOSH) to gather detailed data on the incidence and prevalence of occupational health problems among crop workers. With the addition of these questions, the NAWS sample has increased to 3600 workers per year.

Participation in the NAWS is limited to hired crop workers; workers in animal agriculture are excluded. The NAWS interviews workers employed by either growers or contractors and working in all crops included in Standard Industrial Classification (SIC) code 01. This includes nursery products, cash grains, field crops, fruit and vegetables, and the production of silage or other animal fodder. Landscape workers are interviewed only if they are involved in the production of plants rather than in installation, sales or maintenance. The fieldwork criterion excludes secretaries and mechanics, but includes field packers and supervisors working in the field. Neither unemployed agricultural workers nor H-2A workers are interviewed. Owners of the farm and their unpaid family members are excluded from the sample, as are sharecroppers. Workers in canneries or packing facilities are eligible to be interviewed only if the facility is adjacent to (or on) the farm and at least 50% of its goods are produced by the same employer.

The NAWS uses a multistage sampling procedure, which relies on probabilities proportional to size to obtain a nationally representative sample of crop workers annually. Each year, three survey cycles are conducted in an effort to cover all types of seasonal crop workers. Each of these cycles lasts approximately 12 weeks, beginning in October, February and June. For each cycle there are four sampling levels: the region, the crop reporting district (CRD), the county and the employer. The 12 geographical regions are based on the US Department of Agriculture (USDA) regions, and the CRDs were also created by the USDA as groups of counties within regions with similar agricultural characteristics.

To summarize the sampling method, the NAWS uses the USDA Farm Labor Survey's quarterly estimates of hired and contract labour for each region to allocate the interviews between cycles and regions. A roster of 47 CRDs is selected, using probabilities proportional to size. Of these CRDs, 30 to 35 are selected for each cycle. From these CRDs, counties are selected using probabilities proportional to the counties' hired crop labour expense. Next, farm sites are randomly selected from a list of all farm employers located in the selected counties. The selected farm employers are then contacted to obtain permission to interview the farm workers employed at the site. Finally, interviews are administered to a random sample of workers in the farm worker's quarters.

The allocation of interviews across seasons, counties and growers is based upon the amount of agricultural activity. The more agricultural activity for a given season, county or grower, the more interviews are allocated. This process is intended to produce a representative sample of US hired crop workers.

Sampling method

The USDA *Farm Labor* report divides the continental USA into 17 regions. Florida and California are each categorized as single regions, with the remaining 46 states grouped into 15 regions. Because of budgetary constraints, the NAWS combines several regions for sampling. Only contiguous regions that have similar agricultural characteristics are combined. The level of similarity was determined using various statistical techniques, primarily hierarchical case clustering and multidimensional scaling. The data used were from the *Census of Agriculture* and the *Farm Labor* reports. After the analysis, the 17 regions were reduced to 12 regions for sampling purposes. The following sets of USDA regions are combined: Appalachian I and II, Delta and Southeast, Mountain I and II, the Cornbelt I, II and the Northern Plains. Table 18.1 lists the sampling regions and the states in each region.

Table 18.1. NAWS sampling regions. (Source: US Department of Labor, 2000a.)

NAWS sampling region	USDA region		States in USDA region
	Code	Name	
AP12	AP1	Appalachian I	NC, VA
	AP2	Appalachian II	KY, TN, WV
CBNP	CB1	Cornbelt I	IL, IN, OH
	CB2	Cornbelt II	IA, MO
	NP	Northern Plains	KS, NE, ND, SD
CA	CA	California	CA
DLSE	DL	Delta	AR, LA, MS
	SE	Southeast	AL, GA, SC
FL	FL	Florida	FL
LK	LK	Lake	MI, MN, WI
MN12	MN1	Mountain I	ID, MT, WY
	MN2	Mountain II	CO, NV, UT
MN3	MN3	Mountain III	AZ, NM
NE1	NE1	Northeast I	CT, ME, MA, NH, NY, RI, VT
NE2	NE2	Northeast II	DE, MD, NJ, PA
PC	PC	Pacific	OR, WA
SP	SP	Southern Plains	OK, TX

After establishing regions, the NAWS had to determine the roster of locations for interviews within each region, and USDA CRDs were used as a starting point. Some CRDs covered large areas with intensive agriculture. For example, California's 58 counties were grouped into eight CRDs. However, for the purpose of the NAWS these groupings were too large. Using Census of Agriculture data for the number of farm workers, many CRDs were subdivided into small groups of counties or sometimes individual counties.

Finally, 47 sites were selected from the list of potential sites (CRDs and subdivisions of CRDs). Each region had at least two sites, and the remaining sites were distributed across the regions. Sites were selected using probability proportional to size (pps) sampling, based on crop labour expenditure. To ensure geographical diversity within each region, the potential site list (in geographical order) was divided into as many parts as there were sites to be selected, and one site was selected from each part, using pps. For example, there were to be three sites selected from the Pacific region (Washington and Oregon). The list of potential sites in this region was therefore divided into three parts of estimated equal size, and one site was selected for each part.

The NAWS sampling design also attempts to represent the seasonality of crop work. The seasonal measures are expressed in terms of the percentage of farm work that is done in a specific area during each of the three NAWS cycles. At the regional level, seasonality is measured by the USDA Quarterly Agricultural Labor Survey (QALS); for the CRDs, the seasonality measure is a weighted average of employment levels obtained from unemployment insurance records and seasonal distribution measures provided by Cooperative Extension agents.

For the autumn and summer cycles, sites are selected for all 12 regions. During the winter cycle, however, the seven northern regions are combined into one region and so sample sites are drawn from this region and the remaining five southern regions.

For each cycle, approximately 30 of the 47 sites are selected (without replacement) using probabilities proportional to the CRDs' seasonal farm labour expenses. This was determined by multiplying the total labour expenditure by the seasonality index for each CRD (or sub-CRD). Within each site, one county was selected using pps. This is the county in which interviews start and in most cases the quotas for interviews are filled within the first county. If the quota is not filled, second (or subsequent) counties are selected using pps.

Within each selected county, employers are selected at random using probabilities proportional to the square root of the size of their seasonal workforce, primarily using data gathered by state employment services and provided to the Bureau of Labor Statistics (BLS). The NAWS generates lists of farm employers using information from the BLS, Farm Service Agency and other administrative data sources. These lists are reviewed, updated and supplemented using information from county extension agents, local employment agencies, grower organizations and farm worker service providers.

NAWS interviewers contact the selected employers to secure permission to visit the worksite and interview selected workers. Farm workers are approached for an interview while they are not working (for example, when arriving for work, during lunch, or when leaving). In most cases, interviews are scheduled for a later time, at a location selected by the worker. Individual farm workers are selected and approached at random.

Allocation of interviews within sites

Once interview locations are selected, interviews are allocated according to pps. First, the number of interviews allocated to each region is decided. This allocation is based on the total agricultural wages paid in the region (excluding livestock workers). The allocation uses a compromise between proportional to size and proportional to the square root of the size, to ensure that smaller regions have sufficient sample sizes.

For all cycles, the number of interviews per region is proportional to the size of the seasonal farm labour force in that region, as determined by the National Agricultural Statistical Service (NASS) using information from the QALS. Interviews are then allocated to CRDs proportional to size. After interview allocations are determined for each county, interviewers are provided with Grower Lists that contain randomly selected employers and the number of interviews required for the county. Growers are contacted in order until the interviews for the county are complete. The number of interviews allocated to each employer is proportional to the square root of size.

Limitations of the data

The primary limitation of the NAWS data is that they include only hired workers in crop production, excluding livestock workers. In addition, the refusal rates of both employers and employees are unknown. Since employers have to give permission for the interviews to take place, their refusal could bias the sample of workers.

Other limitations include the issue of the income questions in the NAWS and the relatively large standard errors in the data. The income questions in the NAWS pose a problem because they include only income accruing in the USA, both for the farm worker and for the farm worker's household, and exclude income earned elsewhere. This problem has two dimensions. Firstly, the farm worker may earn income outside the USA that is not reflected in the income statistics in the NAWS. Secondly, there are likely to be other earners in the household who may generate income outside the USA. This could be as wages but also as farm income or other income from self-employment. This limitation is important, because the farm worker's total income and the worker's total household income will be underestimated if income is earned in the origin country (e.g. Mexico) for in-migrants to the USA. At the same time, it should be noted that it is possible with the NAWS data to compare workers' US wages, which is useful for gauging the economic returns to farm work in the USA.

Finally, the standard errors of the NAWS data tend to be high. The NAWS data are often used for national studies but the high standard errors make regional studies using the NAWS problematic (Emerson, 2000; Findeis *et al*., 2000; Huffman, 2000; Thilmany and Grannis, 2000).

Comparing Demographic Characteristics

Comparisons of the demographic characteristics of the samples for the NAWS and CPS show important differences in the survey samples. (It should be noted that the analyses of farm worker images in Chapter 2 of this volume and the statistics presented here will differ slightly, due to differences in the number of observations included in the analyses: Chapter 2 focuses on specific categories of workers and both chapters use weighted data.)

The demographic differences are not surprising, given the differences in the sampling procedures used in the CPS and NAWS. As shown in Table 18.2, the CPS provides a sample in which the majority (over 60%) of farm workers in the USA are non-Hispanic white. This is the case for both the March CPS and the CPS earnings data for all farm workers. Even for crop workers (only) in the CPS earnings data, 44.5% are non-Hispanic white. In contrast, only 14.5% of workers covered by the NAWS are non-Hispanic white. In the NAWS, the vast majority (82%, based on annual average data) of farm workers are Hispanic.

The NAWS sample covers a workforce that is often unauthorized or working illegally in the USA. The population, particularly in the crop production sector, is often migrant and may be out of the USA for part of the year. As a result, it is not surprising that the March CPS sample misses these workers in agriculture. The CPS reflects workers who are more likely to be legal and permanent residents on a farm. The NAWS sample reflects workers who are in many respects 'underground' and on the move.

The demographic differences in the two samples are most evident in differences in race/ethnicity but are observed for other characteristics as well. The NAWS sample, for example, tends to be younger, with about

Table 18.2. Demographic characteristics of survey respondents, NAWS and CPS (weighted data for 1993–1998, annual average). (Sources: NAWS; March CPS; CPS Annual Earnings File.)

			CPS earnings file	
Characteristics	NAWS	March CPS	All farm workers	Crop workers
Mean age (years)	31.0	36.2	36.5	37.1
Gender (% male)	79.8	82.6	81.8	83.5
Age composition (%)				
< 25 years	38.3	25.8	35.4	23.9
25–54 years	56.6	61.2	61.1	62.0
> 54 years	5.1	13.0	13.5	14.1
Education (%)				
Not a high school graduate	81.8	49.9	50.5	60.1
High school graduate	12.8	28.7	28.2	22.7
Beyond high school	5.4	21.4	21.4	17.2
Race (%)				
Non-Hispanic white	14.5	63.3	60.5	44.5
Non-Hispanic black	1.6	4.6	4.1	5.4
Hispanic	82.1	30.3	33.7	47.8
Other	1.7	1.8	1.7	2.4
Marital status (%)				
Never married	43.3	35.6	33.2	32.2
Married	51.9	54.4	54.6	54.0
Other (separated, divorced, widowed)	4.8	10.0	12.3	13.8

38% of survey respondents being under 25 years of age. This compares with roughly 26% of 'farm workers' in the CPS. NAWS respondents are also less likely to be married, in part as a result of the younger age structure. They also have less education, although it should be noted that farm workers in both the NAWS and CPS generally have very low levels of educational attainment. In the CPS, roughly one out of every two workers in the farm production sector has not completed high school, and in the NAWS sample four out of five workers have not attained a high school degree. The results for the crop workers, subset of the CPS earnings microdata show that three out of five crop workers have not completed high school (Table 18.2). Female farm workers tend to have somewhat higher levels of education than males in both the NAWS and CPS data sets, although the female farm workers in the NAWS still tend to be less educated and younger and more likely to be Hispanic than in the CPS data (see Table 18.3 for disaggregated means by gender).

Finally, CPS data show a farm worker population that is more likely to be born in the USA (Table 18.4). In contrast, the NAWS data document a largely immigrant labour force in US crop production, very unlike the CPS. The CPS again misses at least a portion of this segment of the farm worker population.

The NAWS data help to provide insight into the origin and background of the immigrant population working on US farms in crop production. As shown in Chapter 2, 46% of crop workers in recent years of the NAWS have been unauthorized – a significant increase from the 8% in 1989. Of the total number of farm workers in the NAWS over the 1993–1998 period, almost 70% were born in Mexico, with only an average of 23% born in the USA. As shown in Table 18.5, reliance on US workers has declined over this period, with workers from Mexico comprising a higher percentage over the 1990s. The remainder of NAWS workers are from other countries in Latin America, Puerto Rico and Asia, but these percentages are minimal.

Table 18.3. Demographic characteristics of survey respondents by gender, NAWS and CPS (weighted data for 1993–1998, annual average). (Sources: NAWS; March CPS; CPS Annual Earnings File.)

	NAWS		March CPS		CPS earnings file[a]	
Characteristics	Male	Female	Male	Female	Male	Female
Mean age (years)	30.6	32.63	35.9	37.8	36.1	38.2
Age composition (%)						
< 25 years	40.0	31.5	26.5	22.2	26.7	19.8
25–54 years	55.0	63.0	60.9	62.6	60.1	65.2
> 54 years	5.0	5.5	12.6	15.2	13.2	15.0
Education (%)						
Not a high school graduate	84.5	71.1	51.9	40.2	52.6	40.8
High school graduate	11.0	19.9	27.5	34.8	27.5	31.2
Beyond high school	4.5	9.1	20.7	25.0	19.9	28.0
Race (%)						
Non-Hispanic white	11.7	26.0	60.9	74.3	58.1	71.3
Non-Hispanic black	1.5	2.3	5.2	1.7	4.6	2.0
Hispanic	85.3	69.6	32.1	21.9	35.8	24.0
Other	1.6	2.1	1.7	2.1	1.5	2.8
Marital status (%)						
Never married	45.6	34.4	37.8	25.4	35.2	24.3
Married/living together	50.9	55.7	52.8	61.2	52.6	63.2
Other (separated, divorced, widowed)	3.5	9.9	9.3	13.4	12.2	12.5

[a]For all farm workers, including both crop and livestock workers. Statistics for crop workers and livestock workers separately are available from the authors.

Table 18.4. Additional demographic information for CPS farm workers, based on March CPS data (weighted data for 1994–1998, annual average; these questions were not asked in 1993).

Citizenship	All workers	Crop workers only[a]	Full-time workers	Part-time workers
Born in USA (%)	62.0	53.7	55.6	83.6
US citizen (%)	67.4	58.9	61.6	89.0

[a]Based on CPS Annual Earnings File data.

Table 18.5. Additional demographic and related information (percentage of respondents) available in the NAWS Public Access Data (weighted data for 1993–1998).

	1993–1995	1996–1998	Annual average
Place of birth			
USA	27.80	18.12	22.96
Puerto Rico	2.99	1.82	2.40
Mexico	64.53	74.73	69.63
Other Latin America	3.45	3.78	3.61
Asia	0.08	0.41	0.25
Other	1.15	1.14	1.15
Primary language			
English	23.53	13.50	18.52
Spanish	73.26	82.24	77.75
Other	3.22	4.26	3.74
English proficiency			
Speaking			
Not at all	35.60	42.85	39.22
A little	28.96	28.71	28.84
Somewhat	9.32	7.18	8.25
Well	26.13	21.26	23.69
Reading			
Not at all	44.39	50.63	47.51
A little	19.46	21.82	20.64
Somewhat	6.61	6.72	6.67
Well	29.53	20.83	25.18

Not surprisingly, the NAWS shows that Spanish is the primary language of the majority of farm workers and that relatively few crop workers have a good command of the English language. For example, the NAWS data show essentially that two out of five farm workers cannot speak English at all (not even 'a little') and almost half do not read English at all. The NAWS provides important insights into the farm worker population that the CPS does not. For a population that is largely immigrant, the NAWS data are much richer in this regard.

Both the NAWS and the CPS provide data on the individual characteristics of farm workers working in the USA, although clearly the samples are very different. Both surveys also provide data on household or family characteristics for farm workers, but again the samples are quite different. This is in part due to how the 'household' is defined when data are collected. The basic sampling unit of the CPS is the 'household' (living together in the selected housing unit), with data provided in part at the household level. The CPS also includes information on families and individuals within the households in the sample. These data are found on the family and individual (person) files of the CPS. This hierarchical structure of the CPS includes data at the three levels and allows the linking of data across levels. In comparison, the NAWS covers households that may be living in more than one location, i.e. households with members who live and work in different locations but function and consider themselves as a household.

Wages and the Returns to Work

The CPS and the NAWS both ask questions to determine some of the conditions of work, including earned wages and hours worked, for farm workers in the USA. The NAWS asks a number of questions related to wages paid for work and hours, with the questions essentially being asked in two formats: (i) wages and hours last week for those paid hourly; and (ii) the amount of pay received on the last day of work and hours for that work period, for those paid piece-rate. In the first case, the wage is for Task 1, the worker's principal task identified for that time period. Therefore, real hourly wages are available in the NAWS data for Task 1 and are computed for the last pay period (Table 18.6).

From the NAWS data, real hourly wages can be determined both for Task 1 (e.g. preharvest, postharvest, semi-skilled harvest, supervisory, and 'other') and by crop (e.g. fruits and nuts, vegetables, horticultural crops, field crops and miscellaneous). Supervisory workers are paid the highest wages at an average US$9.72 per hour in 1998 dollars over the 1993–1998 period. Work in pre-harvest tasks is generally paid the least, at an average of only US$5.69 per hour. Real hourly wages over the 1993–1998 period range from US$5.76 in fruits and nuts to US$6.18 in horticultural crops (Table 18.7).

The real hourly wage for farm work is also available in the March CPS and in the CPS earnings data. The average real hourly wage over the 1993–1998 period from the March CPS for all farm workers is US$6.70 per hour (Table 18.8), which is higher than in the NAWS. Full-time farm workers earn on average US$6.74 in the CPS earnings data, and part-time workers earn US$6.20 per hour in real wages. In the CPS there is no detailed disaggregation, by type of crop or by principal task, which is not surprising given the coverage of the CPS as compared with the NAWS.

The CPS and NAWS also provide data on total earnings and income, but these questions are not comparable across the two data sets. The CPS provides earnings per week (Table 18.9) and these can be compared with those of workers in other industries – a major advantage of the CPS. For the NAWS, the earnings data include only earnings in the USA, and there are no data on earnings outside the USA. This limits the comparability of the NAWS total earnings and income data to other data sources for workers in other industries and in other types of households to which households farm workers could be compared. If the worker lives only in the

Table 18.6. Average real wages received and hours worked using different survey approaches, NAWS (weighted data for 1993–1998).

	1993–1995	1996–1998	Annual average
Wage (Task 1[a]) (real US$[b])	6.15	5.91	6.03
Hours worked last week at current farm job	37.95	37.88	37.91
Amount received last day of work (before taxes) (real US$[b])	271.30	256.91	264.11
Hours worked during that period	42.56	42.15	42.36
Average wage (real US$[b])	6.34	6.10	6.22

[a]Task 1 refers to the worker's main task in the NAWS.
[b]Based on 1998 dollars, using the CPI for conversion to real dollars.

Table 18.7. Average real wage (real US$[a]) by task (Task 1[b]) and crop, based on the NAWS data (weighted data for 1993–1998).

	Average real wage for task 1		
	1993–1995	1996–1998	Annual average
Task 1			
Pre-harvest	5.75	5.63	5.69
Harvest	6.35	6.20	6.28
Postharvest	6.13	5.94	6.03
Semi-skilled	6.10	5.74	5.92
Supervisor	9.88	9.49	9.72
Other	6.13	6.02	6.07
Crops			
Fruit and nuts	6.03	5.48	5.76
Vegetables	6.22	5.99	6.11
Horticultural crops	6.01	6.36	6.18
Field crops	6.28	5.86	6.07
Miscellaneous	6.09	5.87	5.98

[a]Based on 1998 dollars, using the CPI for conversion to real dollars.
[b]Task 1 refers to the worker's main task in the NAWS.

Table 18.8. Real wages received (real US$[a]) and hours worked by US farm workers, based on March CPS data (weighted data for 1993–1998).

	1993–1995	1996–1998	Annual average
Individual hours, earnings, income			
Hours worked last week at all jobs	38.87	39.50	39.19
Hourly wage[b]	7.14	6.27	6.70
Total wage and salary earnings[c]	12,970	16,023	14,497
Total income[d]	14,159	16,816	15,487
Household earnings, income			
Total wage and salary earnings	27,816	35,820	31,818
Total earnings[e]	31,862	38,914	35,388
Total income	36,639	42,853	39,746

[a]Based on 1998 dollars, using the CPI for conversion to real dollars.
[b]Only for those who answered 'yes' to question: 'Are you paid by the hour on this job?'
[c]Individual-level data (from personal record of CPS).
[d]From person record of the CPS. Includes total wage and salary earnings, total own-business self-employment earnings and total farm self-employment earnings.
[e]Includes household-level wage and salary earnings, own-business self-employment earnings and farm self-employment earnings combined.

USA, this is not a problem. However, since the majority of USA farm workers today spend time outside the USA and may have family workers earning incomes totally outside of the USA, the data in the NAWS only provide a partial picture. The NAWS data are very useful for analysing the average US earnings of farm workers, but are not directly comparable to total earnings and income statistics for other industries in the USA, as the definitions differ.

Receipt of Job Benefits and Public Services

Employment-related benefits and access to public programmes are also questions

Table 18.9. Real wages received and earnings per week (real US$[a]) for US farm workers (weighted data for 1993–1998). (Source: CPS Annual Earnings File.)

	1993–1995	1996–1998	Annual average
US farm workers			
Wage	6.63	6.59	6.61
Earnings per week	279.00	300.00	289.50
Full-time			
Wage	6.73	6.76	6.74
Earnings per week	318.67	343.04	330.85
Part-time			
Wage	6.19	6.21	6.20
Earnings per week	165.33	166.90	166.12

[a]Based on 1998 dollars, using the CPI for conversion to real dollars.

Table 18.10. Distribution of benefits from farm employers to hired farm workers in the USA (% of respondents[a]) based on NAWS data (weighted data for 1993–1998).

Question	1993–1995	1996–1998	Annual average
If you are injured at work or get sick as a result of your work			
Does your employer provide health insurance or provide or pay for your health care?			
Yes	19.84	28.90	24.37
No	66.16	58.19	62.17
Don't know	14.00	12.91	13.45
Do you get any payment while you are recuperating (i.e. worker's compensation)?			
Yes	40.86	55.62	48.24
No	40.35	27.75	34.05
If you are injured off the job (e.g. at home), does your employer provide health insurance or provide or pay for your health care?			
Yes	10.23	5.10	7.66
No	80.91	82.36	82.64
Are you provided with paid holidays and/or paid vacations on your present job?			
Yes	11.51	9.59	10.55
No	83.56	84.80	84.18

[a]Note that the sum of percentages of 'yes' and 'no' responses will not be 100%, because of 'don't know' responses by some survey participants.

examined in the CPS and the NAWS, to varying degrees. The NAWS asks farm workers participating in the survey about the benefits that they receive from their employer, including health care, paid holidays and unemployment insurance (Table 18.10). The NAWS also asks about receipt of social programme benefits by farm worker households. The specific programmes include Food Stamps, Unemployment Insurance (UI), Social Security, WIC (Women, Infants and Children), Medicaid and public health clinic services, among a number of other programmes that are in place in the USA to help the poor and those without employment. Based on the NAWS, very low numbers of farm worker families are receiving benefits from these programmes, despite their poverty status (Table 18.11). In part, this is the result of the relatively high number of workers that are unauthorized, but the picture is very similar even if social programme benefits received by only those workers who are legal are analysed.

The CPS similarly shows relatively low rates of receipt of public benefits or payments by farm worker households (Table 18.12),

Table 18.11. Receipt of benefits from social programmes for all farm worker households in the NAWS (% of respondents). (Source: NAWS Public Access Data.)

Benefits received[a]	1993–1995	1996–1998	Annual average
AFDC[b]	2.56	1.54	2.05
Food Stamps	16.33	10.44	13.39
Disability insurance	1.92	0.83	1.38
Unemployment insurance	19.96	18.20	19.08
Social Security	2.42	1.22	1.82
Veteran's pay	0.26	0.10	0.18
General assistance/welfare	1.71	1.19	1.45
Low-income housing	0.80	0.52	0.66
Public health clinic	1.36	0.25	0.81
Medicaid	13.38	12.73	13.06
WIC[c]	8.41	9.82	9.11
Disaster relief	0.25	0.09	0.17
Legal services	0.31	0.05	0.25
Other	1.51	1.32	1.42

[a]The specific question in the NAWS reads: 'Within the last two years has anyone in your household received benefits from or used the services of any of the following social programs?'
[b]Aid to Families With Dependent Children.
[c]Women, Infants and Children.

Table 18.12. Benefits received by US farm worker households (% of respondents) from public sector or farm employer, March CPS.

Benefits	1993–1995	1996–1998	Annual average
Receipt of benefits or payments by household			
Unemployment compensation	14.90	11.00	12.95
Social Security	13.25	13.60	13.43
Supplemental Security	4.27	3.07	3.67
Welfare (public assistance)	5.79	3.32	4.56
Veteran's pay	1.82	1.25	1.54
Survivor benefits	1.74	0.65	1.20
Disability benefits	1.65	1.46	1.56
Retirement payments	4.16	3.85	4.01
Food Stamps (last year)	14.17	10.43	12.30
Coverage of anyone in household by:			
Medicare	10.61	11.25	10.93
Medicaid	16.23	18.49	17.36
Health insurance	60.99	64.39	62.69
Living in a public housing project that is owned by a local housing authority or other public agency?			
In universe[a]	56.82	56.99	56.91
Yes	2.07	2.19	2.13

[a]'In universe' refers to the portion of the sample that is asked this question in the CPS.

although the receipt of benefits is generally more frequently observed in the CPS than in the NAWS. This is likely a reflection of the differences in the CPS versus NAWS samples (i.e. the presence of a high percentage of unauthorized workers in the NAWS).

Other Survey Data

Finally, some data in either the NAWS or the CPS are not available from the other data source. This is particularly true for the NAWS, which includes a rich set of questions focusing specifically on the farm worker population. The CPS does not provide the in-depth coverage found in the NAWS.

Conclusion

The CPS and the NAWS are compared in this chapter, providing an assessment of the major strengths and weaknesses of the data sets for research on farm workers in the USA. The CPS, like employment-related surveys in other developed countries, provides selected farm worker demographic characteristics, hours of work, wages and receipt of employee benefits, and access to public services. Since these data are available for workers in all industries, not solely agriculture, the CPS has the advantage of allowing comparisons across industries. Therefore, it should be possible to compare the CPS data on farm workers with data on non-farm workers in jobs in the USA with similar skill requirements. This is a major strength.

However, the sampling structure used for the CPS results in a very different population of farm workers than for the NAWS. Farm workers in the CPS are much less likely to be Hispanic, have higher levels of education, tend to be older on average, and have other characteristics quite different from the increasingly solo male Hispanic workforce described in Chapter 2, based on the NAWS. As shown in this chapter, this is true even if only crop workers from the CPS Annual Earnings File are used for comparison.

The NAWS data provide a much clearer view of farm workers in the USA today. A major advantage of the NAWS is that it covers a broader cross-section of farm workers involved in crop work – both those that are authorized and those that are not. Given this advantage, the NAWS is more appropriate for understanding the true status of those who provide hired crop labour to US agriculture.

Acknowledgements

The authors wish to thank Daniel Carroll at the US Department of Labor and Jack Runyan at the US Department of Agriculture for their useful input on the NAWS and CPS data sources.

References and Further Reading

Emerson, R.D. (2000) Changes in the southeastern farm labor market. Paper presented at 'NAWS at 10: A Research Seminar', University of California, Davis, 7 October.

Findeis, J.L., Larson, J., Swaminathan, H. and Wang, Q. (2000) Hired farm workers in the northeast and mid-atlantic states: changes in agriculture, agricultural employment and the conditions of work. Paper presented at 'NAWS at 10: A Research Seminar', University of California, Davis, 7 October.

Huffman, W.E. (2000) Changes in the US and midwestern farm labor markets: NAWS and USDA data. Paper presented at 'NAWS at 10: A Research Seminar', University of California, Davis, 7 October.

Martin, P.L. and Martin, D.A. (1994) *The Endless Quest: Helping America's Farm Workers*. Westview Press, Boulder, Colorado.

Runyan, J. (2000a) Economic Research Service Briefing Room. Farm labour: farm labour data. http://www.ers.usda.gov/briefing/Farm Labor

Runyan, J. (2000b) *Profile of Hired Farmworkers, 1998 Annual Averages*. US Department of Agriculture, Washington, DC.

Thilmany, D. and Grannis, J. (2000) Farm labor trends for the mountain and northwest region. Paper presented at 'NAWS at 10: A Research Seminar', University of California, Davis, 7 October.

US Department of Labor (2000a) *National Agricultural Workers Survey: Statistical Documentation for Fiscal Years 1993–1998*. Washington, DC.

US Department of Labor (2000b) *National Agricultural Workers Survey, Codebook for Public Access Data, Federal Fiscal Years 1993–1998, Cycles 14–31*. Release 2.0, Washington, DC.

Index

Figures in **bold** indicate major references. Figures in *italic* refer to figures and tables.